AF442812

RECENT ADVANCES IN EMBRYOLOGY

RECENT ADVANCES IN EMBRYOLOGY

Vol. 5
Mammalian Embryology

A.P. Diwan
&
N.K. Dhakad

ANMOL PUBLICATIONS PVT LTD
New Delhi-110 002

ANMOL PUBLICATIONS PVT LTD
4374 / 4B, Ansari Road
Daryaganj, New Delhi-110 002

Recent Advances in Embryology
© Reserved
First Edition 1994
ISBN 81-7041-974-3 (Set)

PRINTED IN INDIA

Published by J.L. Kumar for Anmol Publications Pvt. Ltd., New Delhi
and Printed at Efficient Offset Printers, Delhi.

Preface

Embryology is a branch of biology which has a most immediate bearing on the problem of life. It has become an immense field whose confines are difficult to delineate. The focus has been and remain on the embryos, on the gradual emergence of form and structure from what appears to be a very modest beginning. The essence of embryonic development is change transition from one stage to another. Embryo is a fleeting stage, a continum along the axis of time. Embryology is a bank division of biology. It stems the anatomy, the histology and the physiology of adult. To understand it well is to aid in the comprehension of the other biological disciplines.

Though, the authors have taken special care to present a current account, yet they are fully aware about their limitations, and the readers may come across the mistakes of various types. For all types of mistakes they extend their due apology.

The authors have freely consulted other standard books and research papers while preparing the manuscript, so the authors claim no originality of the work.

The authors express their thanks to their friends and colleagues whose continuous inspirations have invited them to bring out this text.

The authors are also thankful to Shri J.L Kumar of M/s Anmol Publications Pvt. Ltd. for bringing out a handsome print of this set in a comparatively short time.

Constructive criticisms and suggestions for improvement of the book will thankfully acknowledged.

Authors

Contents

Reproduction

The Male Reproductive Organs

The primary male organs consist of paired glands called the testes, and these produce the male gametes or spermatozoa. The testes develop in the abdominal cavity adjacent to the adrenal glands but in most mammals they migrate downwards through the body cavity into a special fold of skin called the scrotum, or scrotal sacs. The reason for this is that spermatogenesis will not take place at body temperature, and

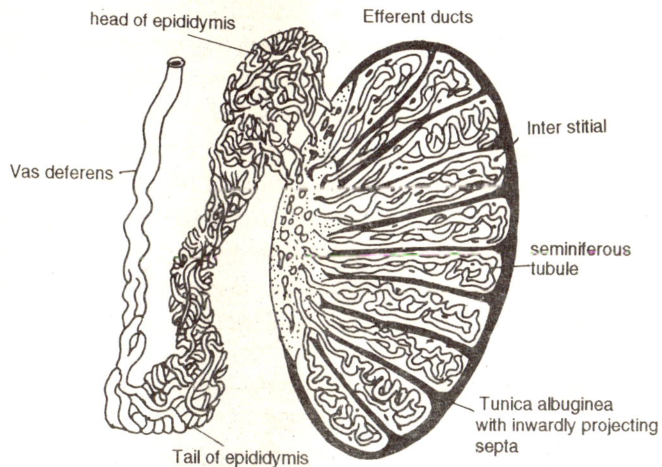

Fig. 1.1.   Diagram of a vertical section of the testis.

within the thin walled scrotum the temperature is several degrees below that of the body cavity. If for some reason the descent does not occur then the animal is infertile. In amphibia, reptiles and in some mammals e.g. the elephant, the testes remain within the abdominal cavity. The testis is guided in its descent by a long cord, the gubernaculum, which extends from the lower pole of the testis to the scrotum, and the gubernaculum anchors the mature testis in the scrotum.

The secondary sexual organs include a variety of ducts and glands which convey the spermatozoa in special secretions to the exterior of the body, so that they can deposited within the female; the secondary sex organs also include those characters of the male which distinguish it from the female e.g. the long mane of the male lion. These secondary sex organs are developed and maintained by means of sex hormones produced within the testis itself. Male sex hormones are called androgens.

Structure of the testis
The testis is a tubular gland surrounded by a fibrous capsule the tunica albuginea. It is divided into several hundred compartments by means of fibrous tissue septa and each compartment contains several tubules, called seminiferous tubules. Each tubule is about 50 cms. long in man and is coiled upon itself, hence the name convoluted seminiferous tubules. All the tubules drain into one border of the testis, into larger collecting tubules which are coiled together in a mass called the epididymis which is applied to the surface of the testis. Between the seminiferous tubules inside the testis, there is connective tissue containing blood vessels and the glandular cells which are called interstitial cells or Leydig's cells and are responsible for the production of the male sex hormone.

Structure of the Tubule
In the adult the seminiferous tubule consists of a basement membrane lined by the seminiferous epithelium; this seminiferous epithelium consists of two types of cell. First the germ cells themselves, secondly the cells of Sertoli which support and nourish the germ cells. The youngest germ cells are those

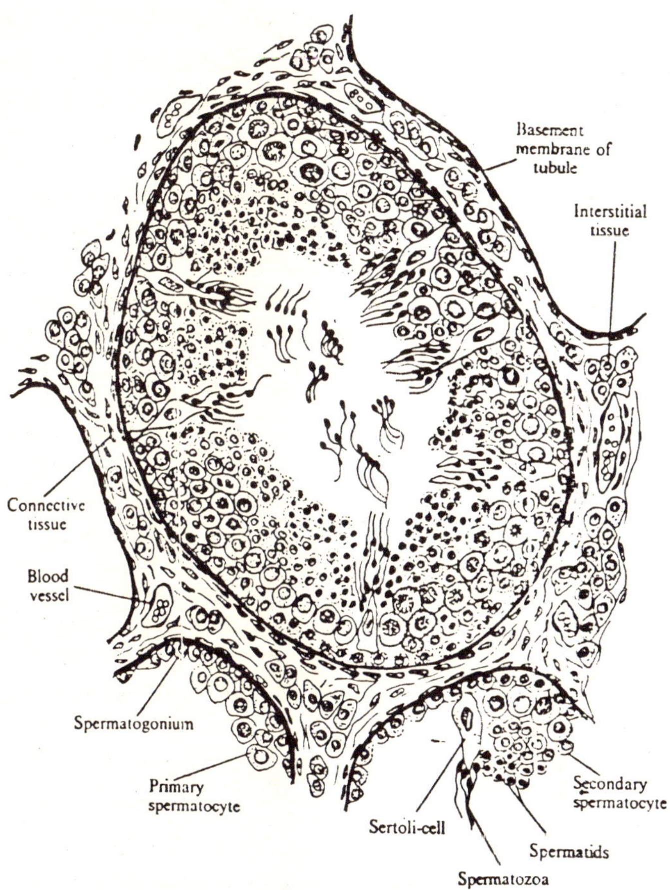

Fig. 1.2.   Transverse section of a seminiferous tubule seen at high magnification showing the seminiferous epithelium and the stages of spermatogenesis.

lying close to the wall of the tubule and these divide to produce cells which pass nearer to the lumen of the tubule where the mature spermatozoa occur. The details of this transformation are described in more detail. The Sertoli cells are slender pillar like cells attached at their base to the basement membrane. At from the lumen of the seminiferous tubules into the larger collecting tubules of the epididymis, whose walls bear a ciliated epithelium which moves the spermatozoa into the vas deferens.

Secondary sex organs

The spermatozoa are conveyed by the vasa deferentia to the base of the bladder where they can pass into the urethra. In structure the vas deferens is a hollow muscular tube, which, by means of muscular contraction, rapidly transports the spermatozoa before they are deposited in the female. Opening

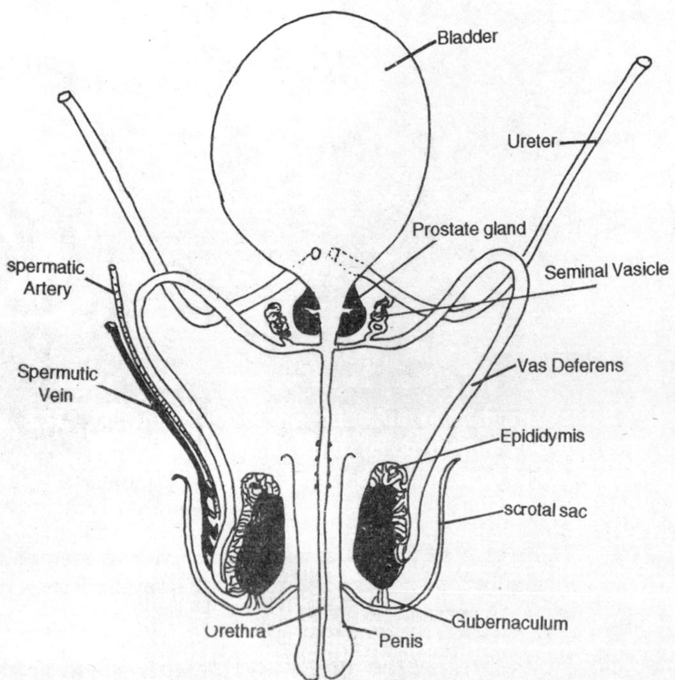

Fig. 1.3.   Diagram of the male reproductive organs.

into the vas deferens before it joins the bladder neck is the duct of the seminal vesicle; the seminal vesicles were once thought to store sperm but their function is to produce a thick secretion to provide the bulk of the fluid in which the sperms are transported. Surrounding the base of the bladder where the vasa deferentia open into the urethra is another gland, the prostate gland, whose secretions are discharged into the urethra together with the spermatozoa and the secretions from the seminal vesicles. The prostate gland produces a thin alkaline secretion; this helps to neutralize any acid urine remaining in the urethra and also to neutralize some of the acid secretions of the female vagina after the sperms have been placed into the female.

These various secretions, spermatozoa and the products of the seminal vesicles and prostate gland are together called semen; the semen is discharged through the urethra which passes through the penis. The penis, through which both urine and semen pass, contains in its walls, sponge-like systems of blood spaces which can become filled with blood so making the penis a more rigid organ so that the semen can be deposited within the female.

THE FEMALE REPRODUCTIVE ORGANS

The primary sex organs in the female consist of paired ovaries situated within the abdominal cavity. The germ cells are liberated from the surface of the ovary into the peritoneal cavity from whence they pass into the secondary sex organs— the Fallopian tubes, the uterus and vagina, leading to the exterior of the body. Paired secretory organs, the mammary glands, are included in the secondary sexual organs of the female, and serve to nourish the newly-born mammal.

Structure of the ovary

The ovary is attached to the wall of the body cavity by a fold of peritoneum. The free surface of the ovary bulges into the peritoneal cavity into which the germ cells are liberated. The ovary is studded with follicles in various stages of development, containing the germ cells. When the follicles are ripe they come to the surface of the ovary where they rupture.

The free surface of the ovary is covered by a thin layer of

germinal epithelium from which the germ cells arise in the embryonic period. In some species of mammal it seems that even in the adult the germinal epithelium can give rise to successive crops of new germ cells which pass inwards to mature in the tissues of the ovary. The timing of the successive crops of new germ cells coincides with the rupture of mature Graafian follicles in which follicular fluid rich in the hormone oestradiol pours over the surface of the ovary. This hormone has been called a 'mitogenic' hormone because of its effect on the germinal epithelium in stimulating cell division and the formation of new crops of germ cells. Beneath the germinal epithelium is a layer of dense connective tissue, the tunica abluginea. Beneath the tunica albuginea the thicker outer part of the ovary, or cortex, contains the follicles in various stages of development. The central part of the ovary or medulla contains a loose connective tissue containing masses of blood vessels.

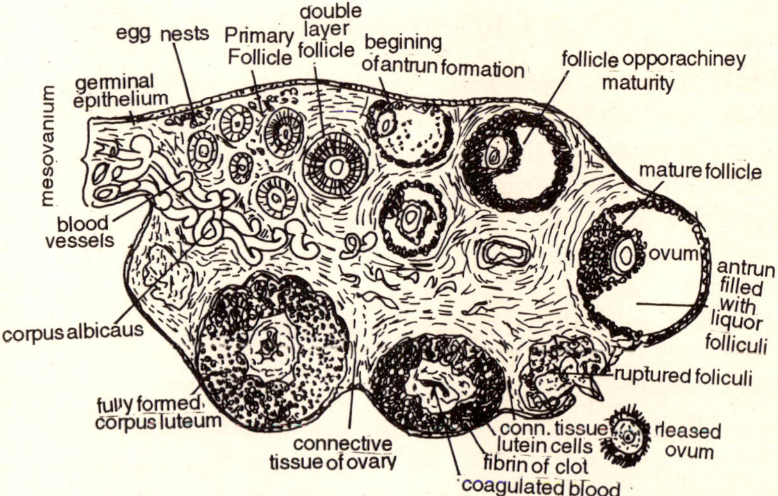

Fig. 1.4.   Mammalian Ovary

The interstitial connective tissue of the ovarian cortex consists of connective tissue fibres and various types of cells. Some of these cells, large polyhedral 'epithelioid' cells are given the name interstitial cells. The number of these cells varies throughout the life of the female mammal. In some mammals with large litters e.g. rodents, there may be enormous numbers of these cells. They may arise from the walls of degenerating Graafian follicles.

Not all of the follicles in the ovary undergo the course of the development into mature Graafian follicles as described. Very many of them undergo a degenerative change called atresia in which there is a hypertrophy of the cells forming the wall of the follicle together with a degeneration of the ovum. The interstitial cells and the cells of the atretic follicles are considered to have an endocrine function.

The oviduct or Fallopian tube

The oviducts are muscular tubes which serve to convey the germ cells from the ovaries to the uterus. The outer end of the tube, nearest to the ovary is expanded, and its edge is split up into fringes, the fibriae, which are closely applied to the surface of the ovary. The lumen of the oviduct is lined by a secretory mucous membrane, in which there are many ciliated epithelial cells. The germ cells are conveyed down the tube to the uterus by means of peristaltic movements of the tube itself and by the effect of the ciliated epithelium.

The uterus

The uterus is a thick-walled muscular structure within which the embryo develops. Its wall has three layers, an outer serous coat, a thick middle coat consisting of interlaced smooth muscle fibres (the myometrium) and an inner vascular mucous layer, the endometrium.

In primitive mammals there are two uteri, each opening into the vagina and this is called the duplex condition and is found in marsupials, many rodents (e.g. rats, mice, rabbit) and bats. In most mammals the distal end of the two uteri is fused to give a biconuate uterus. In higher primates, including man, the two uteri are completely fused together to give a single organ, the uterus simplex.

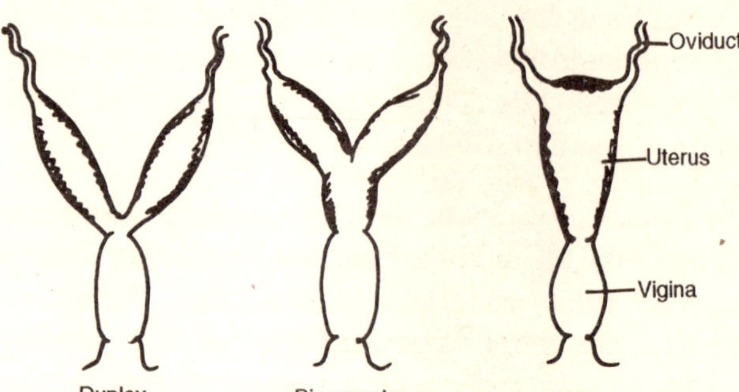

Fig. 1.5.   Types of mammalian uteri.

The vagina is a distensible tube lined by squamous epithelial cells which connects the uterus to the outside world.

GAMETOGENESIS

The process of formation of gametes is called gametogenesis; the formation of eggs is called oogenesis and the formation of sperms is called spermatogenesis. Gametogenesis may be conveniently divided into three stages. The first stage is one in which the cells of the germinal epithelium divide and is called the stage of multiplication. The second stage is one of growth when each of the tiny cells produced in the multiplication stage grows to a larger size. The cell at the end of this stage is called the primary oocyte in the case of the female, and the primary spermatocyte in the case of the male. The third stage in gametogenesis is a period of maturation; during this period very important changes occur in the nucleus with the result that the chromosome number is reduced from the diploid to the haploid number. This reduction in chromosome number takes place in the so-called reduction division of a special-kind of cell division called meiosis. When the primary oocyte divides by the reduction division it does so unequally; the nucleus divides into two equal parts, but almost all of the cytoplasm goes with one half of the nucleus, whereas the remaining half of the nucleus has very little cytoplasm. The latter is called the first polar body. The nucleus with most of the cytoplasm is now called the secondary oocyte and it contains only the haploid-number of chromosomes. IN mammals it is at this stage that the female gamete is released from the ovary; and before this gamete can be considered as fully mature another division of the nucleus has to take place, and this again is an unequal division resulting in the production of a second polar body. The production of this second polar body takes place, in a mammal, when the egg is fertilized by the sperm. We have seen that the development of the female gamete involves three stages, the first of multiplication, the second of growth ending in the formation of the primary oocyte, and the third of maturation involving the production of the secondary oocyte with polar bodies.

The first stage of gametogenesis is complete in the female embryo by the end of intra-uterine life, and she is born with all the oogonia already formed within the ovary. The second stage of growth of the oogonia continues throughout the life of the mammal and we will now look at this growth phase in more detail.

Development of the Graafian follicle

In the cortex of the ovary of the mature mammal are many small clusters of cells called the primary follicles, which have been formed during the embryonic period from invaginations of the germinal epithelium, called sex cords. The primary follicles are very small and there are about 400,000 in the human female at maturity. At birth the first phase of oogenesis, the phase of multiplication has already started producing small collections of germinal cells, called primary follicles. One of the cells in the primary follicle is larger than the rest and is the oogonium, whilst the smaller surrounding cells are called follicular cells. In the second phase of oogenesis, the growth phase, the primary follicle develops and changes occur, in the oogonium as it becomes the primary oocyte, in the follicular cells and also in the connective tissue which surrounds the follicles. The oogonium enlarges, its nucleus gets bigger, and a few yolk granules begin to appear in its cytoplasm. At this stage a well defined shining layer appears around the surface of the oogonium called the zona pellucida. In the primary follicle the oogonium was surrounded by a simple columnar epithelium but as the follicle grows the follicular cells multiply to produce an epithelium which is several layers in thickness. The cells of this epithelium secrete a follicular fluid which accumulates in spaces which begin to appear between the cells.

The follicle by this stage in the human female is about 2 mm. in diameter and is now called the Graafian follicle. The follicle increases in size with the accumulation of more fluid within it and the oogonium is pushed to one side of the follicle where it is attached to the wall of the follicle in a group of columnar cells called the discus proligerus. The cavity of the follicle is lined by a few layers of columnar cells called the membrana granulosa. The connective tissue surrounding the Graafian follicle has become organized into a membrane called the theca (consisting of two layers, the theca interna and externa). Eventually the Graafian follicle may reach a size of 10 mm. in diameter in the human female and bulges from the surfaces of the ovary; by this time the oogonium has grown to its full extent and is called the primary oocyte. The fluid within the follicle is formed at a faster rate than the follicle wall grows and the follicle eventually ruptures.

By the time the follicle has ruptured the primary oocyte has undergone the meiotic division producing the first polar body and is now called the secondary oocyte. When the secondary oocyte is released it is surrounded by a few columnar cells which form the corona radiata, which may have a nutritive function similar to that of Sertoli cells in the male. When the Graafian follicle has ruptured and liberated the oocyte it collapses and the hole left by the departing oocyte becomes plugged with a blood clot. There is now a multiplication of the remaining cells of the follicle, the granulosa and theca cells. The cells enlarge and develop deposits of a yellow pigment called lutein.

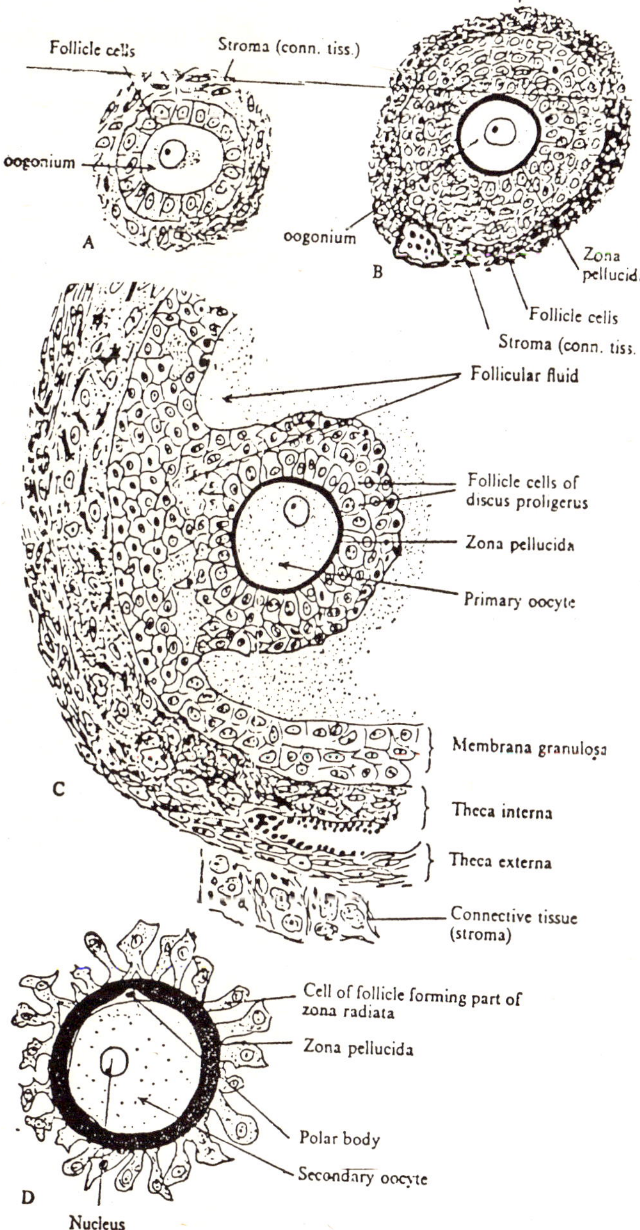

Fig. 1.6. Development of the Graafian follicle. A. Young follicle. B. Older follicle. C. Part of a mature Graafian follicle. D. The maturing ovum liberated by rupture of the Graafian follicle.

The whole structure so produced is called the corpus luteum, a solid ball of yellow pigment cells, which produces hormones which prepare the uterus to receive the fertilized oocyte.

Spermatogenesis, the formation of spermatozoa. Unlike the female, where the germinal epithelium forms the outermost layer of the ovary, in the male the germinal epithelium lines the walls of the seminiferous tubules. The cells nearest to the wall of the tubule, the spermatogonia, are the most primitive, undifferentiated cells of the tubule and they give rise to the other cells by mitotic division. As in oogenesis there are three stages in spermatogenesis.

The first stage of spermatogenesis is that in which the spermatogonia multiply by mitotic division. Some of the products of these divisions pass inwards nearer the lumen of the tubule where they enter the second phase of spermatogenesis, the growth phase. During this phase the spermatogonia become larger, producing the primary spermatocytes. In the third phase each primary spermatocyte undergoes a reduction division producing two secondary spermatocytes containing the haploid number of chromosomes. Each secondary spermatocyte then divides to produce two spermatids. The spermatids do not undergo divis on but a series of changes occur which transform the spermatid into the mature sperm. During this maturation of the spermatids they are attached to the cells of Sertoli.

The Mature Sperm (human). The mature sperm consists of a head, middle piece and tail. The head consists of the condensed nucleus of the spermatid and is a flattened ovoid structure about 5 microns long. The head is capped by a sheath of material called the head cap. In the middle piece of the sperm there is a centriole from which arises a long axial filament which passes through the middle piece and the tail. Surrounding the axial filament in the middle piece is wound a sheath of mitochondrial material, the mitochondrial sheath, which is probably concerned in the respiration of the sperm. In the tail the axial filament is covered by a sheath.

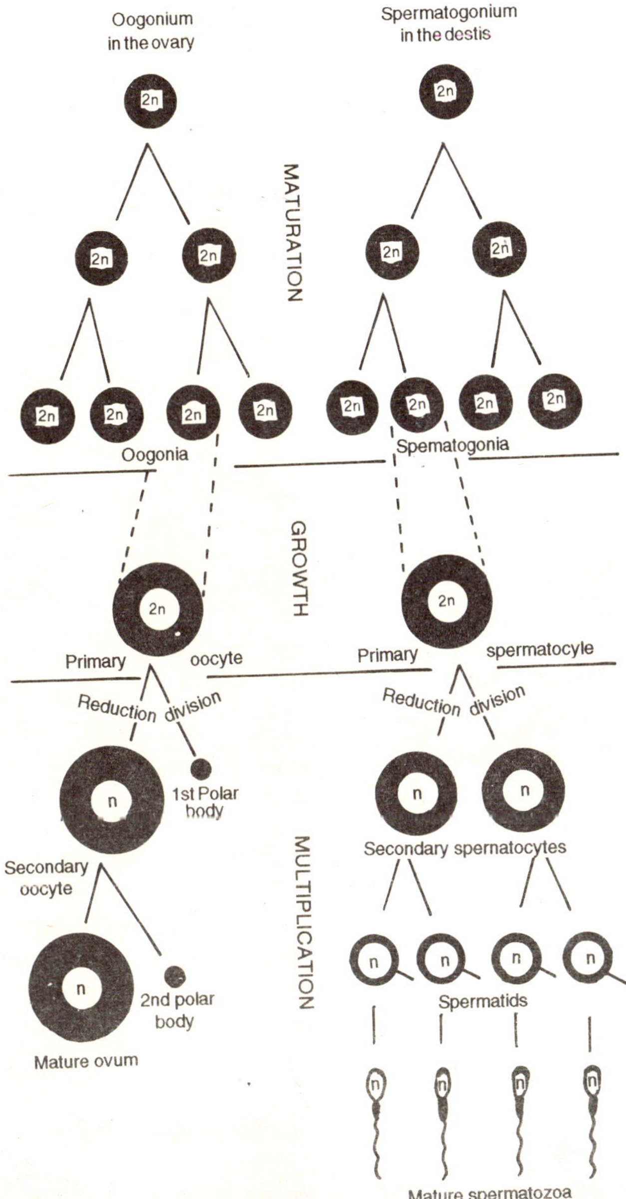

Fig. 1.7. The phases of gametogenesis.

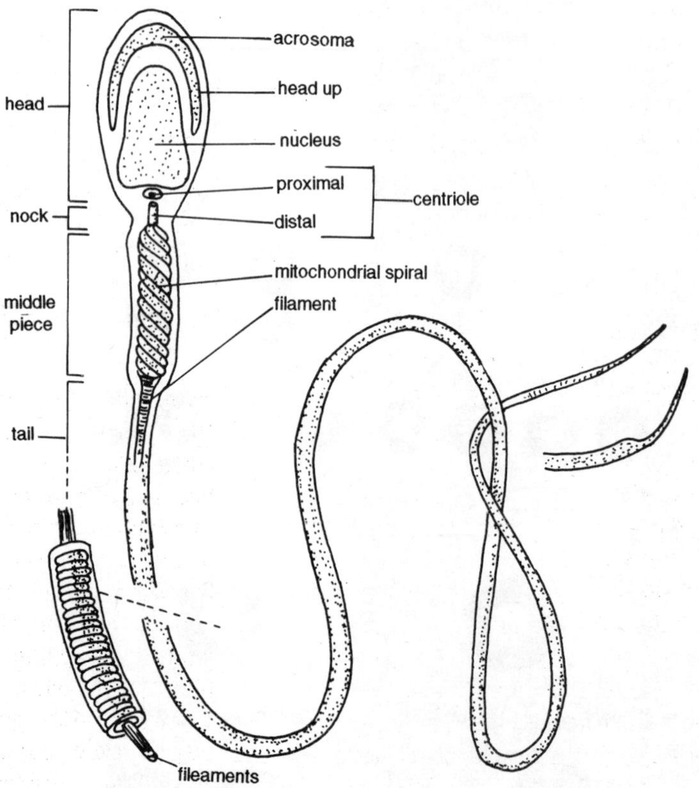

Fig. 1.8.   Structure of a mature sperm.

The spermatozoa remain inactive until they pass from the testis. During their passage from the testis to the penis they are activated by the secretions of the accessory glands. The sperm is capable then of active swimming during which S-shaped waves pass along the tail. The energy for this is derived from the anaerobic breakdown of fructose which is present in the prostate secretions.

THE PHYSIOLOGY OF REPRODUCTION

Hormones and reproduction in the male

In the sexually immature mammal the primary and

secondary sex organs are small and undeveloped. The growth and development of the sex organs is dependents upon the activity of the pituitary gland. The pituitary exerts its effect by the production of hormones called gonadotrophic hormones because of their growth effects upon the gonads. The gonadotrophic hormones are complex protein substances which have been isolated in a relatively pure state; their chemical structure is not yet elucidated. There are almost certainly two gonadotrophic hormone. One of these hormones is called the follicle stimulating hormone (or F.S.H.) because of its effect in the female in stimulating the growth of the follicles in the ovary. F.S.H. stimulates the growth of the seminiferous tubules of the testis and stimulates the activity of the germinal epithelium. The other pituitary hormone is called the luteinizing hormone (L.H.) or interstitial cell stimulating hormone. L.H. stimulates the growth and secretory activity of the interstitial or Leydig cells of the testis. These cells produce certain steroid hormones called sex hormones, because of their effect on the sex organs.

The most important sex hormone in the male is called testosterone. However, there are also female sex hormones or oestrogens produced in the male. The effect of testosterone is

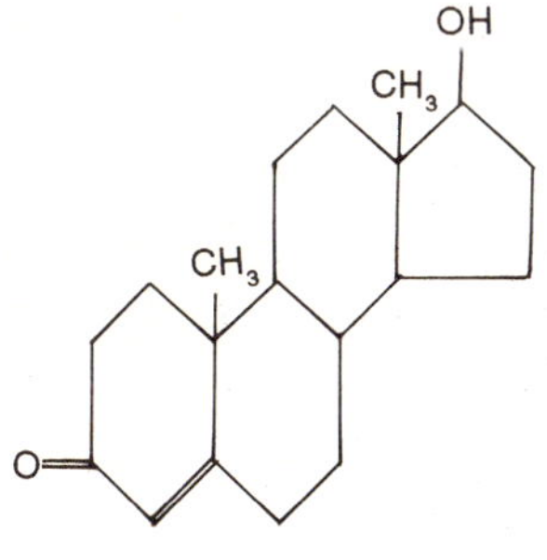

Fig. 1.9. Formula of testosterone.

to promote the growth of the various sex organs - the vas deferens, seminal vesicles, prostate gland, penis etc. It also promotes the development of those other secondary sex characters which vary from one mammalian species to another; in man these include the growth of the beard, enlargement of the larynx with the development of the deeper voice, the male distribution of hair on the body, and the greater development of muscle. The effect on muscle growth is due to the fact that testosterone is what is called a protein anabolic hormone, that is it promotes the retention and incorporation of protein in the tissues.

In the immature animal testosterone will promote the precocious development of the sex organs, and in those species of mammals in which the testes do not descend into the scrotum until sexual maturity it stimulates the descent of the testis. In animals which have been castrated the sex organs gradually atrophy and the administration of testosterone can reverse these charges.

Testosterone has also a direct effect on the pituitary gland and a certain level of testosterone in the blood will inhibit the pituitary from producing gonadotrophic hormones; when the pituitary production of leuteinizing hormone falls then the Leydig cells of the testis stop producing testosterone and the blood level of testosterone falls. With the falling producing of testosterone by the Leydig cells the pituitary gland is now released from the inhibitory effect of testosterone and it begins to produce the gonadotrophic hormones again. By this feed-back mechanism. the secretion of testosterone is controlled.

This effect of testosterone upon the pituitary gland also explains the varying results that experimenters have had in the administration of testosterone to animals. Some workers have found that in some animals testosterone will cause a stimulation of the germinal epithelium of the testis and the production of spermatozoa; because of the appearance of large numbers of dividing cells in the germinal epithelium the hormone has been called 'mitogenic', that is one which stimulates mitotic division within the cells. In large doses however testosterone can depress the growth of the testis, presumably because it inhibits the anterior pituitary gland from producing the gonadotrophic hormones, by the feed-back mechanism.

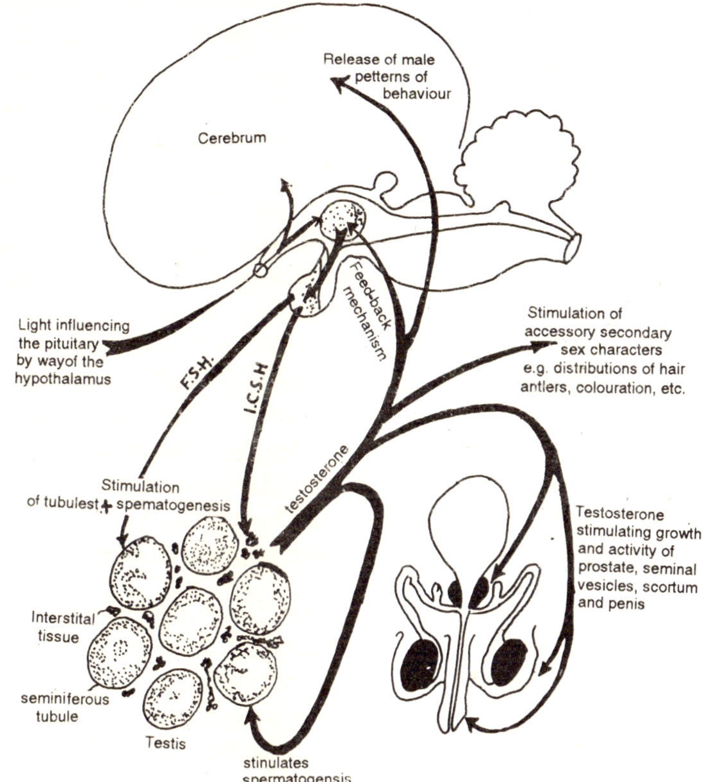

Fig. 1.10. Diagrammatic summary of hormonal control of the sex organs in the male mammal.

Testosterone also appears to be responsible for behaviour changes in animals, and administration of the hormone to sexually immature males results in the appearance of male breeding behaviour. In some species even the female will show a masculine pattern of breeding or sexual behaviour when given testosterone. The aggressiveness of many male mammals can be promoted by giving testosterone. In higher primates, including man, sexual behaviour cannot be controlled so simply, and cultural factors play a much more important role in the determination of the direction of sexual impulses.

HORMONES AND REPRODUCTION IN THE FEMALE

In the mature female mammal sexual activity tends to be an intermittent phenomenon during the breeding season, which is the period when mating can occur. The breeding season may consist of several weeks, or months of the year, and there may be more than one breeding season in the year. During the breeding season itself sexual activity is a cyclic phenomenon, with periods of sexual activity or oestrus ('heat') alternating with periods of sexual inactivity. These cycles of activity stop of course as soon as a pregnancy is started. In some animals, including primates and rodents, these oestrous cycles are continuous throughout the sexual life of the animal and are not restricted to breeding seasons; these animals are said to experience poly-oestrus. In man, because of the unusual feature of menstruation of the oestrous cycles are known as menstrual cycles; but the menstrual cycle is fundamentally similar to the oestrous cycle of other mammals. The variations in breeding activity are best illustrated by referring to particular examples:

Horse. The mating season of the horse is between March and August, although some breeds will mate in the autumn and winter in England. During the mating season, if pregnancy does not occur then there is a regular occurrence of oestrus, each period of oestrus lasting about 20 days. The horse is said to be seasonally poly-oestrous.

Dog. The domestic dog has two breeding seasons in the year, in late Winter and early spring and in the Autumn. During each season there is only one period of oestrous, and the dog is said to be monoestrous.

Golden Hamster. This rodent is polyoestrous, coming into heat at all times of the year, the oestrous cycles recurring every ten days, each cycle lasting about four days.

Roe Deer. Like the dog the Roe Deer is monoestrous. The breeding season is in July and August during which there is only one period of oestrous.

The oestrous cycle. During the oestrous cycle there are wide-spread changes in the structure and behaviour of the female; all the changes that occur are under the control of hormones, produced mainly by the anterior pituitary gland and the ovary. The aim of these changes is to mature an ovum and prepare the uterus to receive and nurture the ovum if it is fertilized by a sperm.

In the early phase of an oestrous cycle the ovary is activated by the secretion of follicle-stimulating hormone (F.S.H.) from the anterior pituitary gland. The ovary responds by the progressive growth of one or more Graffian follicles, depending upon the species of mammal. Small amounts of luteinizing hormone (L.H. is said to have a synergistic effect. The ripening Graafian follicles secrete a steroid sex hormone called oestradiol which has widespread effects upon the secondary sex organs, and also has an effect on the secretory activity of the pituitary gland. Oestradiol is only one of a number of substances isolated from the ovary, blood and urine of the female mammal, all of which have some effect on the sex organs; the name oestrogens have been used to describe this class of substances, although oestradiol is the most potent naturally occurring oestrogen.

Oestradiol stimulates the growth of the uterus; the myometrium increases in thickness because of growth of its individual cells, its vascularity increases and the endometrium thickens, becoming more vascular as its secretory glands grow in length. The Fallopian tubes and vagina are also stimulated. In the sexually inactive phase or anoestrus, the epithelial lining of the vagina is thin, only one or two cells thick, but after stimulation by oestrogen the epithelium thickens and cornifies, and flattened cornified squames appear in the vaginal secretions. The mammary glands also increase in size under the influence of oestrogen, which stimulates the growth of he duct system.

Fig. 1.11. Diagram illustrating the changes in various structures

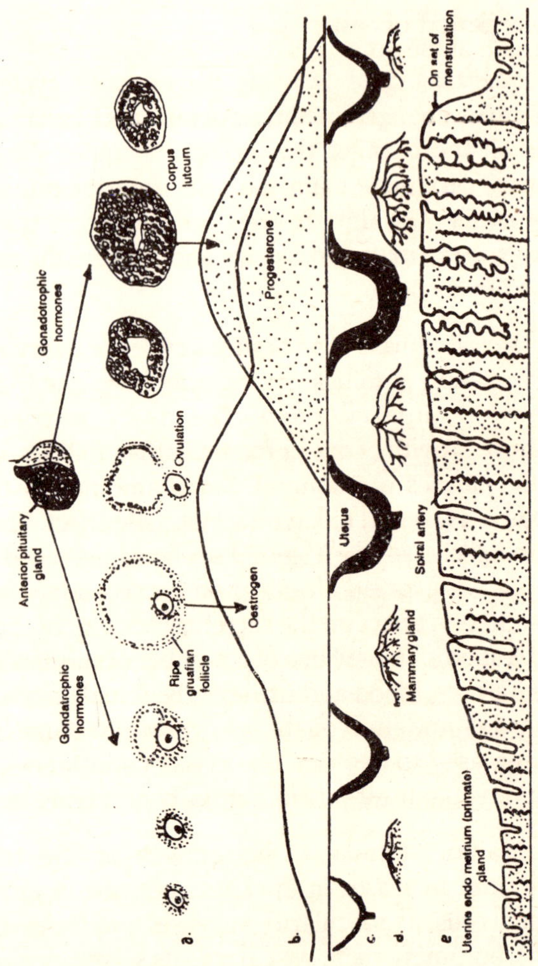

throughout the oestrous (or menstrual cycle). a Shows the development of the Graafian follicle, ovulation and the corpus luteum. b. Shows the blood levels of oestrogen and progesterone. c. Shows the progressive development of the uterus. d. Shows the development of milk forming tissue in the mammary glands. Early there is a growth of the duct system, and later in the progesterone phase the development of glandular elements. e. Shows the development of the primate endometrium with its peculiar feature of menstruation.

The rising level of oestradiol in the blood has important

effects upon the pituitary gland. It inhabits the formation of F.S.H. and at the same time stimulates the production of further amounts of L.H. which brings about ovulation and the formation of the corpus luteum. This first phase of the oestrous cycle, as described above, is called the *follicular phase,* because of the growth of the Graafian follicles, or the oestrogen phase, because of the importance of this hormone in this part of the cycle. In the uterus the follicular phase is associated with growth, both of the myometrium and endometrium, and this phase of uterine change is called the proliferative phase. The follicular phase finishes at ovulation when the egg escapes from the ovary and the remains of the Graffian follicles grow to produce special glandular structures under the influence of the pituitary luteinizing hormone (L.H.), called *corpora lutea.*

We now enter the *luteal phase* of the oestrous cycle. This is often called the progesterone phase, because of the importance of this hormone at this time. The corpora lutea are large yellow pigmented bodies studded in the ovarian cortex. They are formed by the proliferation and growth of cells in the wall of the ruptured Graafian follicle. Under the influence of another pituitary hormone called luteotrophin or lactogenic hormone the corpora lutea secrete a hormone called progesterone, in addition to small amounts of oestradiol. Pugesterone produces further changes in the endometrium of the uterus the increase in thickness and vascularity of the uterus progresses as the glands of the endometrium become tortuous and begin to per-secretions into the cavity of the uterus. Because of these glandular changes this phase of uterine activity is called the *secretory phase.* Progesterone also has effects on the mammary glands which haw already been primed by the effects of oestradiol; now, glandular element begin to appear around the ends of the duct systems of the mamma glands.

We have seen that the time of maximal oestradiol activity is at the time of ovulation and after this time the level of oestrogen production gradually falls, as progesterone comes to play a more important par in the cycle. It is at the time of maximal oestradiol production that the female mammal is most willing to receive the male and this is the true period of 'heat.' Mating and fertilization usually occur about this time. If fertilization of an ovum does not occur then corpora lutea gradually disintegrate and retrogressive changes occur in the secondary sex organs. The falling level of blood oestrogen in the luteal phase the cycle, together with some inhibitory effect of progesterone on the anterior pituitary gland, are responsible for a gradual decline in the production of L.H. and therefore the corpora lutea degenerate.

Menstrual cycle

In the human female when the corpus luteum disintegrates at the end of the luteal phase of the cycle, there is a complete breakdown of the hypertrophied endometrium; blood and broken-down tissues are discharged from the vagina and this constitutes menstruation. In the human the menstrual cycle lasts about 28 days. In the first half of the cycle there is the follicular phase, culminating in ovulation at about the fourteenth day. In the second half of the cycle, the luteal phase, the corpus luteum is formed and the endometrium enters the secretory phase, and in the absence of fertilization the luteal phase is ended by the appearance of the menstrual flow. Following menstruation only fragments of endometrium are remaining and these lie in crypts in the myometrium. From these fragments the entire endometrium is reformed during the proliferative phase of the next cycle. The primate endometrium undergoes this almost complete breakdown because of a peculiarity in its blood supply. When the supply of progesterone is waning as the corpus luteum degenerates at the end of the luteal phase, certain spiral arteries of the endometrium go into such intense spasm that the tissues supplied by them die and undergo degenerative changes. The whole of the dead endometrium is then sloughed off from the uterine wall together with blood.

The oestrous cycle and pregnancy

If during the luteal phase of the oestrous cycle an ovum is fertilized and settles in the uterine cavity then the retrogressive changes in the secondary sex organs do not occur, nor does the corpus luteum degenerate. We have seen that in the absence of pregnancy the corpus luteum degenerates; it does this because of the decline in the production of pituitary gonadotrophic hormones. As soon as there is a union established between the fertilized ovum and the uterine wall increasing amounts of gonadotrophins appear in the maternal blood and this maintains the structure and function of the corpus luteum. The gonadotrophic hormone is produced by the placenta, the organ which unites the mother and foetus, and through which is receives its nourishment. The placenta also produces large amounts of oestrogen and progesterone and gradually replaces large amounts of oestrogen and progesterone and gradually replaces the corpus luteum as a source of these hormones, so that later in the pregnancy one can remove both ovaries from the female mammal without disturbing the pregnancy. The function of the large amounts of sex hormones produced by the placenta is to promote further growth of the uterus, vagina and mammary glands.

During pregnancy, growth of further Graafian follicles and ovulation is prevented because the large amounts of oestrogen produced by the placenta inhibit the anterior pituitary gland from producing follicle-stimulating hormone.

Reproduction and the environment

The reasons why animals tend to breed at relatively restricted times of the year have been studied and classified under two headings, internal physiological mechanisms on the one hand, and external environmental factors on the other. An internal physiological 'clock' cannot be the sole factor in determining periodic breeding; seasonal breeding has an adaptive significance in that young are produced at favourable times of the year, and obviously a rigid internal mechanism would, through the course of time, fail to adapt the animal to changing climatic conditions.

It appears that animals have become adapted to respond to certain environmental factors which herald the oncoming favourable season. In temperate latitudes many animals have as it were harnessed their breeding behaviour to the length of day and respond to an increase in day length by development of the sex organs, so that the young will develop in a favourable season with warmth and an adequate food supply. The first study of the effect of light on sexual cycles was made on the Canadian bunting, a bird which normally breeds in Spring. By giving these birds extra periods of light in the Autumn it was possible to cause development of the testes, even at a time when the temperature animals, reptiles, fish, birds and mammals it has been possible to cause the development of the sex organs out of the breeding season. The light exerts its effect by way of the eyes and connections with the hypothalamus and anterior pituitary gland. Not only is the increase in day length of importance but in some autumn breeding animals the shortening days may also act as the stimulus to sexual development.

When some animals with fixed breeding seasons in temperate latitudes are transferred from the Southern to the Northern Hemisphere, after a However, animals imported from tropical countries tend to continue differences in daylight to which the species inhabiting temperate zones respond. Thus may be due to the fact that owing to the comparative uniformity of conditions in their own countries they have never acquired the capacity to respond to variations in light intensity or duration, characteristic of many animals living under seasonally changing conditions. Thus Java deer when imported to England continue to produce young in late Autumn, as they are believed to do in Java, a condition which is abnormal for any deer inhabiting temperate countries. But even in tropical climates animals may have special breeding seasons; if reproduction were continued throughout the year the competition for food for the young might be too intense for survival of the species. A further advantage of seasonal breeding may be the synchronization of the male and female sexual cycles by the fact that they are both adapted to the same environmental factor. What these environmental factors are in tropical climates in uncertain. That breeding cycles in some tropical animals are harnessed to some environmental factor and not dependent upon an internal rhythm seems certain; a tropical insectivorous bat lives throughout the period of daylight until about ten minutes before sunset in dark and almost thermo-static caves and yet it was found that in one year of observation no pregnancies occurred until a few days at the beginning of September. It seems impossible that such synchronization of the sexual cycles of such a group of bats could be achieved by an internal physiological mechanism.

The homiothermic vertebrates (warm blooded) have become almost independent of the temperature of their surroundings and are able to carry out breeding at any time of the year. They do, however, tend to breed in Spring in temperate latitudes to ensure a favourable environment with adequate food for the young. But even in homiothermic animals temperature may play some role in the timing of breeding behaviour. If Autumn weather is warm and food supply is good, sexual behaviour in the robin and other birds may be pronounced and may even lead to reproduction in a few birds, although they normally only breed in Spring. In this Autumn breeding phase sexual activity is developing hen daylight is decreasing.

We have now seen something of the way in which sexual cycles are controlled by the endocrine organs, and the way in which the endocrine organs, by way of the hypothalamus of the brain, are synchronized with external environmental factors so that the young are cared for under favourable conditions.

THE PLACENTA

The uterus has been preparing to receive a fertilized ovum throughout the oestrous cycle. In the follicular phase of the cycle there is growth both of the myometrium and endometrium and an increase in the blood supply of the uterus. The endometrium is thickened, new blood vessels grow and its glands increase in length. In the luteal phase of the cycle these changes continue and the endometrium takes up secretary character. The glands become tortuous and their epithelium becomes active, secretions passing into the uterine cavity.

The early development of the fertilized egg occurs during its passage through the Fallopian tube, since fertilization is usually achieved high up in the Fallopian tube. The fertilized egg or zygote undergoes divisions to produce a ball of cells called the morula. Then morula than differentiates into an outer layer and an inner cell mass producing the blastula. The cells of the outer layer form the trophoblast or trophoblastic ectoderm. This enters into the formation of the chorion, the outermost, covering of the developing zygote. This outer layer is called *trophoblast* because it enters into the formation of placenta, the organ concerned in nourishing the foetus. The inner cell mass of the blastula contains the cells from which the embryo will develop.

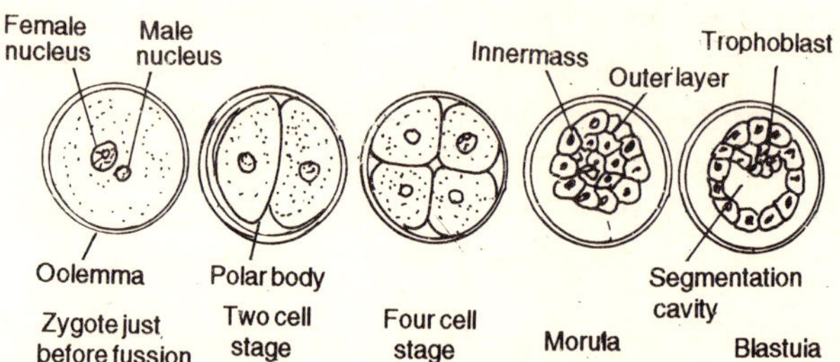

Fig. 1.12.   Diagrams showing the early development of the fertilized egg of the mammal up to the blastula stage.

During its passage down the Fallopian tube the nutrition of the zygote is dependent on the small amounts of food stored in the protoplasm. When the morula reaches the uterine cavity it is bathed in the secretions of the uterine glands, and these secretions probably have some nutritive function. In some species the blastocyst lies in the cavity of the uterus, and in contact, by means of its trophoblast, with the endometrium all over its surface. In some other species the blastocyst becomes attached to the endometrium on one surface, and projects freely into the uterine cavity or its other serface. In other types the blastocyst sinks into the endometrium and becomes surrounded completely by maternal tissues; this type of implantation of the blastocyst is found in man and is called the interstitial type. In order to understand the varied types of placenta found in mammals it is necessary to examine in some detail the formation of the various foetal membrane which take part in the formation of the placenta. There are four foetal membranes concerned in the adaptation of the foetus to life in the uterus, the amnion and chorion, formed from the original embryonic body wall, and the yolk sac and allantois, parts of the original gut of the embryo.

We had left the development of the embryo at the blastula stage, consisting of an inner cell mass and an outer layer, the trophoblast; a two layered vesicle is produced by a growth of endoderm cells around the inner layer of the trophoblast (enclosing the yolk of the egg, when this is present) Only a part of the wall of the blastula is destined to form the embryo and this is called the embryonic area. This embryonic area gradually sinks into the centre of the blastula as it becomes covered over by the amniotic folds. In man the embryo is not covered by the amniotic folds and there is merely a hollowing out of the cells of the embryonic area to produce a cavity, the amnion. The amniotic folds meet and fuse above the embryo and because of the double walled nature of the folds the embryo becomes to be surrounded by two membranes, an outer chorion and an inner amnion. The chorion thus becomes the membrane which is in contact with the endometrium of the uterus, and projections called chorionic villi grow out from its surface to make a more intimate contact with the maternal tissues. The amniotic membrane surrounds a fluid filled cavity, the amniotic cavity, in which the embryo floats.

Fig. 1.13.   Development of the late blastula. Note the embryonic area gradually sinking into the blastula in A, B and C, gradually

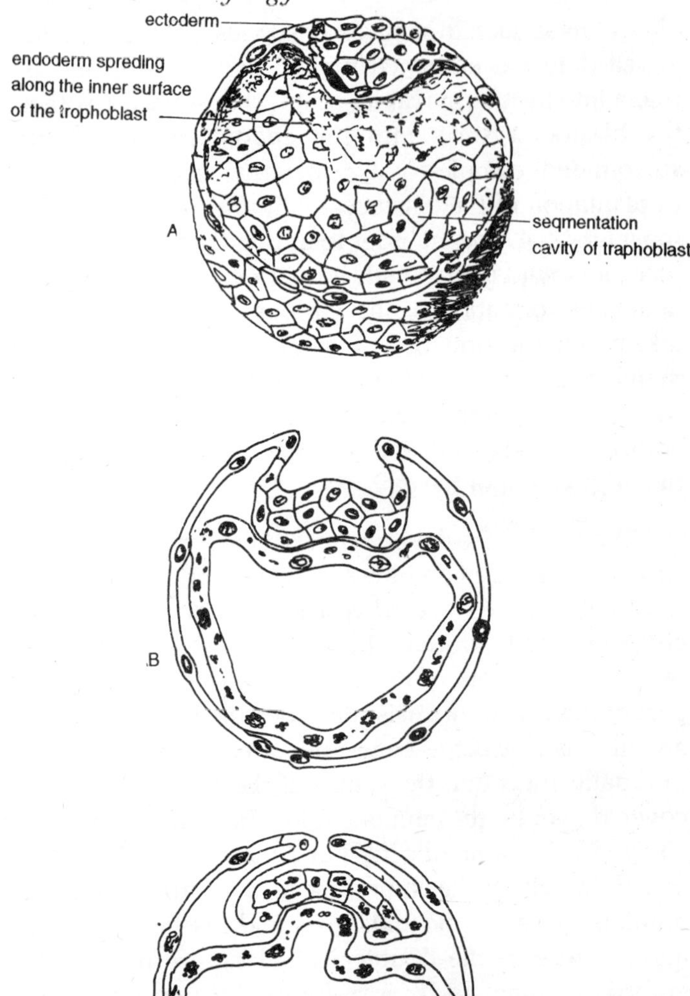

becoming covered by the amniotic folds. The endoderm is shown progressively spreading over the inner surface of the trophoblast.

We have seen above how the endoderm grows round the

inner surface of the trophoblast to produce a two layered vesicle. In the heavily yolked eggs of birds and reptiles this layer encloses the yolk of the egg and is called the yolk sac. The primitive gut of the embryo, a groove on the under surface of the embryonic area, is open into this yolk sac but later as the gut is folded off it is separated from the yolk sac except for a narrow passage, the yolk sac stalk. Later mesoderm grows out from the embryo between ectoderm and endoderm; this process provides a double layer of mesoderm in the amniotic folds and a layer of mesoderm between the ectoderm and endoderm of the yolk sac. It is in this layer of mesoderm that the embryonic blood vessels develop. In the heavily yolked eggs of birds and reptiles the inner layer of the yolk sac becomes vascular in order to absorb the nourishment from within the yolk sac. But in mammals there is very little nourishment within the yolk sac and it is from outside that it has to look for nourishment. Thus the outer layer of the trophoblast becomes highly developed

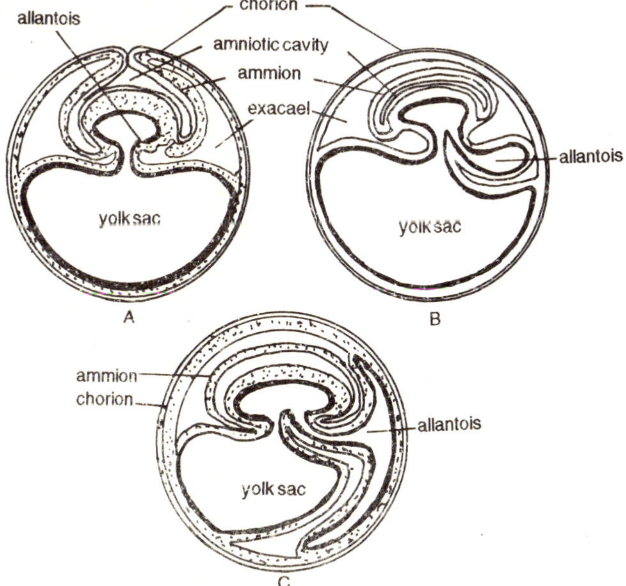

Fig. 1.14. Showing the development of extra embryonic structure.

and supplied with blood vessels from the mesoderm. In primate embryos, including man, the yolk sac is a rudimentary structure only. In Marsupial mammals which have a less developed placenta the yolk sac is a more important organ, and is the major organ for nourishment of the embryo. In rodents the yolk sac becomes partly vascularized; the non vascularized area becomes eroded away so that the cavity of the yolk sac opens into the uterine cavity. The inner wall of the yolk sac is applied directly to the endometrial lining of the uterus, producing what is known as an inverted yolk sac placenta.

In most mammals the chief absorbing surface occurs on a structure called the allantois. This is a blind ended outgrowth of the hind gut carrying mesoderm on its surface as it grows into the extra embryonic coelom. In the eggs of reptiles and birds the allantois functions as a bladder for the storage of excretory products, but it also serves to transport oxygen which diffuses through the shell, through the blood vessels of the allantois to the developing embryo. In mammals the allantois has become progressively more important in the nourishment of the foetus, and blood vessels of the allantois pass into the villi of the chorion which make connection with the lining of the uterus.

The processes which grow out from the trophoblast (i.e. extra-embryonic ectoderm of the chorion) are called trophoblastic or chorionic villi. The villi are connected to the embryonic blood vessels through the yolk sack or allantois, depending upon the species of mammal. Not all of the surface of the trophoblast is covered by villi and the arrangement of the villi varies from species to species. In the horse and the pig the villi are diffusely arranged over the surface of the trophoblast producing a diffuse placenta. In the sheep and cow the villi are localized in patches called cotyledons producing the cotyledonary placenta. In carnivores the villi are restricted to a band encircling the embryo producing a zonary placenta. In man (also rodents and insectivores) the villi are at first scattered over the whole trophoblast and later become limited to a disc shaped area producing a discoidal placenta.

Types of placental union with the uterine wall

There is a great variation in the intimacy of contact between the foetal and maternal tissues in the placenta of mammalian species. On the foetal side of the placenta there are three layers of tissue, the chorion, the mesenchymal tissues and the endothelium of the 'capillary vessels. On the maternal side there are uterine secretions, the endometrial epithelium, connective tissues and the endothelium of the blood vessels. There are thus at least six possible layers of tissue separating the foetal and maternal blood. In the diffuse placenta found in the pig and horse these six layers which separate the maternal and foetal blood streams persists, producing what is called an epithelio-chorial placenta.

This description of the placental structure of the pig and the horse give a rather exaggerated picture of the separation of maternal and foetal blood. The foetal cells of the chorion are low cuboidal in shape and foetal capillaries ramify close to the chorion and may actually penetrate between the chorionic cells to form intra-epithelial plexuses. The capillaries at places displace the cuboidal cells and compress their cytoplasm into thinned out plates. Thus over a large part of the placental surface only a very thin plate of epithelial cytoplasm separates the foetal and maternal blood. In other mammals the barrier between maternal and foetal blood is reduced by the erosion of the layers of the uterine wall by the trophoblastic villi. In the syndesmochorial placenta of cattle and sheep the endometrial epithelium is eroded and the epithelium of the trophoblastic villi is in contact with the uterine connective tissue and in the endotheliochorial placenta of the cat and dog the uterine tissues are further eroded and the epithelium of the trophoblastic villi is separated from the maternal blood only by the endothelium of the maternal capillaries. In insectivores, rodents and man the maternal capillaries are eroded so that the trophoblastic

Fig. 1.15. Diagrammatic representation of different placental types.

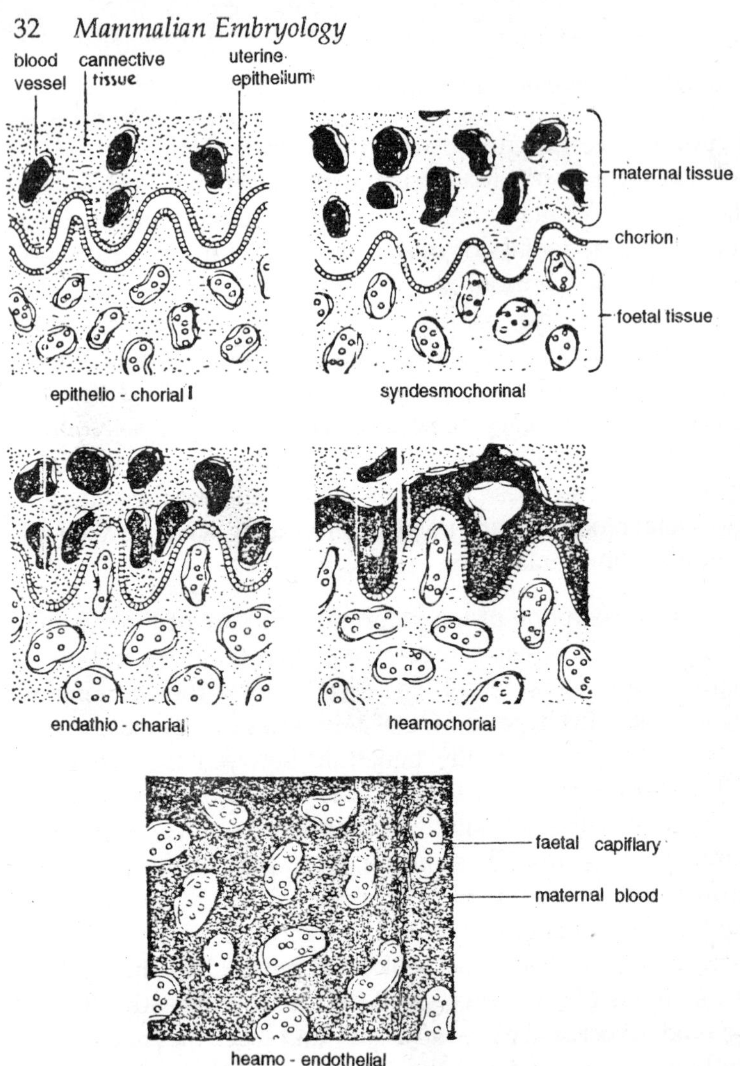

Maternal blood is shown in red.

villi are bathed in lakes of maternal blood. The epithelial cells of the chorion are provided with many minute projections called microvilli which interdigitate with maternal tissue and provide a very intimate form of contact between the two. The long processes of foetal tissue, called trophoblastic villi, run vertically from the surface of the placenta into maternal tissues. Between these trophoblastic plates are the maternal capillaries with a thick endothelium. These capillary endothelial cells show, in electron micrographs, an interesting structure which may well be related to the way in which some substances are transferred by the placenta between maternal and foetal blood. The cytoplasm of the maternal capillary endothelium is spongy, containing many vacuoles. In the cytoplasm between these vacuoles are mitochondria and Golgi material. The vacuoles are sometimes seen to connect directly with the lumen of the capillary and it seems probable that the vacuoles are formed by invagination of the surface membrane of the cell. Thus substances included in the vacuoles may be transferred by these means from maternal blood to foetal tissues without having to pass through the 'cytoplasm' of the endothelial cells. The basement membrane of these endothelial cells varies a great deal in thickness, being absent in some places so that the plasma membrane of the maternal endothelial cells may be in direct contact with the cells of the foetal trophoblastic villi, The irregular basement membrane may represent erosion by the trophoblastic villi.

In the human placenta the trophoblast consists of an inner layer of cells (Langhans cells) lying next to the foetal connective tissue and blood vessels and an outer sycitial layer which is bathed by the maternal blood. The Langhans cells become flattened and many disappear as pregnancy advances. The cytoplasm of the syncytium is foamy and many ovoid nuclei are scattered through it. The foaminess of the cytoplasm is due to the presence of many vacuoles, each with a granular outline, the granules probably being of R.N.A. Synthesis of foetal protein may occur at these sites until the foetal liver takes over this role.

In addition to these smaller vesicles there larger vacuoles, usually near to the outer surface of the syncytium under the microvilli which occur here. These vacuoles may arise by invaginations of the surface membrane of the cell and may be concerned with the 'absorption' of maternal plasma into the cell by the process of pinocytosis. The large number of microvilli increases the surface area of the cell for absorption or secretion.

In the rabbit, guinea pig and rat there is even more intimate contact since the epithelium and mesenchymal tissues of the chorionic villi disappear leaving only the endothelium of the trophoblastic capillaries separating maternal and foetal blood, producing the haemoendothelial placenta.

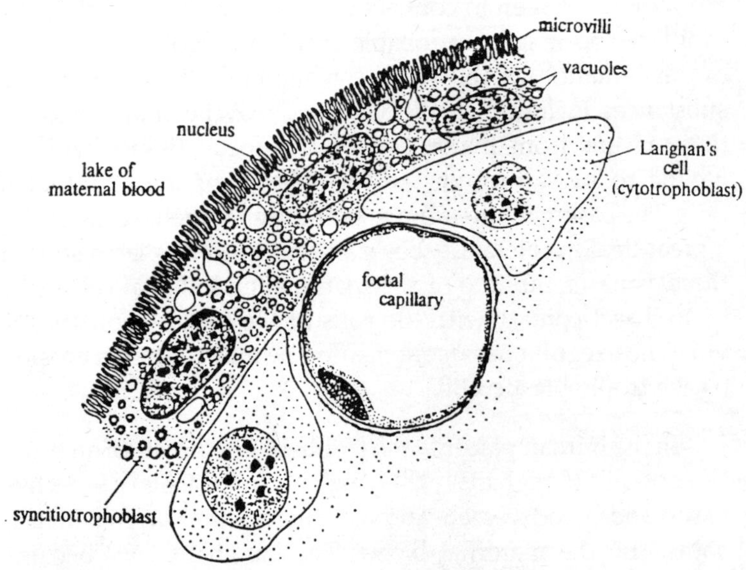

Fig. 1.16. Diagram of a small area of human placenta made from electron micrographs. Note the microvilli of the syncytio-trophoblast projecting into a pool of maternal blood. Some of the vacuoles of the syncitio-trophoblast are shown opening between the bases of the microvilli.

Type of placenta, efficiency and length of gestation period. In the ungulates with epithelio-chorial or syndesmochorial placentae the barrier between maternal and foetal circulations is much greater than in the other placental types. Nutrition of the foetus in ungulates is assisted by secretions of the uterine glands called uterine milk. It is difficult to correlate the efficiency of the placenta with the type of placenta; even in one species the placental structure is not constant throughout development, and in some species e.g. rat, there are additional structures to the chorio-allantoic placenta in the form of the yolk sac placenta.

SUMMARY OF TYPES OF PLACENTA

Placental type	Epithelio-chorial	Syndesmo-chorial	Endothelio-chorial	Haemo-chorial	Haemo-Endothelial
Maternal tissues					
Endothelium	+	+	+	-	-
Connective tissue	+	+	-	-	-
Epithelium	+	-	-	-	-
Foetal tissues					
Chorionic epithelium	+	+	+	+	-
Mesenchyme	+	+	+	+	-
Endothelium	+	+	+	+	+
Examples	Horse Pig	Cattle Sheep	Dog, Cat	Insectivores Man, Lower Rodents	Rat Rabbit

+ indicates presence
- indicates absence

However there is a general tendency for animals possessing a more highly developed placenta (i.e. fewer layers) to have a shorter gestation period, although the young are born in a helpless condition. This feature is illustrated in the following list of gestation periods :

Mouse. gestation period 19 days. Haemoendothelial placenta. Young very immature at birth.

Rat. gestation period 21 days. Haemoendothelial placenta. Young helpless and blind.

Dog. gestation period 63 days. Endothelio-chorial placenta. Young are helpless and blind.

Cat.   gestation period 65 days. Endothelio-chorial placenta. Young are helpless and blind.

In the above examples the young are in a very similar condition at birth, although the gestation period of the cat and dog is three times as long as that of the mouse and rat. This difference is explained on the grounds of a more efficient placenta is rats and mice.

Guinea Pig. gestation period approx. 68 days. The young are born active with eyes open. There is an accessory yolk sac placenta throughout the gestation period. The additional period of gestation compared to the mouse and rat is used to further the development of the young.

Pig.   gestation period 119 days. Epitheliochorial placenta. This is a primitive type of placenta and a long gestation period is necessary.

Thus it is seen that the more primitive type of placenta is associated with a longer gestation period, the guinea pig being an exception to this general rule perhaps because the young are so well developed by the time they are born.

PHYSIOLOGY OF THE FOETUS

The placenta and metabolism of the foetus

Through the placenta the foetus obtains its nourishment; water, carbohydrates, proteins, fats, mineral salts, vitamins and oxygen, and through the placenta pass the waste products of the foetus, including carbon-dioxide and urea, to be excreted eventually by the lungs and kidneys of the mother. The blood vessels of the foetus and mother do not join one another and all these substances must diffuse across the barrier between the two sets of blood vessels. There are three devices in the placenta which promote the passage of materials across the barrier between the two sets of vessels. We have already seen two of these devices: first the large surface area of the placenta provided by the branching villi and secondly the reduction in the thickness of the placental barrier with varies from species to species. The third device consists in the arrangement of maternal and foetal blood vessels so that the maternal blood vessels containing blood at relatively high pressure are first adjacent to foetal vessels containing blood at low pressures, whilst foetal blood vessels containing blood at relatively high pressure lie adjacent to maternal vessels containing blood at low pressure: this arrangement encourages an efficient transfer of substances between the two series of blood vessel.

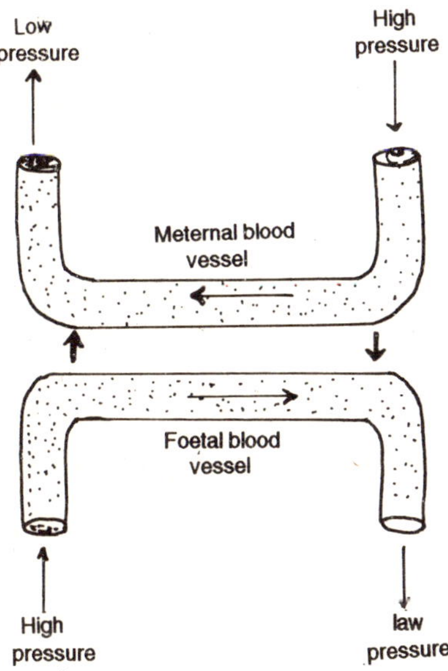

Fig. 1.17. Diagram to illustrate the principle of the arrangement of blood vessels in the placenta. The two short thick arrows indicate the direction of the exchange of materials.

Respiration of the foetus

The need for an increasing supply of oxygen by the foetus is somewhat anticipated by the maternal part of the placenta in that the maternal vascular bed in the uterus grows more rapidly than the foetal contribution to the blood vessels of the placenta. Thus, at an early stage in gestation the maternal blood leaving the placenta still contains large amounts of oxygen. As the gestation period progresses the maternal blood leaving the placenta becomes progressively more de-oxygenated; the placenta reaches its maximum size at a time when the foetus is still growing rapidly, and at the end of the gestation period the supply of oxygen is critical.

The foetus is adapted in two main ways to obtain oxygen efficiently from the placenta. First there is the special arrangement of foetal blood vessel which has already been described above. Secondly, in many species foetal haemoglobin has somewhat different properties from adult haemoglobin, in that it takes up oxygen with a greater avidity and gives it up less readily. foetal haemoglobin can take up oxygen at partial pressures at which maternal haemoglobin would give up oxygen. The foetus is a relatively inactive creature and can exist with haemoglobin which retains its oxygen more avidly than does adult haemoglobin. But once the animal is born it may need to use oxygen as rapidly as does an active adult, particularly in those species such as the horse where the young are very active at birth. It has been found that in some animals, e.g. the goat, there is a gradual alteration of the properties of foetal haemoglobin produced towards the end of the gestation period to approximately those of the adult, so that the animal will be better adapted to an active life breathing air.

Carbohydrate metabolism.

As in the adult, the foetus uses glucose as an important source of energy. Glucose can pass across the placenta from mother to foetus in all animals. There may be persistent differences between the blood sugar level of the foetus and mother which may imply that the transfer of glucose across the placental membranes is not merely one of diffusion. In most mammals the foetal blood sugar tends to be lower than that of the mother but in some epithelio-chorial placentae, e.g. pig, the blood sugar is higher on the foetal side.

In early foetal life the maternal side of the placenta acts as a temporary liver for the foetus in that it holds stores of glycogen at a time when the foetal live contains little or no glycogen. Later in the gestation period when the foetal liver has developed stores of glycogen, less in stored in the placenta. From experiments it has been found that the foetus exercises a strict glycogen economy, drawing upon glycogen stores only in emergencies.

Lipid metabolism.

The foetus usually has good stores of fat. This may come from several sources. It may be able to synthesize fat from carbohydrates and amino acids. Some lipids pass across the placenta and become available for the synthesis of fat, but the mechanism of transfer across the placental membranes is not understood. Most of the fat fed to the mother and stored in her tissues does not pass unchanged across the placental barrier. If the food fat is stained with the dye Sudan III than the mother's stores of fat are intensely stained red, although none appears in the foetus. Further evidence which indicates that fat is not transferred directly across the placental membranes is the different chemical constitution of foetal and maternal fat.

Protein metabolism.

Foetal protein may be derived directly by transfer of intact protein molecules from the mother or by transfer of amino acids. Antibody proteins (gamma globulins) are known to be transferred intact from mother to foetus in some species (e.g. rabbit) and this confers on the foetus a passive type of immunity. In other species much or all of the antibody proteins are received by way of the colostrum, the first pale secretion of the mammary glands of the mother. Amino acids probably do not pass across the placenta by simple diffusion and active transport mechanism may be involved.

Endocrine Function of the Placenta

In addition to serving the nutrition of the foetus the placenta is an endocrine organ producing several hormones in large quantities. Gonadotrophin is produced from the chorion and is called chorionic gonadotrophin. In the human this hormone is produced in such large quantities by the end of the second month of pregnancy that a specimen of urine, in which the hormone is excreted, will cause the growth of the ovaries and ovulation when injected into a sexually immature mouse. Modifications of this effect are used in the early diagnosis of pregnancy. When the urine is injected into a female toad called Xenopus, eggs are shed into the surrounding water within 24 hours. Even mature male amphibians have been used in pregnancy tests and injections of chorionic gonadotrophin are followed by a release of sperms, which have to be identified in the urine microscopically. In addition to gonadotrophin the placenta also produces large amounts of oestrogen and progesterone. The exact significance of these various hormones is not well understood but they undoubtedly play their part in the development and maintenance of the sex organs during the pregnancy and prepare the mammary glands for their function after the birth of the young. The high local concentration of progesterone at the placental site, by increasing the membrane potential of the uterine smooth muscle cells renders them less excitable. Thus the uterus is mechanically inactive and the foetus is safeguarded.

The circulation of the Foetus and Modifications at Birth

The circulation is adapted to life in utero, in which the placenta is the organ of nutrition and respiration; the foetal lungs and gastro-intestinal tract do not function as the respiratory and nutritive organs in utero and their blood supply is correspondingly small. But immediately at birth the placenta ceases to have these functions and there must be a rapid adjustment of the organism to an independent life. The lungs are rapidly converted from semi-solid organs with a small blood supply into air filled organs with a large blood supply and are responsible for the gaseous exchange of the whole organism. The gastro-intestinal tract takes over the nutritive functions of the placenta as the young animal begins to feed. In order to make these transformations possible there must be special devices within the cardio-vascular system of the foetus so that blood which once went to the placenta can now be diverted to the lungs and gastro-intestinal tract.

Foetal oxygenated blood is returned from the placenta by way of the umbilical vein which passes to the liver where it joins the hepatic portal vein. Some of the oxygenated blood goes to the liver but most of it is shunted away through a connection of the umbilical vein with the inferior vena cava called the ductus venosus, a feature only of the foetal circulation. In the inferior vena cava the oxygenated blood mixes with venous blood draining from the lower part of the body. When this blood reaches the heart most of it is shunted from the right auricle through an opening in the septum between the two auricles, into the left auricle. The opening in the septum, called the foramen ovale, is a special feature of the foetal heart and it censures that the oxygenated blood returning from the placenta is diverted away from the right ventricle and lungs to supply the head (i.e. central nervous system) and upper limbs of the foetus by way of the left ventricle and aorta. This flow of blood has been confirmed using X-ray studies of the foetus after radio-opaque material has been injected into the circulation. The blood is diverted from the right auricle into the left auricle by a valvular arrangement in the wall of the right auricle.

The venous blood which drains into the right auricle from the head and neck of the foetus by way of the superior vena cava passes through the right auricle into the right ventricle, from which it passes out into the pulmonary artery. But since the foetal lungs are not functioning as organs of gaseous exchange, only a small amount of blood is needed, sufficient to meet the metabolic needs of the tissues in the lung and much of the blood in the pulmonary artery passes directly into the aorta by way of a connection called the ductus arteriosus. This blood supplies the lower part of the body, and much of the it passes into the umbilical arteries (branches of the internal iliac arteries) supplying the placenta with deoxygenated blood.

There are thus four special features of the foetal circulation:

1.   The placental circulation.
2.   The ductus venosus which shunts blood from the

umbilical vein away from the liver into the inferior vena cava.

3. The foramen ovale through which oxygenated blood returning to to the heart via the inferior vena cava passes into the left auricle and so to supply the upper part of the body.

4. The ductus arteriosus which shunts blood from the pulmonary arch into the aorta, so by-passing the lungs.

It will be seen that the right ventricle of the foetus, pumping blood to the lungs and to the lower part of th body and placenta, performs more work than the left ventricle.. This position is reversed after birth.

At birth there are dramatic changes in this circulation. The umbilical vessels become increasingly irritable towards the end of the gestation period and with the physical stimuli of birth these vessels go into spasm, thus excluding the placental circulation from the foetus. This means that when the mother bits through the umbilical cord to free the young animal it does not bleed to death. The ductus venosus also goes into spasm so that blood in the hepatic portal vein now has to pass through the tissues of the liver and cannot be directly diverted into the inferior vena cava. With the expansion of the lungs at birth larger amounts of blood pass into the lungs from the plumonary arch, and the wall of the ductus arteriosus contracts so that blood in the pulmonary arch can no longer be diverted into the aorta. With the increased supply of blood to the lungs there is an increasing amount of blood returning to the left auricle and the pressure of blood in the left auricle rises. This rise in pressure pushes a loose flap of tissue against the foramen ovale, closing the connection between the left and right auricles. The adult type of circulation is now achieved. The ductus arteriosus, ductus venosus and umbilical vessels, initially closed by muscular spasm are gradually permanently obliterated as fibrous tissue grows in the lumen of the vessel, and the flap of tissue which is closing the foramen ovale gradually fuses with the septum..

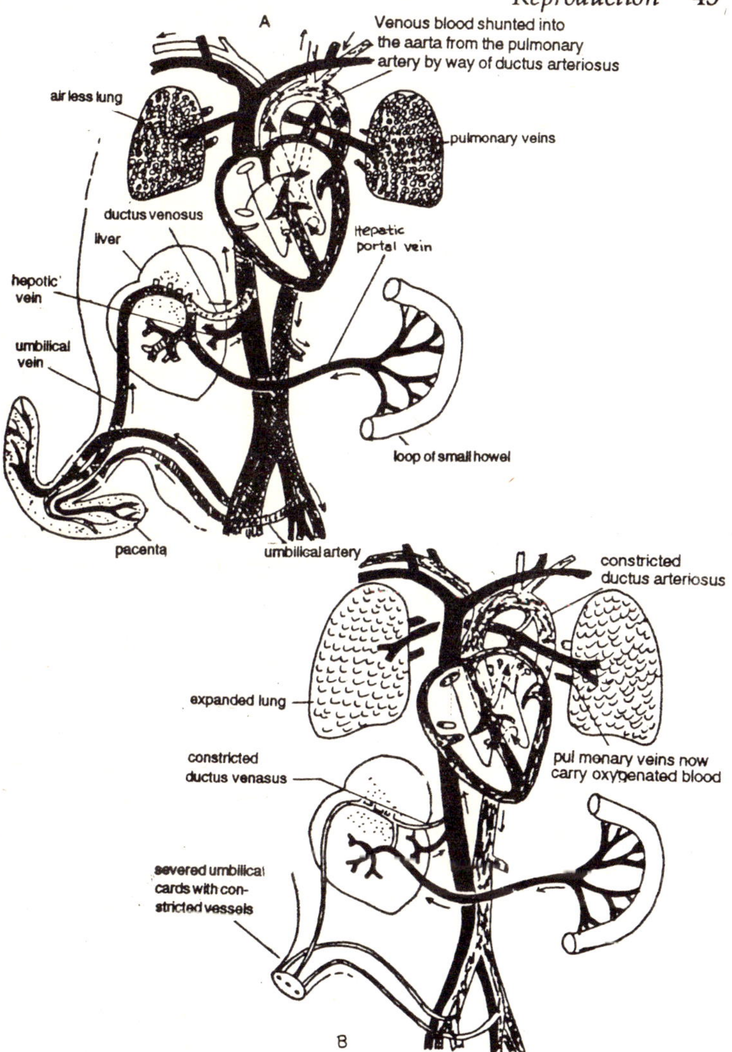

Fig. 1.18. *A. diagram showing the circulation of blood in the foetus before birth.* Red indicates oxygenated blood and superimposed shading indicates the degree of contamination by de-oxygenated blood. *B. The changes in the circulation which occur after birth.* The umbilical cord has been severed and the remains of the umbilical vessels, the ductus venosus and ductus arteriosus have closed down. The blood in the two sides of the heart is now separated by the closure of the foramen ovale. The change in the oxygen content of the blood is indicated by alteration of the superimposed shading in A.

Parturition.

At the end of the gestation period parturition occurs when the young are expelled from the uterus into the outside world by means of powerful intermittent contractions of the myometrium. What initiates the process of parturition is not known but it has been suggested that the declining production of the hormone progesterone, which has an inhibitory effect on uterine motility, sensitizes the uterus to the posterior pituitary hormone oxytocin, which stimulates intermittent powerful contractions of the uterus. In cases of slow labour associated with weak uterine contractions, injections of oxytocin are used to produce more powerful uterine contractions.

In those species in which there is an intimate mingling of foetal and maternal tissues in the placenta the uterine contractions not only propel the young through the birth canal but they serve to separate placenta from the uterine wall, and after birth of the young the persistent contraction of the uterine muscles serve to close the blood vessels which have been torn upon during the separation of the placenta.

Lactation

The mammary glands have been prepared for their function during the pregnancy by the effect of the sex hormone produced by the placenta. As mentioned previously oestrogen causes a growth of the duct system of the glands and progesterone initiates the development of the glandular elements around the ducts. Btu the glands are unable to produce milk during pregnancy because the large amounts of oestrogen inhibit the pituitary gland from producing a hormone vital for lactation, the lactogenic hormone. Towards the end of pregnancy the oestrogen production by the placenta gradually falls, and after parturition, when the placenta is expelled, the level of oestrogen in the body falls still further Lactogenic hormone is now produced by the anterior pituitary gland and lactation starts soon after parturition. The first pale secretions called colostrum are rich in proteins, and in some species they contain antibodies, which serve to protect the young for some months until they have been broken down by the body.

During the lactation period milk is secreted by the glandular cells of the mammary gland continuously. Milk, however, only passes along the larger ducts that open onto the nipple during suckling. When the infant mammal grasps the nipple in its mouth stimulation of sense organs in the nipple cause a reflex liberation of the hormone oxytocin from the posterior pituitary gland. This hormone reaches the mammary gland by way of the blood stream. On reaching the mammary gland the hormone stimulates certain contractile cells surrounding the glandular cells causing an emptying of milk into the larger ducts within open onto the nipple. Milk now appears from pores on the nipple and the young, by compressing the dilated ducts which underly the nipple increase the flow of milk.

2

Development of Rabbit

Mammals evolved from reptiles. Some mammals still lay their large eggs like most modern reptiles and birds. These animals are called the *oviparous animals* and such condition of reproduction is called *oviparity*. Most mammals give birth to young ones directly and in them the development is internal and takes place inside the body. Such animals are the *viviparous animals* and the stage is called *viviparity*. The viviparous animals are those animals in which the eggs are internally fertilized, the fertilized eggs are retained in a specialized part of oviduct called *uterus*, and the embryological development of these eggs take place inside the uterus. The developing embryo of such a viviparous animals often establishes a direct connection with the uterine wall and draws its nutritional, respiratory, excretory and other metabolic requirements from the maternal body. All mammals, except prototherian mammals which are oviparous (*e.g., Ornithorhynchus*) or *ovoviviparous* (a reproductive condition of egg-lying animals, in which the eggs though develop inside the maternal body, but are not supplied with maternal nourishment, *e.g., Echidna*), are viviparous. The viviparity is an adaptation to adverse environmental conditions such as cold, dryness and seasonal changes. It permits the mother freedom to move about, to feed, and to migrate, a freedom which is not enjoyed by a bird parent during its task of incubation. Viviparity is also found in some elasmobranchs, s teleosts and reptiles. In most of these animals, however, the egg is supplied with abundant yolk, but develops within organs of its mother's body, until hatching.

EVOLUTION OF VIVIPARITY IN MAMMALS

Palaentological evidences indicate that the mammals are derived from ancestors which were closely related to the early reptiles. It is expected, therefore, that they must have had large yolky eggs and their developmental history must have resembled with that of the reptiles. At some stage of their evolution some of the mammals switched from oviparity to vivipary. In such cases embryo received sufficient supply of nourishment from the mother while it was retained in the uterus, and the yolk supply of the egg became unnecessary. As a result, there was a progressive decline in the yolk supply and eventually it disappeared altogether. In this respect, there is a distinct graduation in the three sub-classes of the class Mammalia.

1. Sub-class Prototheria

Mammals of this sub-class are mostly *oviparous* as they lay eggs that develop outside the maternal body, *e.g., Ornithorhynchus*. However, some of them are *ovoviviparous, i.e.,* through developing in the maternal body, they are not supplied with nourishment, *e.g., Echidna*. The eggs, therefore, have a large amount of yolk. They are meroblastic and develop in a reptilian pattern (*Flynn* and *Hill*, 1939).

2. Sub-class Metatheria (the Massupials)

The quantity of yolk in the egg is insignificant and it becomes unnecessary for the development of the embryo. Nevertheless, it is ejected at the beginning of cleavage. Although the developing embryos receive nourishment from the mother in the uterus, but the adaptations for this procedure are not much advanced and the young are born poorly developed.

3. Sub-class Eutheria (Placental mammals)

The microlecithal egg is devoid of yolk right from the very beginning. Therefore, the embryos of these mammals develop inside the uterus and draw their nutrition from the maternal body through a 'physiological bridge' the *placenta*.

THE SPERM

It is microscopic. It is formed of a head, a middle piece and a tail. The head is pear-shaped. It has a nucleus and an acrosome. The nucleus and acrosome are surrounded by a plasma membrane. The middle piece is connected with the head by a neck. The middle piece contains two centrioles and the mitochondria. The tail is long. Its central core is occupied by an axial filament.

THE EGG

The egg is spherical in shape. it is microscopic and is about 0.11 mm. in diameter. The smaller size of the egg is due to the absence of yolk. Since there is no yolk, the egg is said to be *alecithal*. When the egg is released from the ovary it is surrounded by three membranes namely, corona radiata, a middle zona pellucida and an inner plasma membrane. The corona radiata is formed of follicle cells. The zona pellucida corresponds to the vitelline membrane of lower vertebrates. There is a fluid-filled space between zona pellucida and the surface of the egg and it is called *perivitelline space*.

Mammalian eggs are Secondarily Alecithal

The egg of Amphioxus has lesser amount of yolk (alecithal) like that of mammals. The absence of yolk or the presence of lesser amount of yolk is a primitive condition as it is exhibited by the primitive chordate. The amount of yolk increases in the case of the animals which are highly evolved. Thus fishes, amphibians, reptiles and birds are provided with a large amount of yolk in their eggs. The amount of yolk reaches its maximum in birds. Among the mammals, the amount of yolk in prototheria is as high as in birds. But the eutherian mammals (man, rabbit, pig etc.) possess only a negligible amount of yolk (alecithal). Thus the egg of highly advanced mammals resembles the egg of primitive chordate, *Amphioxus*. The alecithal condition of mammalian egg is due to the internal development of the embryo. The mother provides all the necessary materials for the development of the embryo. That is why yolk is not deposited inside the mammalian eggs. Thus the mammalian eggs are not *primarily alecithal* but only *secondarily alecithal*. The Amphioxus-egg is primarily alecithal. The alecithal egg of mammals is derived from the megalecithal egg of reptiles and birds. In the course of evolution, the mammalian eggs became alecithal only because of internal development.

The fact that mammalian eggs are derived from megalecithal eggs is further confirmed by the pattern of development especially gastrulation. The gastrulation of mammalian egg is exactly like that of megalecithal eggs of reptiles and birds and not like that of alecithal eggs of *Amphioxus*. Again like megalecithal eggs, the mammalian eggs develop an *yolk sac*.

FERTILIZATION

Fertilization takes place in fallopian tube. During copulation, the penis is inserted into the vagina; the sperms thus transferred by the male, travel up the oviduct and reach the fallopian tube to fertilize the eggs. Complete sperm enters the ovum; the tail, however, soon degenerates. Soon after the entry of the sperm, the egg completes its maturation. The fertilized egg passes down the oviduct into the uterus where it becomes attached to the uterine wall. A vascular organ, the *placenta* is formed through which developing foetus gets the nourishment from the maternal blood. The period of *question* (= period of carrying the young within the uterus) is normally one month. When fully formed the uterus expels the young ones by muscular contraction at birth. Rabbit is viviparous and the young-born possesses all the organs fully formed. The Female rabbit gives two young ones at a time. At birth the young one is hairless and its eyelids are coherent.

CLEAVAGE

The fertilized eggs or rabbit and other entherian mammals are *microlecithal*, minute in size, *i.e.,* ranging in diameter from 0.08

to 0.15 millimetre. Cleavage of the eggs of rabbit is of *holoblastic* type, but despite a microlecithal condition the blastomeres tend to show size differences from the start. Moreover, the numbers of blastomeres do not increase by a regular doubling sequence but tend to show arithmetical progression. The cleavage process of rabbit culminates in blastula by undergoing following stages:

First cleavage

The first cleavage is *meridional* or vertical and takes place within 22 to 24 hours after mating or 10 to 12 hours after fertilization. This results in the production of two unequal blastomeres. The smaller one is called *micromeres* and the larger is called *mega* or *macromere*. The blastomeres are held together by the zona pellucida.

Second cleavage

The second cleavage is also called *meridional* or *vertical*, but at right angles to the first one. The four cell stage is reached about 24 to 32 hours after fertilization. Although total, the cleavage is *irregular* as the synchronization of untosis in the blastomeres is lost very early. The macromere divides first and thus a three cell stage is produced. This is followed by the division of the micromere. Thus a four-cell stage is formed.

Further cleavage divisions are irregular. The *eight-cell stage* is found after 32 to 41 hours of mating. One member of the larger blastomeres of the four-cell stage egg divides, forming a five-cell condition, which is followed by the division of second larger blastomere, producing a six-cell stage. After a short period, one of the smaller cell divides and thus, a total of seven blastomeres is formed. The last cleavage is followed by the division of the other smaller cell, producing a eight-cell stage. The mitotic spindles of each of these cleavages are formed at right angles to one another, thus, showing an independence and asynchrony of cleavage. The rate of mitotic divisions increases and at about 65 to 70 hours after mating, the 16-cell stage is reached.

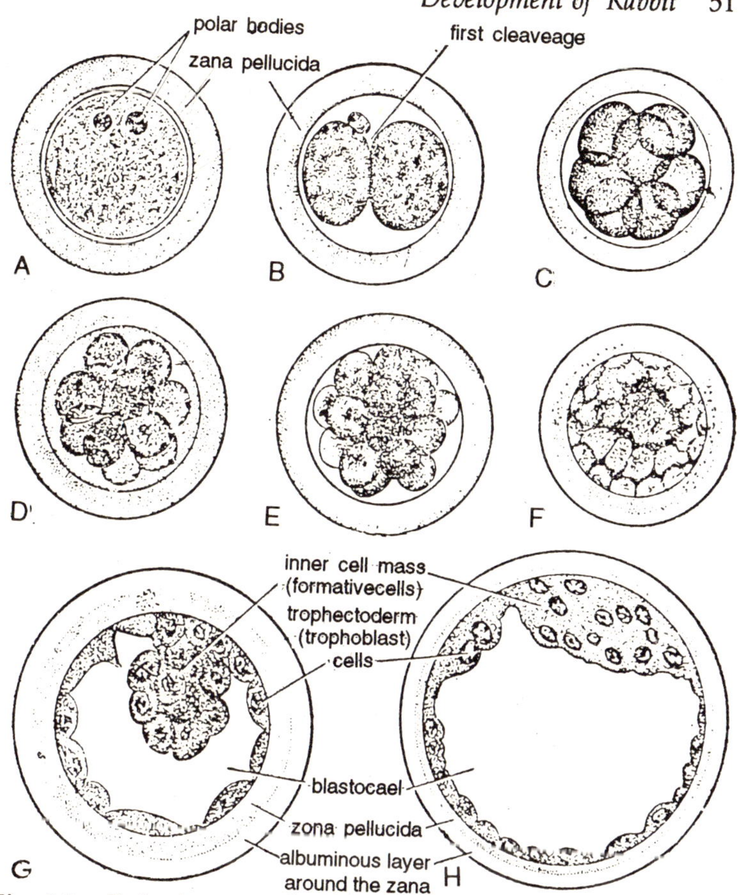

Fig. 2.1. Early cleavages of rabbit.

MORULA

Cleavage results in the production of a solid mass of cells. This mass of cells appears like a bunch of mulberry. Hence this stage of the embryo is known as *morula*. It has no cavity inside. It is formed within 65 to 70 hours after mating. The morula consists of two types of cells, the micromeres and macromeres. The macromeres became flattered and form a covering. It is called *trophoblast* or *trophectoderm*. It is nutritive in function. The interior of morula is occupied by the micromeres.

They are called *inner cell mass*. The inner cell mass alone develops into the body proper of the embryo.

BLASTOCYST

Soon a cavity appears inside the cell mass on one side; this cavity gradually increases. This results in the shifting of the central mass to one side which eventually becomes attached to the outer layer at the animal pole. The central mass is now called the *inner cell mass* while the cavity is falsely called the blastocoel, filled with the fluid imbibed from the secretion of the uterine wall. This stage can be referred to us *blastula* and is called *blastocyst* or *blastosphere* in mammals. By the further divisions of the inner cell mass, the blastocyst, grows to the size of 0.28 mm. diameter with inner cell mass attached at *embryonic knob* at the animal pole. From the embryonic knob, the embryo arises. The general outer layer of cell which encloses the blastocoel and the knob is called the *trophoblast*. It soon establishes relations with uterine wall and helps in the nutrition of the developing embryo. The cell of the trophoblast overlying the embryonic knob, are called the *cells of Rauber*.

IMPLANTATION

The embryo reaches the uterus in the blastocyst stage. Then it becomes attached to the uterine wall. The process by which the blastocyst is attached to the uterine wall is called *implantation*. Implantation is really the adhesion of the trophoblast of the embryo to the uterine wall. The adhesion occurs selectively where there is maternal blood vessels beneath the uterine epithelium. By this attachment the maternal blood diffuses into the embryo.

Types of implantation

In mammals there are three types of implantation. They are (*a*) central or superficial implantation(*b*) eccentric implantation and (*c*) interstitial implantation.

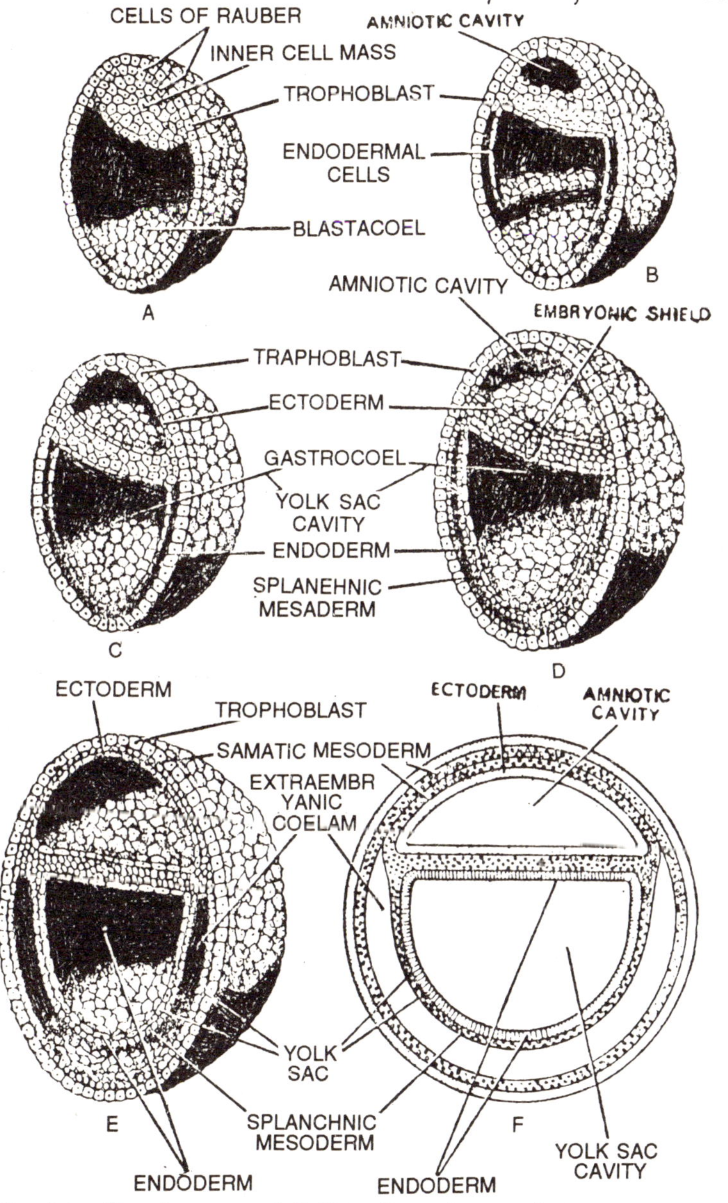

Fig. 2.2. Development of Rabbit, Gastrulation of the mammalian embryo. The inner cell mass forms the amniotic cavity by delamination. The embryonic shield is located between the amniotic cavity and the archenteron.

(*a*) Central or superficial implantation

In central implantation the blastocyst remains within the cavity of the uterus; the trophoblast (outer layer of blastocyst) makes a superficial attachment to the uterine wall (epithelium), *e.g.*, Ungulates, pig, carnivores, monkey etc.

(*b*) Eccentric implantation

Here the blastocyst remains in between the folds of the uterine epithelium. Then the fold s grow over so that the blastocyst is no more in the cavity of the uterus, *e.g.*, rat ; squirrel etc.

(*c*) Interstitial implantation

Here the blastocyst burrows deep into the uterine wall *e.g.*, hedgehog, guinea pig some bats, man etc.

GASTRULATION

Gastrulation begins either after implantation or earlier. There are two steps as in chick. They are

1.   The development of endoderm and
2.   The development of primitive streak and mesoderm.

1. The development of endoderm

The endoderm or hypoblast is formed by delamination. The innermost layer of cells of the inner cell mass (cells facing the blastocoel) splits off from the remaining cells. The delaminated cells multiply rapidly and gradually extend downward along the inner surface of trophoblast. They meet and fuse on the midventral side enclosing a cavity. This cavity is called *yolk sac* (archenteron) because it is homologous to the yolk-sac of birds. In birds this sac contains yolk; but in mammals there is no yolk. Even then, the yolk-sac develops as an ancestral memory. After the formation of hypoblast, the inner cell mass is called *embryonic knob*.

As the hypoblast is being formed, the cells of Rauber covering the inner cell mass disintegrate and the inner cell mass is exposed to the exterior. At the same time, the embryonic knob proliferates rapidly and the cells arrange themselves in the form of a disc called *embryonic disc* or *germ disc*. The embryonic disc is connected with the trophoblast around its edge. The upper layer of cells of the embryonic disc develops into the embryonic ectoderm and the trophoblast is now called *trophectoderm.*

Fate map

The organ-forming areas (fate-map) are arranged in the embryonic disc as in chick.

2. Development of primitive streak

The second step in gastrulation is the development of primitive streak and the subsequent formation of mesoderm. After the establishment of the endoderm, the embryonic disc at one side proliferates cells at a rapid rate and hence the cell number increases in this region. This results in a definite increase in thickness of the embryonic in this region. The thickened area is crescentic in shape when it first appears. It is called the primitive streak. The side, where the primitive streak appears, represents the caudal end of the embryonic disc. The opposite end represents the cephalic end where the head develops later. Soon the primitive streak elongates cephalo caudally. This is followed by the over all lengthening of the embryonic disc. The primitive streak consists of primitive plate, primitive ridges, primitive groove, primitive pit and Hensen's node as in chick.

As the primitive streak is growing, its cells migrate downwards into the blastocoel. These are the mesenchyme cells. Once inside, they migrate laterally as well as forwards between ectoderm and endoderm. Here these cells arrange into two sheets on either side of primitive streak. They form the mesoderm. The mesoderm cells migrate outside the germ disc also. The mesoderm which is found inside the germ disc is called *embryonic mesoderm* and that lying outside the germ disc is called *extraembryonic mesoderm.* The extraembryonic mesoderm develops into the extra embryonic structures like foetal membranes. The mesoderm undergoes the same type of changes as in chick.

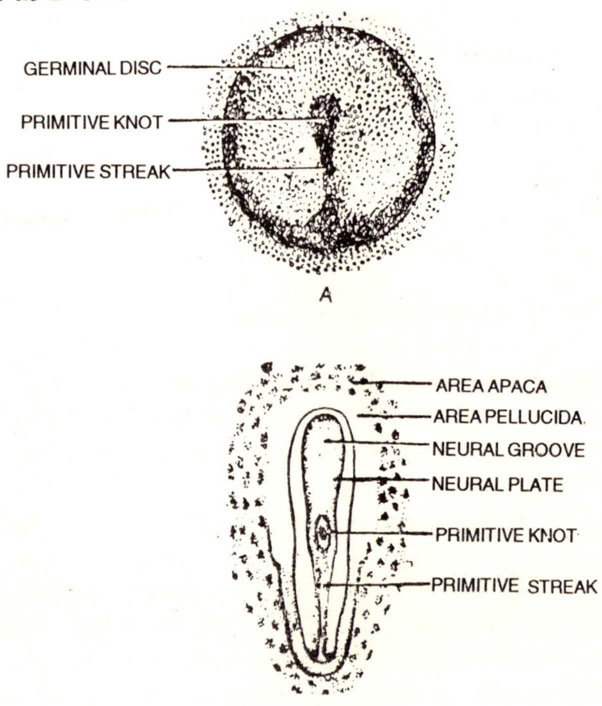

Fig. 2.3.   Development of mammals. A—Appearance of the primitive streak and groove in the blastodisc; B—Surface view of an early embryo.

The cells migrating into the blastocoel from the Hensen's node, arrange themselves in the form of a rod in front of Hensen's node. The anterior portion of this rod is called *prechordal plate* and the posterior portion is called *notochord*.

The ventral side of notochord is in close contact with the endoderm. Sometimes, in this region, the endoderm disappears completely and the roof of archenteron is formed of notochord only. The posterior end of notochord makes contact with the Hensen's node. As the primitive streak recedes pact wards,

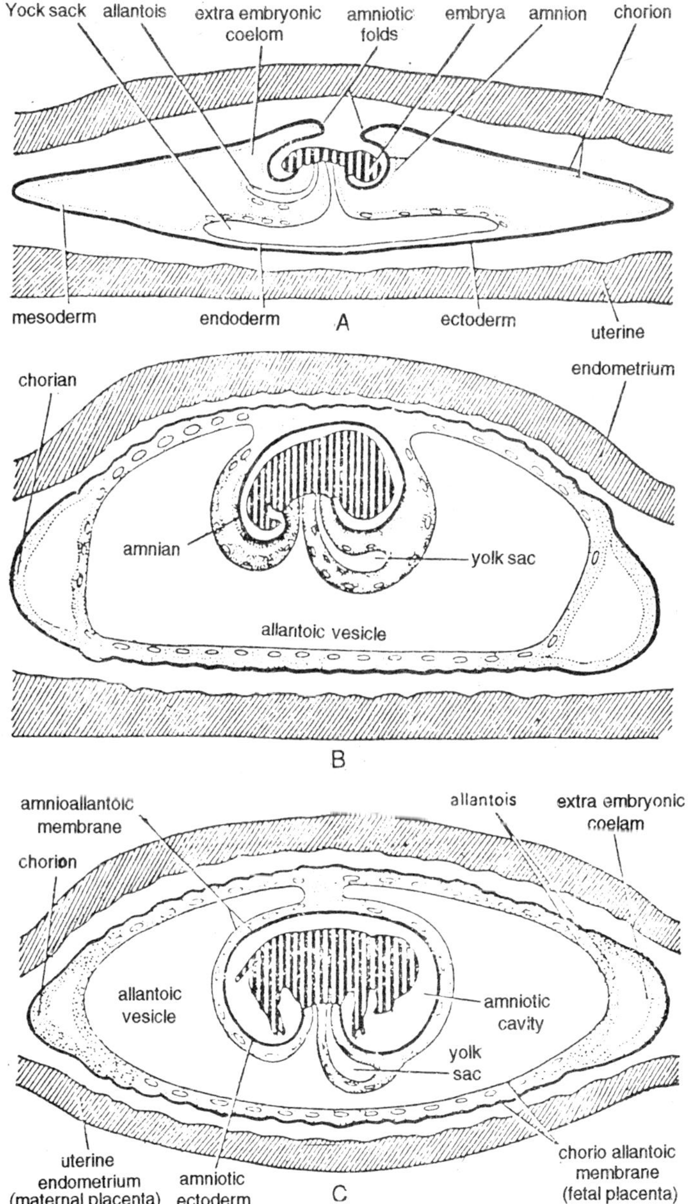

Fig. 2.4.   Development of extra-embryonic membranes of pig. A-B—Early stages. C—Fully developed extra-embryonic membranes.

the notochord stretches backwards. When the primitive streak disappears its place below the epiblast is taken by the notochord. The neural tube, noto-chord, mesoderm and coelom-develop as in chick.

FOETAL MEMBRANES OF RABBIT

The mammalian membranes including rabbit develop four *extra embryonic membranes* or *foetal membranes* to cover and the protect the embryo. These are *anmion, chorion, allantois* and *yolk sac*. The structure, development and function of these membranes are more or less similar to those of chick and slight modification.

AMNION AND CHORION

Amnion and chorion develop as upward projecting folds of somatopleure (somatic mesoderm and trophectoderm) called *amniotic folds*. Two types of amniotic folds develop. They are named according to their positions. They are the amniotic head fold, and the amniotic *tail fold*. First of all the amniotic fold appears as a transverse fold behind the tail. It is called *amniotic tail fold*. It grows over the embryo dorsally and then grows forwards. Later a transverse fold appears in front of the head. It is called amniotic head fold. It grows upwards and the backwards over the embryo. Finally the head fold and the tail fold meet and fuse together. This fusion takes place near the anterior end because the tail fold develops first and grows forward rapidly. The place of their final fusion is marked by a temporary connection called *seroamniotic connection* or *seroamniotic raphe*. After the union of the folds, the inner and outer layers separate. The inner layer becomes the amnion and the outer layer becomes the *chorion*. For further development refer to the chick.

Functions

1. The embryo is freely immersed and bathed in the amniotic fluid. Hence the amniotic fluid functions as the *artificial swimming pool of the embryo*.

2. As the amniotic fluid is more viscous and gelatinous, it functions as an efficient shock absorber. Thus the soft embryo is protected from external as well as internal mechanical pressures.

3. The amnion prevents the adhesion of embryo to the other parts.

4. The amnion provides facilities for the free movement of the embryo.

5. The chorion provides psysiological exchange between the embryo and the mother. It develops finger-like outgrowths called *villi*. The villi dip into the corresponding depressions called *crypts* developed in the uterine wall. This enhances the absorption of nutrition from the maternal side.

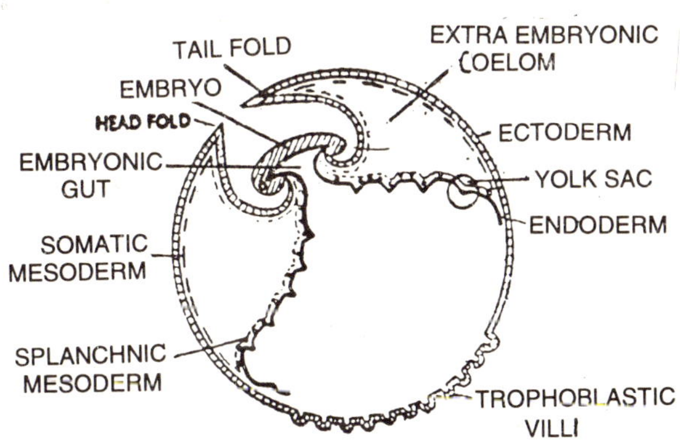

Fig. 2.5. Diagram showing formation of amnion and chorion.

YOLK SAC

There is no yolk in mammalian eggs. Hence the yolk sac is not necessary for mammalian embryos. Even then the yolk sac develops, as it developed in the ancestors (reptiles)). Hence the development of yolk sac in mammals is a mere repetition of an ancestral character. So the yolk sac in mammals is empty.

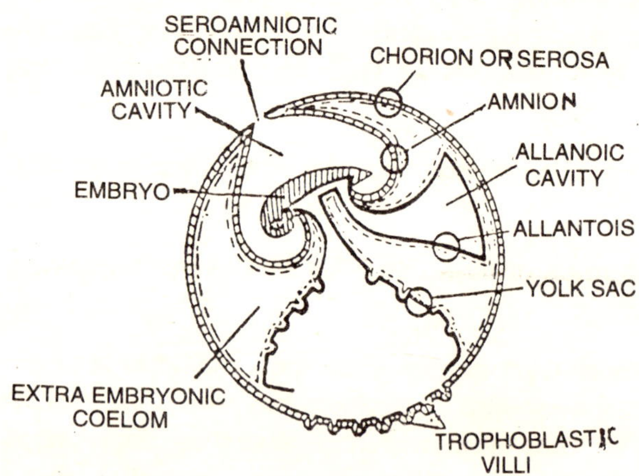

Fig. 2.6.   Diagram, showing formation of yolk sac placenta.

The yolk sac initiates its development, when gastrulation begins. In the blastocyst stage, the inner cell-mass liberates some cells into the blastocoel. These cells arrange themselves to form a flatened layer below the inner cell-mass. This layer forms the *endoderm* or the hypoblast. As more and more cells are added to this layer, the endoderm grows downward along the inner surface of the trophoblast. The space enclosed by endoderm is called the yolk sac. In the early stage the yolk sac is lined by the endoderm only. But when the mesoderm develops, the splanchnic mesoderm grows in close association with the yolk-sac endoderm. The mesoderm will not extend to the lower part of the yolk sac. The upper part of the yolk sac is highly vascularised and contains vitelline arteries and veins. The lower part of the yolk sac is not vascularised.

The connection of yolk sac with the embryo becomes narrowed into a tubular stalk called *yolk stalk*. Later the yolk sac gradually decreases in size and becomes shrivelled.

The yolk sac absorbs uterine milk and the milk is transported by the vitelline circulation to the embryo as nourishment. Thus the yolk sac is nutritive in function. In addition, it assists respiration. Blood cells are also produced by splanchnopleure of the yolk sac.

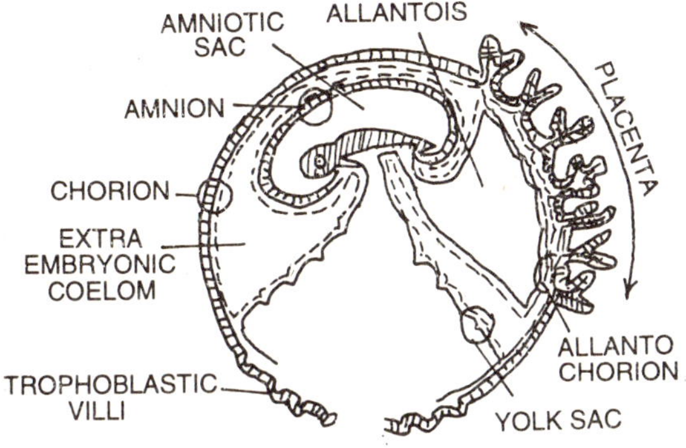

Fig. 2.7.   Diagram, showing formation of placenta.

ALLANTOIS

The allantois of rabbit and other mammals is composed of splanchnopleure (endoderm + splanchnic mesoderm). It arises from the hind gut as a small sac which gradually increases in the extra-embryonic coelom, *i.e.*, space between the amnion and chorion. Soon, it grows over the embryo outside the amnion where it enlarges. Later on, mesodermal component of allantoic wall give rise to the blood vessel system of the allantois and fuses with mesoderm of chorion to form of *allantoic chorion*.

Functions of Allantois

With the acquisition of viviparity in mammals, the original function of the allantois as a urinary bladder becomes altogether lost. As the maternal organism supplies the embryo with all the necessary nutrition, so also it takes over the removal of the waste products (CO_2, urea, etc.) of metabolism from the embryo. However, allantois of mammals supplies oxygen and nutrients to the embryo, which are taken from maternal blood circulation.

The nutrition, respiration and excretion of the developing embryo and foetus

When the developing blastocyst reaches the uterus, the

lower region of the trophoblast develops villi which penetrate into the uterine wall. The uterine wall at this time becomes more vascular *i.e.*, its blood vessels dilate. At this stage the blastocyst is nourished by the secretion of the uterine wall and by the food absorbed from it by the trophoblast villi. With the formation of foetal membranes, larger finger-like villi arise from the allanto-chorion and the embryo sinks deeper into the decidua formed by the multiplication of the cells of the uterine epithelium. Thus a firm connection develops between the foetus and the mother through *placenta*. The placenta is partly maternal and partly embryonic. The chorionic villi soon become more vascularized to form *placental villi* containing circulation which is continuous with that of embryo via allantoic blood vessels and capillaries connected with the extra-embryonic arteries and veins. The tissues of the uterine wall in contact with these villi break to form lacunae through which maternal blood flows to bathe the villi. This facilities the diffusion of food and oxygen from the maternal blood to that of foetus while the waste products and carbon dioxide produced by foetus diffuse in the reverse direction. Thus we can say that the foetus utilizes the digestive system, lungs and kidneys of the mother.

Soon the heart of the foetus becomes four-chambered with the pulmonary circulation still wanting as the lungs are expended. Therefore the blood returning to the right auricle does not pass to the lungs but is largely short-circuited into the left auricle through *foramen ovale*. This aperture lies in the partition that separates the two auricles. From left auricle the blood goes to the left ventricle and then to the general circulation.

At the time of birth, while the foetus is passing, the foetal membranes burst with the result the young burned rabbit possesses no enclosure of foetal membranes. The young rabbit takes the first breath, expands the lungs thereby establishing the pulmonary circulation. In the meantime foremen ovale and ductus arteriosus close and thus a true double circulation is established.

3

Development of Pig

The maturing eggs of the pig are expelled from the ovary during the period of heat, following which they become promptly fertilized.

As soon as the furtilization is completed the development begins. cleavage, blastulation and even the gastrulation occurs in a typical mammalion pattern.

At the completion of germ-layer formation the embryonic disc possesses a typical primitive streak. This is quickly followed by the appearance of neural folds and then mesodermal somites. The stages immediately succeeding correspond to those of three-day chick embryos, but are complicated by flexion and spiral twisting this makes sections difficult for the beginner to interpret. However, in embryos about 6 mm. long the twist of the body has disappeared sufficiently so that its structure may be studied to better advantage. At this time the state of development is generally comparable to that of a four-day chick.

The fetal membranes of the pig stand somewhat intermediate between those of the chick and man. The amnion, chorion and allantois develop very much as in the chick. The yolk sac for a time grows rapidly, but its functions are soon transferred to the allantois which fuses with the chorion; the two enter into a contact-relation with the lining of the uterus and constitute a placenta, which is the organ of fetal respiration, nutrition and excretion. The development and relations of these extra-embryonic structures are described.

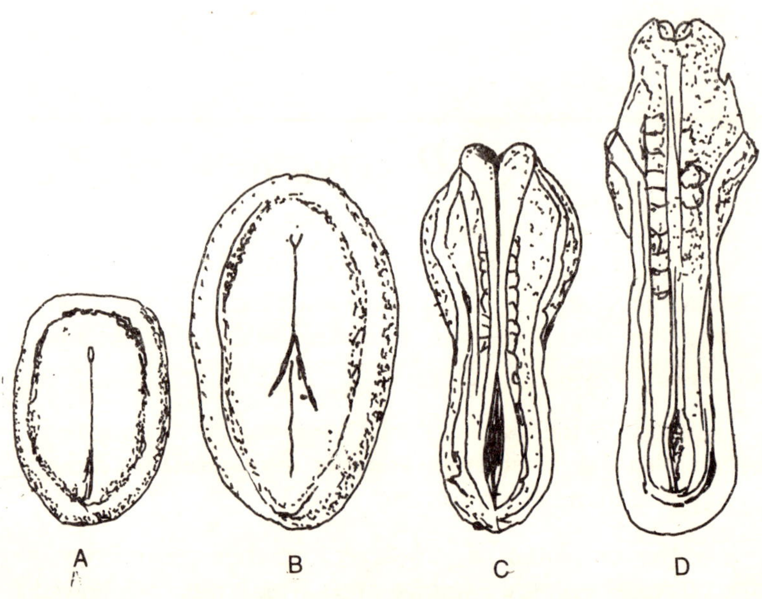

Fig. 3.1.   Early pig embryos, in dorsal vie. Blastoderm at twelve days, with primitive streak and knot. B, Blastoderm at thirteen days, with primitive streak and neural groove. C, Embryo of fourteen days, with seven somites. D, Embryo of fifteen days, with 11 somites.

A. Anatomy of A Six mm. Pig Embryo

The general structure of 6 mm. pig embryo is illustrated in Figs. 3.3-3.4. This should be compared with the chick embryo of four days and the 5 and 8 mm. human embryo. Some familiarity with the gross anatomy of the 6 mm. pig embryo will make the detailed study of the 10 mm. pig embryo, which follows easier to understand.

B. Anatomy of a Ten mm. Pig Embryo

This is the most instructive single stage of later development. Nearly all the important organs are represented, and yet the embryo is not so complex as to confuse unduly a beginner. Embryos between 8 and 14 mm. long may be used satisfactorily

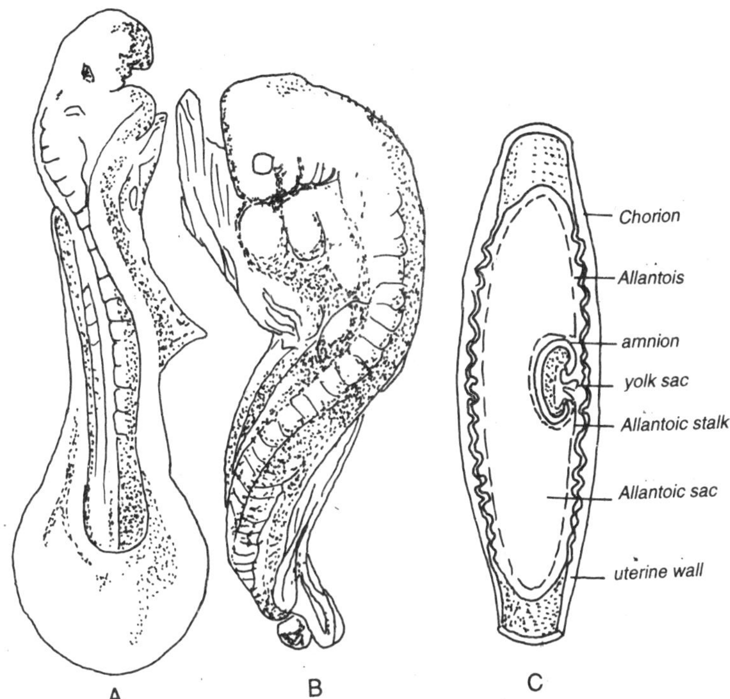

Fig. 3.2.   Early pig embryos, in dorsal view. A, At sixteen days, with seventeen somites. B, At seventeen days, approximately 4 mm. long, with about full number of somites. C, Hemisection of uterus, showing fetal membranes and placental relations.

in conjunction with the descriptions that follow. At this period a human embryo is slightly further advanced than a pig embryo of equal size, but at corresponding stages of development they are fundamentally alike.

External Form

The *head*, which is relatively large on account of the dominance of the brain, makes a right-angled bend at the *cephalic flexure*. On the under surface of the head are the *olfactory pits,* now drawn into elongate grooves and bounded by *lateral* and *median nasal processes*. The lens of the *eye* is prominent as it lies beneath the ectoderm, surrounded by the optic cup. At the sides of the head are four *branchial arches,*

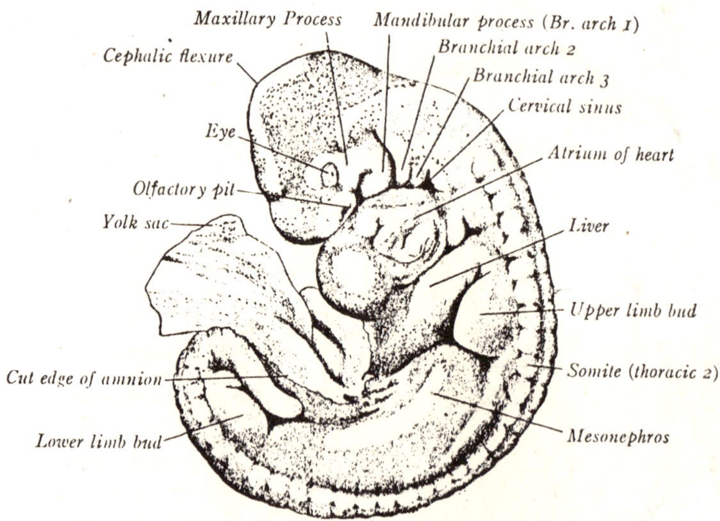

Fig. 3.3.   Pig embryo of 6 mm., with amnion removed.

separated by three *branchial grooves*. The first branchial arch of
each side forks ventrally into two parts. The smaller *maxillary
processes* show signs of fusing with the median nasal processes
to form the upper jaw, while the larger *mandibular processes*
have united already into the lower jaw. Next caudad is the
prominent second, or hyoid arch. Small tubercles, which will
combine into the *auricle* of the external ear, bound the first
branchial groove; the groove itself will become the *external
acoustic meatus*. The third branchial arch is still visible in the
future neck region, but the fourth arch has sunk into the
cervical sinus; both disappear at a slightly later stage.

At the *cervical flexure* the head is bent at right angles to the
body, thus bringing the ventral surface to the head close to the
trunk; it is probably owing to this flexure that the third and
fourth branchial arches buckle inward to give rise to the *cervical
sinus*. Along its dorsal surface the trunk curves convexly, but
this feature is not so prominent as at 6 mm. The reduction in
trunk curvature results from the increased size of the heart,

liver and mesonephroi. These organs are plainly indicated through the translucent body wall, while the position of the septum transversum may be noted between the heart and the liver. The *limb buds* are growing rapidly; the upper limbs are at a level between heart and liver. The *umbilical cord* is relatively large; it attaches at the lower end of the trunk. Dorsally the *somites* occur in serial order; toward the tail they become progressively smaller. Paralleling them and extending in a curve between the bases of the limb buds is the *mammary ridge;* on this thickened band of ectoderm will differentiate the mammary glands. The *tail* is long and tapering. Between its base and the umbilical cord is the *genital tubercle.*

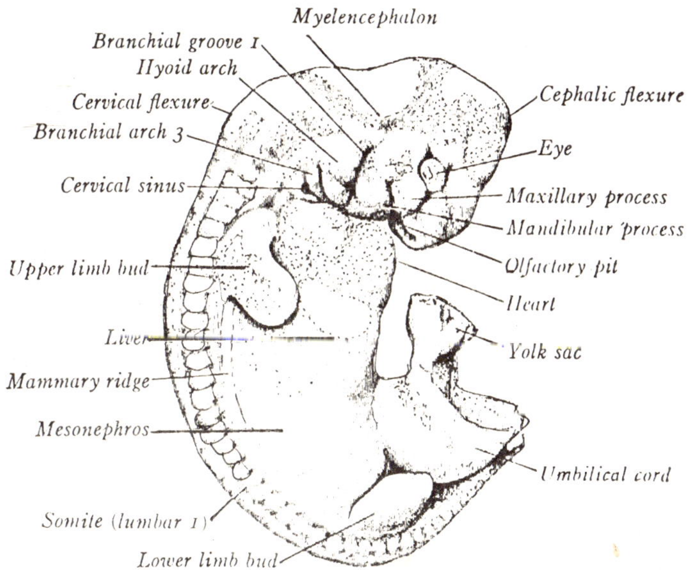

Fig. 3.4.   Pig embryo of 10 mm., viewed from right side. X 7.

Nervous System

Brain and Spinal Cord. Five distinct regions of the brain can be distinguished : (1) The *telencephalon* exhibits its rounded lateral outgrowths, the *cerebral hemispheres.* Their cavities, the *lateral ventricles,* communicate by interventricular foramina with the third ventricle. (2) The *diencephalon* shows a laterally flattened cavity, the *third ventricle.* From the ventrolateral side of the diencephalon pass off the optic stalks, while an evagination of the midventral wall (the infundibulum) will produce the neural lobe of the *hypophysis.* (3) The *mesencephalon* never subdivides, and its cavity becomes the *cerebral aqueduct* leading caudad into the fourth ventricle. (4) The *metencephalon* is separated from the mesencephalon by a constriction, the *isthmus.* Dorsolaterally it becomes the *cerebellum,* ventrally the *pons.* (5) The elongate *myelencephalon* is roofed over by a thin and non-nervous ependymal layer. Its ventrolateral wall is thickened and still gives internal indication of the *neuromeres.* The cavity of the metencephalon and myelencephalon is the *fourth ventricle.*

The *spinal cord* begins without specific demarcation and extends into the tapering tail. Just beneath the hind-brain and spinal cord lies the *notochord.*

Cranial Nerves

Of the twelve pairs of cranial nerves, all but the olfactory and abducent are represented in where they occur in the order listed : (1) The *olfactory nerve* is not grossly demonstrable at this stage. (2) Fibers of the *optic nerve* are growing brainward within the optic stalk, cut through in this illustration. (3) The *oculomotor nerve,* a motor nerve to four of the eye muscles, takes origin from the ventrolateral wall of the mesencephalon and passes downward between the two parts of the bent brain. (4) The *trochlear nerve,* motor and destined for the superior oblique muscle of the eye, really arises from the ventral wall of the mesencephalon but emerges dorsally at the isthmus. The next eight pairs of nerves pass off from the rhombencephalon; four of these are rostral to the otocyst in the metencephalon and four lie caudally in the myelencephalon. (5) The *trigeminal nerve* is conspicuous because of its large *semilunar ganglion* and three branches (ophthalmic, maxillary and mandibular rami) which carry motor impulses to the jaw muscles and bring sensory impulses from the head. (6) The *abducent nerve* originates from the ventral brain wall and passes to the eye, where it will innervate the external rectus muscle. (7) The *facial nerve* is mixed, sensory and motor; it bears the *geniculate ganglion* and divides into chorda tympani, facial and superficial petrosal rami in the order named; most of the nerve has to do with the motor innervation of the face, whereas the sensory supply goes to the tongue. (8) The *acoustic nerve* arises just rostral to the otocyst; it bears the *acoustic ganglion* which will send sensory fibers to the internal ear. Caudal to the otocyst is the *glossopharyngeal nerve,* showing a proximal *superior* and a more distal *petrosal ganglion;* its sensory and motor fibers innervate both tongue and pharynx. (10) The *vagus nerve* is mixed in function and has both a *jugular* and a *nodose ganglion;* its fibers innervate chiefly the viscera. (11) The *accessory nerve* has motor fibers which take origin both from the lateral wall of the myelencephalon and from the spinal cord as far caudad as the sixth cervical ganglion; an internal branch accompanies the vagus, while the external branch is distributed to the sterno-mastoid and trapezius muscles. (12) The *hypoglossal nerve* arises by five or six rootlets from the ventral wall of the myelencephalon; it is purely motor and supplies the muscles of the tongue.

It should be noted that the fifth, seventh, ninth and tenth cranial nerves pass into the four branchial arches in the order named. This primitive relation is maintained in the adult when the nerves innervate the derivatives of these arches.

A nodular chain of ganglion cells extends caudad from the jugular ganglion of the vagus. These have usually been interpreted as *accessory vagus ganglia.* They may, however, be continuous with *Froriep's ganglion* which sends sensory fibers to the n. hypoglossus. In pig embryos of 15 mm. this chain is frequently divided into four or five ganglionic masses, of which occasionally two or three (including Froriep's ganglion) contribute fibers to the root fascicles of the hypoglossal nerve.

Spinal Nerves

Each nerve has a single *spinal ganglion*, from which the sensory *dorsal root* fibers are developed. The motor fibers take origin from the ventral cells of the neural tube; they form the *ventral roots* which join the dorsal roots in the common *nerve trunk*. In the region of the upper and lower limb buds the spinal nerves unite and give rise respectively to the *branchial* and *lumbosacral plexuses*.

Autonomic System

Paired masses of ganglion cells, located dorsolateral to the descending aorta, are the forerunners of this division of the nervous system. These segmental masses are *autonomic ganglia* and, as ganglionated chains, become the *sympathetic trunks*. Fibers pass from the spinal cord to the ganglia and these fibers comprise a branch of the spinal nerve known as its *communicating ramus* (r. communicans).

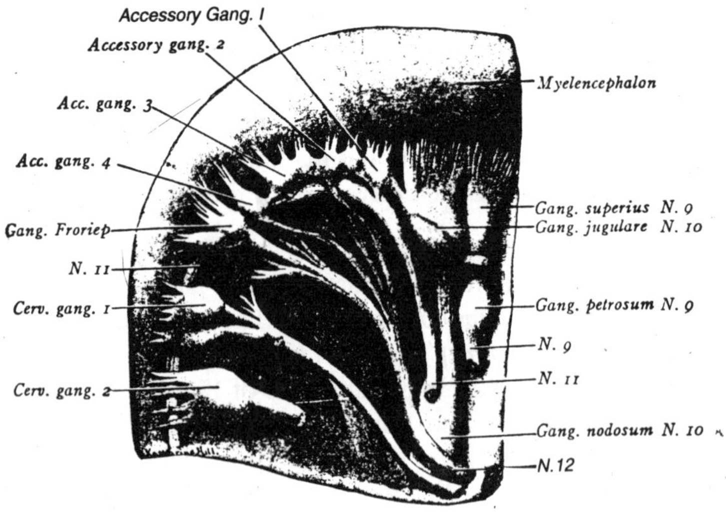

Fig. 3.5.   Dissection of postotic cranial nerves and ganglia of 15 mm. pig embryo.

Sense Organs

The *olfactory pits* are deep fossae, flanked by the median and lateral nasal processes. The stalked *optic cups* are prominent and the *lens vesicles* have detached from the ectoderm. Each *otocyst* is a compressed oval vesicle with a tubular *endolymph duct* growing from its medial side.

Digestive and Respiratory Systems

Mouth and Pharynx. The pharyngeal membrane disappeared at a considerably earlier stage and the *stomodeum*, or ectodermal mouth, then becomes continuous with the pharynx. From the dorsal wall of the stomodeum, *Rathke's pouch* (epithelial hypophysis) extends as a long, stalked sac which forks at its end near the primordium of the neural lobe. The floor of the mouth and pharynx is occupied by the *tongue* and *epiglottis*. From the mandibular arches arise paired *lateral swellings* that become the body of the tongue. Lying between these thickenings is the transient *tuberculum impar*. The thyroglossal duct, which formerly opened just caudal to the tuberculum impar, is already obliterated; the *thyroid gland* itself, composed of branching epithelial cords, is now located in the midplane between the second and third branchial arches. A median ridge, named the *copula*, unites the second arches and represents the primitive root of the tongue; it connects the tuberculum impar with the epiglottis which develops from the bases of the third and fourth branchial arches. On each side of the slit-like *glottis* is an *arytenoid fold* of the larynx.

The pharynx is flattened dorsoventrally; it is broad at the oral end. Opposite the third branchial arch the pharynx bends sharply in conformity with the cervical flexure. The paired *pharyngeal pouches* are large, and each bears a dorsal and a ventral wing. The first pouch on each side persists as the *auditory tube* and *tympanic cavity;* the 'closing plate' between it and the first branchial groove forms the *tympanic membrane,* while the ectodermal groove becomes the *external acoustic meatus.* The second pouches are destined largely to disappear; about each develops a *palatine tonsil.* The dorsal wing of each tubular third pouch forms a *parathyroid gland*; the ventral wings differentiate into the *thymus*. The fourth pouches are smaller; their dorsal wings give rise to another pair of parathyroids, while the ventral wings are rudimentary. A tubular outgrowth, just caudal to each fourth pouch, is sometimes regarded as a fifth phayngeal pouch; it forms an *ultimobranchial body*.

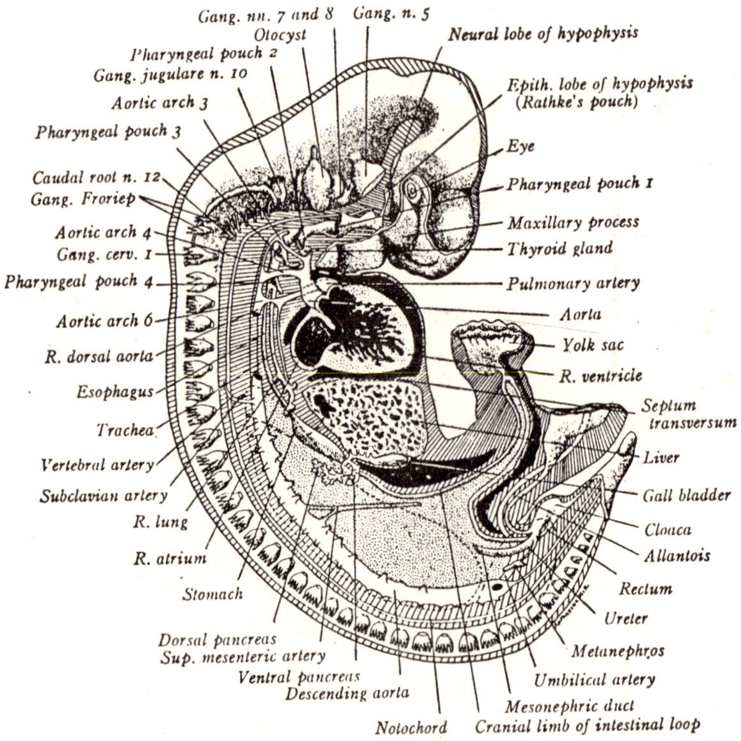

Fig. 3.6.   Reconstruction of 10 mm. pig embryo viewed from right side.

Larynx, Trachea and Lungs. The *larynx* and *epiglottis* are appearing and the *trachea* is a definite tube. Terminally the trachea bifurcates into *primary bronchi*; each of these has already divided again into secondary bronchial buds which indicate the two lobes of the *left lung* and the middle and lower lobes of the *right lung*. From the right side of the trachea itself appears another bud which, in the pig, represents the upper lobe of the right lung.

Esophagus and Stomach. The *esophagus* extends as a narrow tube past the lungs, whereupon it dilates into the laterally flattened *stomach*. The entire stomach has rotated about its axis so that the original dorsal border, now the convex greater curvature, lies to the left and the primitive ventral border (lesser curvature) to the right. At this stage the rotation is incomplete.

Intestine

The pyloric end of the stomach opens into the *duodenum* which also shows the effect of stomach rotation; the stem of the *hepatic diverticulum,* originating from it ventrally, now lies to the right. The diverticulum itself has differentiated into various things. From its tip has become the four-lobed *liver,* filling in the space between the heart, stomach and duodenum. One of the several ducts now connecting the liver with the parent diverticulum will persist as the *hepatic duct.* The main stem of the diverticulum is the *common bile duct,* while a side sacculation is the cystic duct and gall bladder. The *ventral pancreas* springs from the common bile duct near its point of origin. It is directed dorsad and caudad, to the right of the duodenum. The *dorsal pancreas* arises a little more caudally (man, cranially) from the dorsal wall of the duodenum; its larger, lobulated body grows dorsad and cephalad. The two glands will interlock into a single organ; it the pig it is the duct of the dorsal pancreas that persists as the functional duct.

Beyond the duodenum the intestine is thrown into a loop, which extends well into the umbilical cord and connects with the *yolk stalk* there. Owing to rotation in the entire loop, the cranial limb of the intestine lies to the right, the caudal limb to the left. The small intestine (jejunum and ileum) extends as far as a slight enlargement on the caudal limb of the loop (Fig. 3.). This is *caecum* which marks the beginning of the *large intestine* (colon and rectum). The cloaca is now subdividing into the *rectum* and *urogenital sinus*.

Coelom and Mesenteries. The coelom is a continuous, communicating system which includes the single *pericardial* and *peritoneal cavities*, still connected by paired *pleural canals*. Between the heart and liver is a prominent partition, the *septum transversum*; the liver is broadly fused to this septum which will comprise much of the diaphragm. The double sheet of splanchnic mesoderm that serves as the primitive dorsal mesentery receives special names at different levels. Where it suspends the stomach it is known as the *mesogastrium,* or *greater omentum.* Then in order come the *mesoduodenum, mesentery* proper of the small intestine, *mesocolon* and *mesorectum.* The mesentery proper and some of the mesocolon follow the intestinal loop out into the umbilical cord. The ventral mesentery is limited in extent. It persists as the *lesser omentum* between stomach and liver, encloses the liver, and continues as the *falciform ligament* between the liver and ventral body wall. A saccular recess, between the caval mesentery and liver on the right and the stomach and its mesenteries on the left, is the *vestibule* and *omental bursa.* It opens through a narrowed *epiploci foramen.*

Urogenital System. The *mesonephroi* are large and complex in the pig. Along the middle of their ventromesial surfaces *genital ridges* have become prominent. In a ventral dissection the course of the *mesonephric ducts* can be traced along the ventral margins of the mesonephroi and into the *urogenital sinus.* The allantois is a conspicuous, stalked sac which communicates with the ventral part of the urogenital sinus.

The *metanephroi,* or permanent kidneys, lie far caudad between the roots of the umbilical arteries. At the present stage each consists of a tubular epithelial portion, surrounded by a mass of condensed mesenchyme. The epithelial tube has budded off the mesonephric duct, near its ending; proximally there is a slender duct, the *urter*, while a distal dilatation is the *renal pelvis*. From the pelvis grow out later the calyces and collecting tubules of the kidney. Encasing the pelvic primordium is a layer of condensed mesenchyme, derived from the lower nephrotomes and destined to differentiate into the secretory tubules, or *nephrons*.

Heart. This organ lies within the pericardial cavity. There are two *atria* and two thicker-walled *ventricles*. In addition, a small chamber, the *sinus venosus*, receives all the blood returned to the heart and directs it into the right atrium while the *bulbus cordis* still serves as a common arterial outlet.

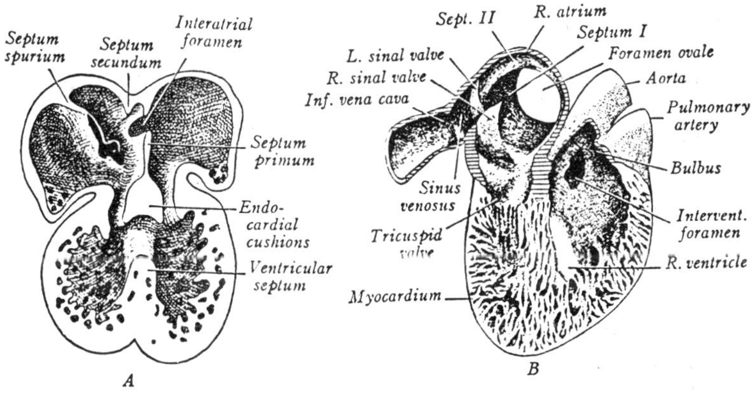

Fig. 3.7.   Dissections of heart of pig embryos. B, At 12 mm., with right wall removed.

The entrance from the sinus venosus into the right atrium is a nearly sagittal slit, guarded by right and left *valves of the sinus venosus (B)*. Dorsally the two valves join and continue a short distance as the temporary *septum spurium (A)*. Somewhat later the sinus largely loses its identity by merging with the right atrium, although its middle part does persist as the *coronary sinus*. The dorsal wall of the left atrium is receiving a single *pulmonary vein.* The two *atria* are incompletely partitioned by the *septum primum* which contains an opening, the *interatrial foramen.* On the right side of this partition a second sickle-like fold, the *septum secundum,* is forming. It also becomes an incomplete septum which bears an opening, known as the *foramen ovale.* After birth these two septa, together with the left valve of the sinus venosus, will fuse to complete the final atrial septum.

In a slightly younger embryo the atria and ventricles communicated through a common canal, bounded by two thickenings named *endocardial cushions.* At the present stage the two cushions have joined midway, have received the septum primum, and now subdivided this passage into two *atrioventricular canals.* About the right canal the endocardium is already undermined and in the process of forming the *tricuspid valve* (b); similarly, on the left is the developing *bicuspid valve.* The two *ventricles* are separated by a *ventricular septum;* it is complete except for the *interventricular foramen* which connects the left ventricle with the bulbar part of the right (A,B). The *bulbus cordis* separates distally into *ascending aorta* and *pulmonary artery,* but proximally it is still undivided (b). The ventricular walls are thick and spongy, forming a meshwork of muscular trabeculae separated by sinusoids. Until later, when coronary vessels are developed, the heart receives all its nourishment from the blood circulating in the sinusoids.

Vascular System. Arteries. The aortic-arch system is still represented, although somewhat modified and in process of transforming into its permanent derivatives. The first two pairs of arches have disappeared. The *ascending aorta* continues into the third and fourth pairs of arches. The third pair and the extensions of the dorsal aortae into the head constitute continuous channels, to be known as the *internal carotid arteries.* Near their bases arise the *external carotid arteries,* which extend into the region of the lower jaw. The fourth aortic arch is largest, and on the left side it will form the permanent *arch of the aorta.* The sixth (fifth?) aortic arches connect with the pulmonary trunk, and from them small *pulmonary arteries* pass to the lungs; the left arch continues until birth as the *ductus arteriosus.*

The paired *dorsal aortae* unite opposite the eighth somites and continue caudad as the median *descending aorta.* The aorta shows dorsal, lateral and ventral branches. The dorsal branches pass upward between the somites, and accordingly can be called *intersegmental arteries.* From the seventh pair, which is located just where the dorsal aortae combine, the *subclavian arteries* pass off to the upper limb buds, and *vertebral arteries* run cephalad into the head. The latter vessels are formed by longitudinal anastomoses between the first seven pairs on each side, after which the stems of the first six atrophy. Under the brain the vertebrals are continuous with the unpaired *basilar artery;* the latter connects with the internal carotids beneath the diencephalon. Lateral branches of the descending aorta supply the mesonephroi and genital ridges. Ventral branches form the *coeliac artery* to the stomach region, the *superior mesenteric artery* (primitive vitelline) to the small intestine and the *inferior mesenteric artery* to the large intestine. The *umbilical arteries* (to the allantois and placenta) belong in this ventral series, but they now arise laterally from secondary trunks which persist as the *common iliacs.* Beyond this point the aorta narrows into the diminutive *caudal artery* extending into the tail.

Veins. Three sets of plexuses, which are the forerunners of the *dural sinuses,* occur alongside the brain. They drain into the *anterior cardinal veins,* now becoming the *internal jugular veins.* After receiving the newer *external jugular veins* from the mandibular region, the anterior cardinals open into the *common cardinal veins* (ducts of Cuvier). The latter vessels also receive the *sub-clavian veins* from the upper limb-buds and the *posterior cardinals* from the lower body. They empty into the sinus venosus. The *posterior cardinal veins* are the oldest veins caudal to the level of the heart. They course dorsal to the mesonephroi and drain the mesonephric sinusoids. However, the posterior cardinal veins are already beginning to decline, and midway along their lengths an interruption occurs; for this reason only the cranial halvescommunicate with the common cardinal stems.

Considerable diversion of blood from the posterior cardinal veins has been brought about by the development of *subcardinal veins* along the ventromesial surfaces of the mesonephroi. These vessels arose as longitudinal channels in a mesonephric plexus that was originally tributary to the posterior cardinal veins. Connections between the post- and subcardinal systems of each side still exist, while the two subcardinals also communicate by a prominent anastomsis across the midplane of the body. Their drainage is now shifting into the just organizing inferior vena cava. A fairly prominent *ventral vein* of the mesonephros follows the ventral border of this organ, but it soon disappears.

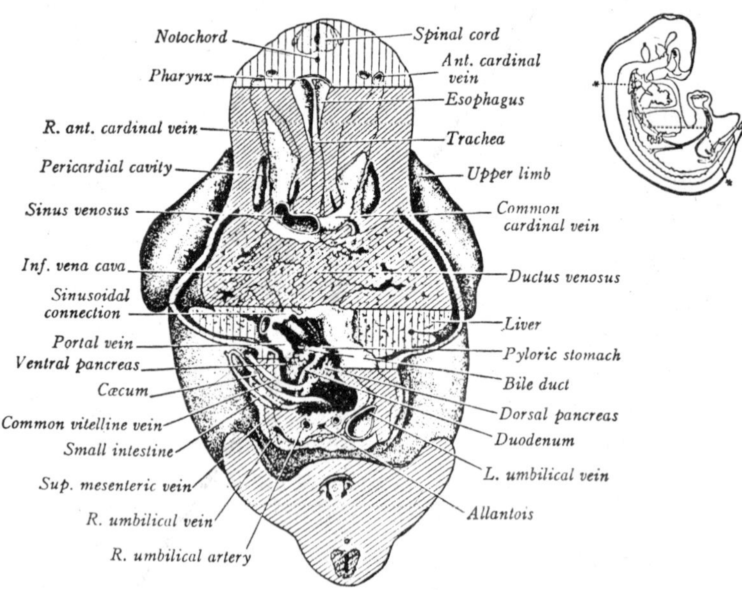

Fig. 3.8.   Ventral reconstruction of 10 mm. pig embryo, especially to show umbilical and vitelline veins.

The *inferior vena cava* is becoming established at this stage. It has a compound origin, some of which is now identifiable. In the mesonephric region the larger right subcardinal is an important component. More cephalad a vein has developed in a specialized portion (caval mesentery) of the mesogastrium. This vessel connects the subcardinal with the hepatic (vitelline) sinusoids). The blood-flow through the sinusoids is already consolidating into a definite channel, and this is the hepatic part of the inferior vena cava. It empties into the common hepatic vein (primitive right vitelline), which constitutes the stem of the vena cava.

The *umbilical veins* follow the allantoic stalk back from the placenta. In the umbilical cord they have merged into a common vessel, but they separate again on entering the embryo where they course in the ventrolateral body wall of each side to the level of the liver. The left umbilical is the larger of the two, and it alone persists in older fetuses. Cephalad of the liver the original stems that connected with the sinus venosus have disappeared, and umbilical blood is now routed through the liver in sinusoidal channels. Such an enlarged fetal passage, connecting the left umbilical with the inferior vena cava, is the important *ductus venosus*.

Distally the two *vitelline veins* are fused. Passing inward from the regressive yolk sac, they course cephalad of the intestinal loop. In the pancreas-region the left vein receives the *superior mesenteric vein* which is a new vessel arising in the mesentery of the intestinal loop. Above this junction a cross anastomosis and a continuation of the right vein make a new channel which is known as the *portal vein*. It gives off branches to the hepatic sinusoids, which have arisen much earlier from a breaking up of the vitelline veins in this region, and connects with the left umbilical vein within the liver. Beyond the sinusoids the vitelline vessels are retained as *hepatic veins* and the stem of the inferior *vena cava*.

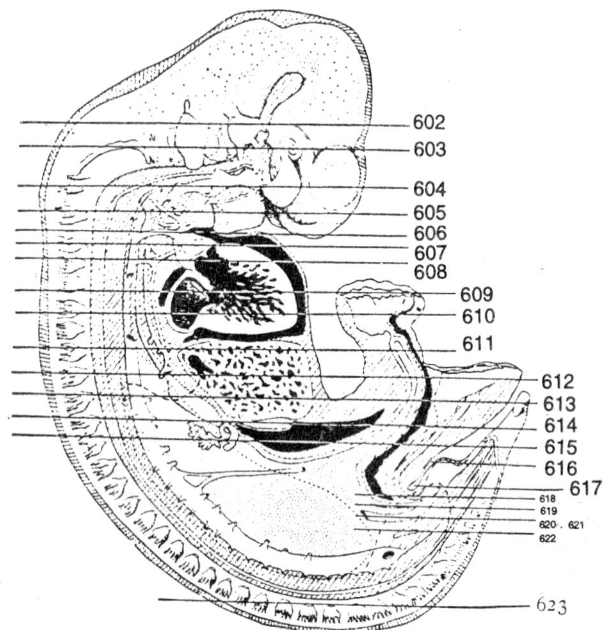

Fig. 3.9. Reconstruction of 10 mm. pig embryo.

TRANSVERSE SECTIONS OF A TEN MM. PIG EMBRYO

The more important levels, are illustrated and described. These are useful for the identification of organs, but the student must interpret his sections with reference to the dissections and reconstructions. All sections are drawn from the cephalic surface; accordingly, the right side of the embryo is at the reader's left.

Section through Cephalic Flexure. Since the head is flexed, the sections first encountered pass through the *mesencephalon* and *metencephalon.* At a slightly lower level the metencephalon also becomes continuous with the thin-roofed *myelencephalon,* but presently the mid-brain gives way to the *diencephalon* as the brain becomes cut twice. Several important structures should be identified in the mesenchyme between these two portions of the brain. In the midplane, but nearer the metencephalon, is the single *basilar artery;* ventrolateral to the diencephalon are the paired *internal carotids.* These three vessels unite at the location of the future *arterial circle* (of Willis). About halfway between the midplane and the lateral wall appear branches of the *anterior cardinal veins* and the *oculomotor* and *trochlear nerves.* Of the two nerves, the trochlear is smaller and slightly more lateral in position; in some series it extends only a slight distance.

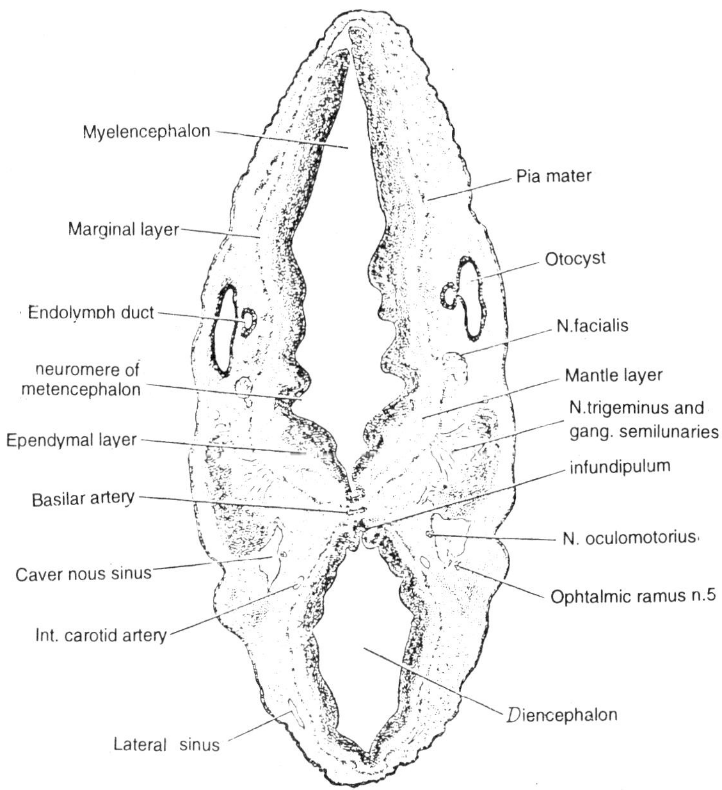

Fig. 3.10. Transverse section through semilunar ganglia and otocysts of 10 mm. pig embryo.

Section through Infundibulum, Semilunar Ganglia and Otocysts

The brain is sectioned twice. At the bottom of the section is the *diencephalon,* cut transversely; its cavity is the *third ventricle.* Midventrally the diencephalon gives off the *infundibulum* which furnishes the neural lobe of the *hypohysis.* The *metencephalon* and *myelencephalon* are sectioned frontally; there is no clear demarcation between the two. Their walls bear the prominent scalloping of the *neuromeres,* while the common cavity is the *fourth ventricle.* The wall of the entire neural tube now shows a differentiation into three layers : (1) an inner *ependymal layer,* densely cellular, next the central canal; (2) a middle *mantle layer,* of nerve cells and fibers; and (3) an outer *marginal layer,* chiefly fibrous. A thin, vascular layer surrounds the brain wall as the primitive *pia mater.*

The interval between the two portion of the brain contains several structures, sectioned transversely. Next the metencephalon is the unpaired *basilar artery;* ventrolateral to the diencephalon are the paired *internal carotid arteries.* Near the latter the *oculomotor nerves.* In this embryo the trochlear nerves had not grown down to this level. On the left side is a part of the ophthalmic branch of the *trigeminal nerve.* Tributaries of the *anterior cardinal veins* occur, the largest rostral to the semilunar ganglia. This is a portion of the *cavernous sinus,* while the stem of the *middle dural plexus* is just caudal to the semilunar gaglion; alongside the diencephalon the *lateral sinus* is cut.

Near the beginning of the hind-brain are the large *semilunar gangila;* from their medial sides nerve fibers of the *trigeminal nerves* join the brain wall. This ganglion, situated at the pontine flexure of the metencephalon, constitute one of the most important landmarks of the embryonic head. Slightly caudad lie the *facial nerves;* the left is cut as it leaves the brain wall. Midway along each side of the hind-brain will be seen the apex of an *otocyst,* and mesial to it, the *endolymph duct;* on the left side the latter is cut near its origin from the otocyst.

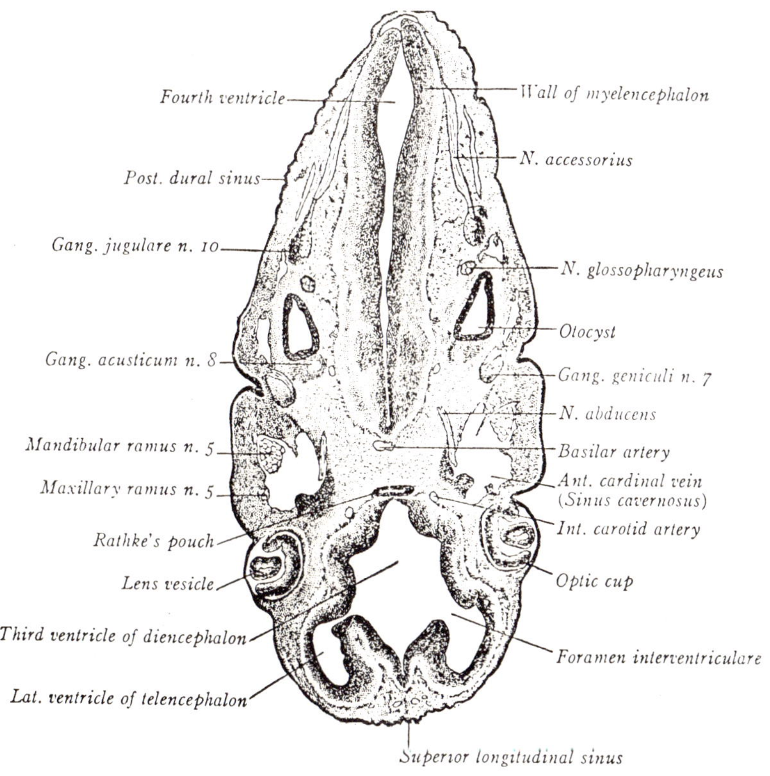

Fig. 3.11.   Transverse section through cerebral hemispheres and eyes of 10 mm. pig embryo.

Section through Cerebral Hemispheres and Eyes. This level shows some important new features. The *diencephalon* is now continuous with the *telencephalon*. The latter consists of a mesial region which has evaginated paired *cerebral hemispheres*; their cavities, the *lateral ventricles*, connect through the *interventricular foramina* with the *third ventricle* of the diencephalon. Close to the ventral wall of the diencephalon is a section of the epithelial lobe of the hypophysis *(Rathke's pouch)*, near which are the *internal carotid arteries.* Lateral to to the diencephalon are the *optic cups,* sectioned just caudal to their stalks. The double wall of the optic cup comprises the *retina;* the thin outer layer is the pigmented epithelium, the inner and thicker coat is the nervous layer. The *lens* is now a closed vesicle, distinct from the overlying *corneal ectoderm.*

The irregular vascular spaces are tributaries of the *anterior cardinal veins.* The largest space is the *cavernous sinus,* in the vicinity of the fifth nerve. The upper half of the section contains portions of the *posterior dural plexus,* while the two small vessels between the cerebral hemispheres at the bottom of the section represent the *superior longitudinal sinus.*

By working above and below the present level all the cranial nerves and ganglia, as well as the central connections and peripheral courses of these nerves, will be observed. In Fig. 3., transverse sections of the maxillary and mandibular branches of the *trigeminal nerve* are seen, while the *abducent nerve* is sectioned longitudinally as it passes from the under surface of the hind-brain toward the eyes. On the rostral side of the otocyst occur the *geniculate ganglion* of the *facial nerve* and the *acoustic ganglia* of the *acoustic nerve.* The *otocyst* is a sharply defined, epithelial sac which lies at the junction of the metencephalon and myelencephalon and makes a convenient landmark in identifying ganglia and nerves. Caudal to the otocyst, the *glossopharyngeal nerve* and the *jugular ganglion* of the *vagus nerve* are cut transversely while the trunk of the *accessory nerve* is sectioned lengthwise as it curves forward from the level of the spinal cord.

Section through Mouth, Tongue and First and Second Pouches. The tip of the head, with parts of the *telencephalon* and *olfactory pits,* is now separate from the rest of the section. Since the level last described, Rathke's pouch has opened into the ectodermal *stomodeum* between the jaws; the present section is at the actual *oral opening,* bounded by the *maxillary* and *mandibular processes* of the *first branchial arches.* At a slightly lower level, the mandibular processes merge in *lateral lingual swellings* which will become the body of the *tongue*. With the disappearance of the pharyngeal membrane, the stomodeum and entodermal mouth cavity have become continuous. The *pharynx* shows ventral portions of the *first* and *second pharyngeal pouches*, destined to be utilized as auditory tubes and tonsillar fossae, respectively. Opposite the first pouch, externally, is the *first branchial groove,* or future external acoustic meatus. A shaving has been cut from the *tuberculum impar* of the *tongue* as it rises above the floor of the pharynx; the upper (caudal) end traces into connection with the second arches which unite as the *copula.* The *facial nerves* of the *second branchial arches* are cut across, but since the previous level the *trigeminal nerves* have ended in the maxillary and mandibular processes of the first arches.

The *mylencephalon* is sectioned close to its continuation into the spinal cord; *Froriep's ganglion* and some of the *accessory nerve* are included. Between the myelencephalon and the pharynx are seen on each side the several rootlets of the *hypoglossal nerve,* fibers of the *vagus* and *accessory nerves,* and the *petrosal ganglion* of the *glossopharyngeal nerve* which stands in relation to the third arch. Mesial to the petrosal ganglia are the *dorsal aortae,* and lateral to the vagi are the *anterior cardinal veins.* In the midplane is a bit of the *notochord,* cut lengthwise. The *basilar artery* still lies beneath the myelencephalon, but a short distance caudad it is replaced by the paired *vertebral arteries.*

Section through Third Pouches and Olfactory Pits. The tip of the head is now small and includes on either side the open *olfactory pits,* lined with thickened ectodermal epithelium. Each pit is bordered by a *lateral* and a *median nasal process,* which assist in the formation of the nose and upper jaw. In preceding sections the epithelium within the pit forms a superficial plate that is the *oro-nasal membrane;* it ruptures later and produces a *primitive choana.* The first three pairs of *branchial arches* show; the first is fused as the *mandible,* and the third is slightly sunken in the *cervical sinus.* The dorsal wings of the *third pharyngeal pouches* extend toward the ectoderm of the *third branchial grooves;* attached to these wings are prominent *parathyroid* primordia. The ventral wings are large, epithelial sacs that can be followed into succeeding sections; one is shown here as a separate drawing. They give rise to the *thymus.* The floor of the pharynx is sectioned through the *epiglottis.*

Ventral to the pharynx are portions of the *third aortic arches* (internal carotids) and the solid cords of the *thyroid gland. (External carotid arteries* arise at a slightly higher level, close to the ventral origins of the third aortic arches; they can be traced for a variable distance into the substance of the lower jaw). Beneath the thyroid and in the mandible are portions of the *external jugular veins.* Dorsally the section passes through the *spinal cord* and the first pair of *cervical ganglia.* Between the spinal cord and pharynx, laterally, are the *anterior cardinal veins,* the *hypoglossal nerves* and the *nodose ganglia* of the *vagi.* Lateral to each ganglion is the external branch of the *accessory nerve* and mesial to the ganglia are the small *dorsal aortae.* The *notchord* is conspicuous in the midplane.

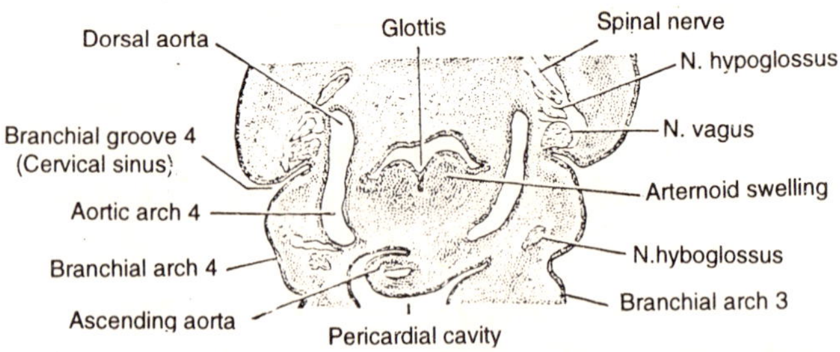

Fig. 3.12.　Transverse section through glottis and fourth aortic arches of 10 mm. pig embryo.

Section through Glottis and Fourth Aortic Arches. At this level the first three branchial arches have been passed and the cephalic border of the heart is coming into view. The illustration includes only the region of the *fourth branchial arches,* and the fourth pair of *aortic arches* which course through them. The left aortic arch will become the permanent *arch of the aorta,* the right the stem of the right *subclavian artery.* The *ascending aorta* connects with these aortic arches a little cephalad in the series. The *fourth branchial groove* opens into the *cervical sinus* which is seen here. This duct extends for some distance before coming in contact with the fourth pharyngeal pouch. The section cuts across the *pharynx,* the *glottis* (entrance to the larynx), and its bordering *arytenoid swellings.* In addition to a spinal nerve, section of the *vagus* and *hypoglossal nerves* are encountered.

Section through Fourth Pouches and Larynx. The head has been passed and the section is now dominated by the *heart,* lying within its *pericardial cavity.* The tips of the *atria* are sectioned as they project at the sides of the *bulbus cordis.* The bulbus is dividing into the aortic stem and the pulmonary trunk. The small section of the *ascending aorta* traces cephalad into the third and fourth aortic arches and caudad into the ventricle. The *pulmonary trunk* is cut twice; its distal portion can be followed caudad into connection with the sixth aortic arches.

The crescentic *pharynx* is continued laterad as the small *fourth pharyngeal pouches.* Each gives origin to a dorsal wing (*parathyroid*), encountered a few sections cephalad in the series and shown here as a separate drawing. At the level of the main section the pharynx is also continuous with a saccular *ultimobranchial body.* From the midventral wall of the pharynx arises the solid epithelial plate of the *larynx.* A section of the *vagus nerve* is located between the dorsal aorta and the *anterior cardinal vein* of each side. Ventral to the anterior cardinals (soon to be called the *internal jugular veins)* are small *external jugular veins.* The left *dorsal aorta* is larger than the right in anticipation of its conversion into the permanent descending aorta of this level.

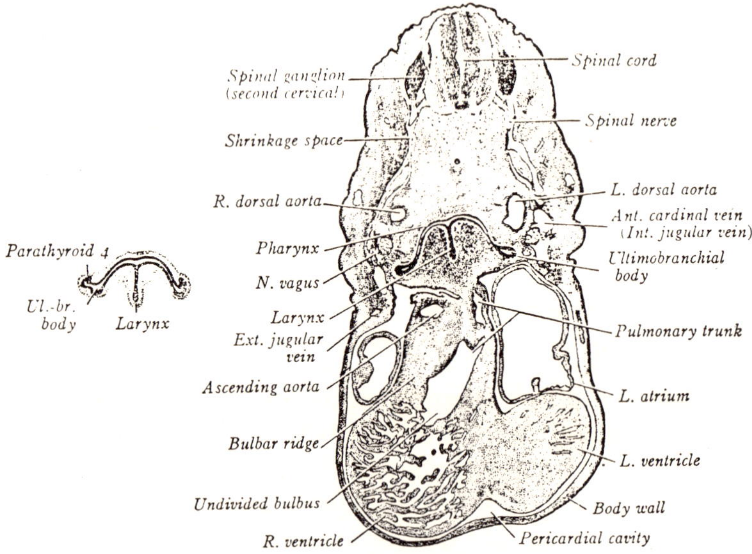

Fig. 3.13.   Transverse section through fourth pharyngeal pouches, ultimobranchial bodies and larynx of 10 mm. Pig embryo.

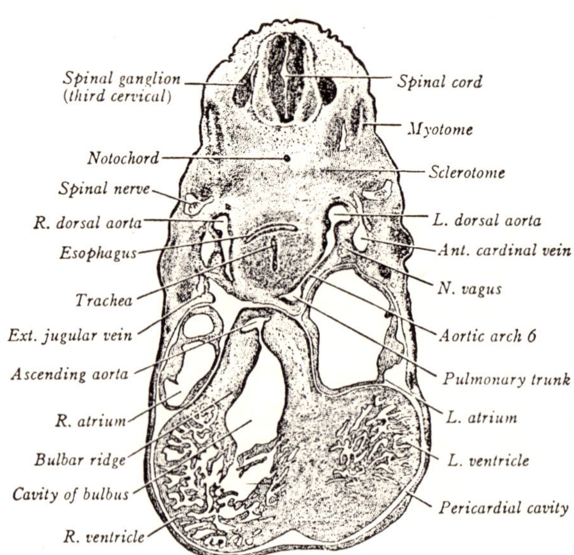

Fig. 3.14.   Transverse section through pulmonary arches and bulbus cordis of 10 mm. pig embryo. X 23.

Section through Pulmonary ARches and Bulbus Cordis. The heart is little changed from the last level. However, the *ascending aorta* communicates with the *bulbus*, while the latter shows on each side the thickened internal *ridges* that will progressively meet and continue the separation of the aortic and pulmonary trunks. The *sixth aortic arches* connect the dorsal aortae with the main *pulmonary trunk*, whose origin was traced in the preceding section. On the left side of the embryo the arch is complete; it represents the *ductus arteriosus*, which remains patent until birth. From these pulmonary arches small *pulmonary arteries* may be traced caudad in the series toward the lungs. The *esophagus* is now separate from the *trachea;* both are cut through their extreme cephalic ends. Ventrolateral to the spinal cord are somewhat diffuse *myotomes*, while *sclerotome* masses surround the notochord. The *vagus nerves* are prominent. At about this level the external jugular veins join the *anterior cardinals.*

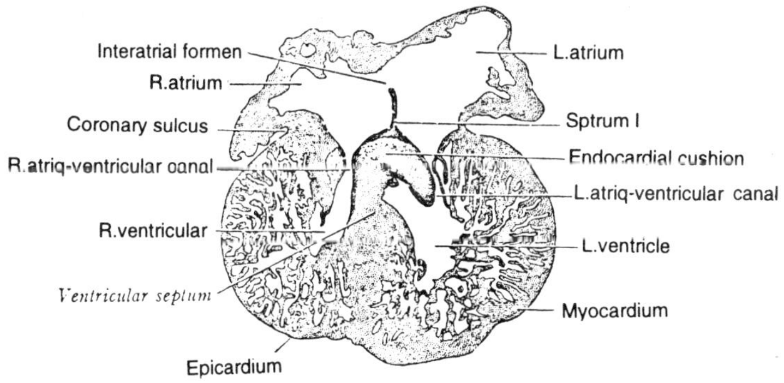

Fig. 3.15. Transverse section through heart of 10 mm. pig embryo.

Section through Heart and Interatrial Foramen. Only the heart is figured, taken from a total section much like. All four chambers are shown. The *Atria* are partially separated by the *septum primum*, which is incomplete because of the *interatrial foramen;* this foramen will remain open until birth. A septum secundum is not seen satisfactorily in embryos of this size. Each atrium communicates with the ventricle of the same side through an *atrio-ventricular canal.* Between these openings is the fused portion of the *endocardial cushions.* At an earlier stage these cushions were double, but they have fused midway and thus divide the originally single canal into two; they will also help in the formation of the *bicuspid* and *tricuspid valves.* The atria are marked off externally from the ventricles by the *coronary sulcus.* Between the two ventricles is the *ventricular septum;* a little higher in the series of sections the thin region (above leader) gives way to a temporary *interventricular foramen.* The ventricular walls are thick and spongy, forming a network of muscular *trabeculae* surrounded by blood spaces, or sinusoids. This muscular layer constitutes the *myocardium.* It is lined by an endothelial layer, the *endocardium,* while the entire heart is surrounded by a layer of mesothelium, the *epicardium,* or visceral pericardium.. The latter sac is continuous with the parietal pericardium, which lines the body wall in the region of the heart.

Section through Common Cardinals and Sinus Venosus. The section is also marked by the large *heart* and the bases of the upper *limb buds.* Dorsal to the atria are the *common cardinal veins.* The right vein empties into the *sinus venosus;* the left crosses the midplane and connects with the sinus at a lower level. Just above the plane of this section the right common cardinal has received the right *subclavian vein* from the limb bud; the left subclavian is still separate. The sinus venosus drains into the right atrium through a slit-like opening in the dorsal and caudal atrial wall. The opening is guarded by the right and left *valves of the sinus venosus,* both of which project into the atrium. The *septum primum* appears like a complete membrane at this level, which is a little caudal to the interatrial foramen and the atrioventricular canals. The septum joins the fused *endocardial cushions,* as does the *ventricular septum* from below.

The *esophagus* and *trachea* are tubular. Around the epithelium of both are condensations of mesenchyme from which their fibrous and muscular layers are to be differentiated; laterally in this mass lie the *vagus nerves,* unlabeled in the illustration. Ventral to the trachea are the *pulmonary arteries.* The left *dorsal aorta* is large and is here continuing the arch of the aorta caudad; the right dorsal aorta at this level forms a part of the right subclavian artery. dorsolateral to the aortae are *autonomic ganglia.* The condensation of mesenchyme about the notochord foreshadows a future *vertebra.*

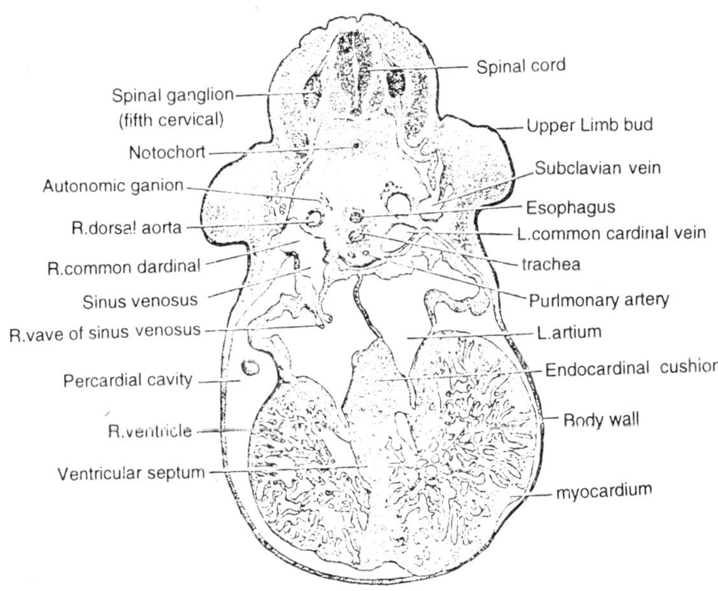

Fig. 3.16. Transverse section through common cardinals and sinus venosus of 10 mm. pig embryo.

Section through Branchial Plexus and Tracheal Bifurcation. Other distinctive features are the presence of upper *limb buds* and the *liver.* The seventh pair of cervical *spinal nerves* is cut lengthwise in diagrammatic fashion. The spindle-shaped ganglion is associated with the *dorsal root* whose fibers grow out from its cells; the fibers of the *ventral root* arise from the middle, cellular layer of the cord and join the dorsal root in the common *nerve trunk*. On the right side, a short *dorsal ramus* unites with similar rami of several other nerves to form the *brachial plexus*, a part of which is seen. Adjacent to the esophagus are the cut *vagus nerves*, which follow this tube laterally.

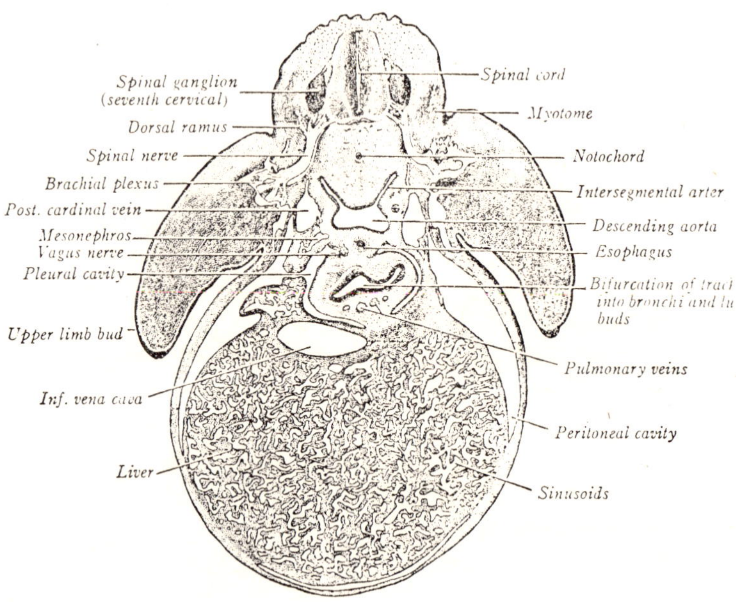

Fig. 3.17. Transverse section through brachial plexus and tracheal bifurcation of 10 mm. pig embryo.

The *descending aorta* shows its manner of origin from paired vessels. From the seventh pair of *intersegmental arteries*, which arise dorsally from it, the *subclavian arteries* are given off two sections caudad in the series. Traced cephalad these seventh intersegmentals become continuous with the *vertebral arteries.* The latter lie mesial to the stem portions of the spinal nerves. In some embryos they are imperfectly developed at this stage. Lateral to the aorta are the *posterior cardinal veins,* easily traced to the common cardinals of the previous figure.

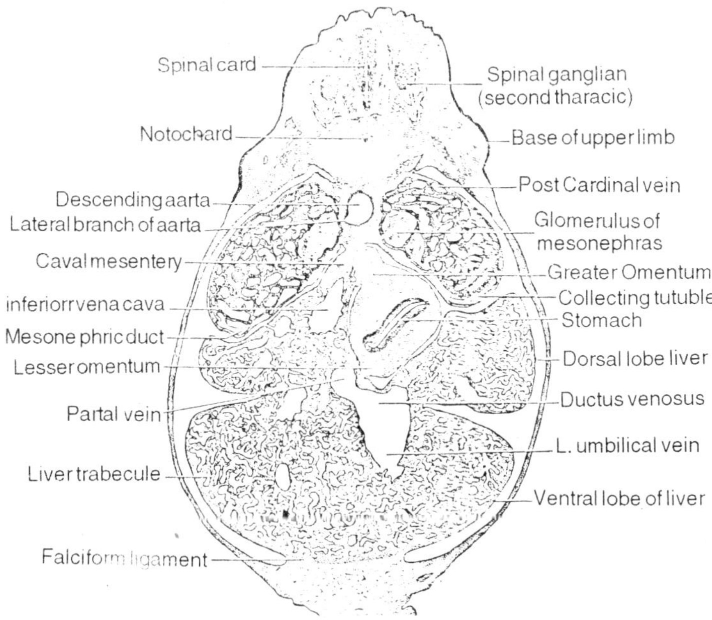

Fig. 3.18. Transverse section through stomach and liver of 10 mm. pig embryo.

Beneath the *esophagus* the trachea branches into short *primary bronchi.* These continue laterad into buds which represent the upper left and middle right lobes of the *lungs.* The upper right bud (unpaired) comes off the trachea a little cephalad in the series of the paired lower lobes are more caudad. Crescentic *pleural cavities* bound the bulging pulmonary tissue laterally.

On the embryo's left this cavity is separated from the peritoneal cavity by a *pleuro-peritoneal membrane,* and here the parietal and visceral pleura assume typical relations. The broad mesial mass of mesenchyme containing the esophagus and lungs is the *mediastinum.* Ventral to the bifurcating trachea are sections of the *pulmonary veins,* which can be traced into the left atrium at a slightly higher level. The liver, with its close network of trabeculae and sinusoids, is large and nearly fills the *peritoneal cavity.* A little cephalad in the series its attachment to the septum transversum is seen better. In the present section the *inferior vena cava* is leaving the liver and entering the septum. This portion of the vena cava is, in reality, the *common hepatic vein* (stem of the primitive right vitelline).

Section through Lungs, Upper Limbs and Autonomic Ganglia. The *limb buds* are ectodermal sacs stuffed with dense, undifferentiated mesenchyme. Ectodermal thickening, at the tips, exert control over their progressive organization. Flanking the now circular *descending aorta* are the cranial ends of the *mesonephroi,* while above the aorta is a pair of *autonomic ganglia;* the *communicating rami* connecting the spinal nerve trunks do not show but can be seen in neighboring sections. The esophagus is just beginning to dilate into the *stomach,* and at this level the *vestibule* of the omental bursa appears as a crescentic slit to the right and below it. The mediastinum of higher levels has changed into a typical mesentery of the lower esophagus (*mesoesophagus*) and stomach (*mesogastrium,* or *omentim*). The *lungs* are sectioned through their paired lower lobes. Both *pleural cavities* still communicate freely with the peritoneal cavity. In the right dorsal lobe of the liver is located more of the intrahepatic portion of the *inferior vena vaca;* this particular segment is organizing from enlarged hepatic sinusoids. Near the midplane is the large *ductus venosus,* which traces into union with the vena cava a short distance cephalad. The *posterior cardinal veins* are coming into intimate relation with the *mesonephroi;* these temporary kidneys are out near their cranial ends.

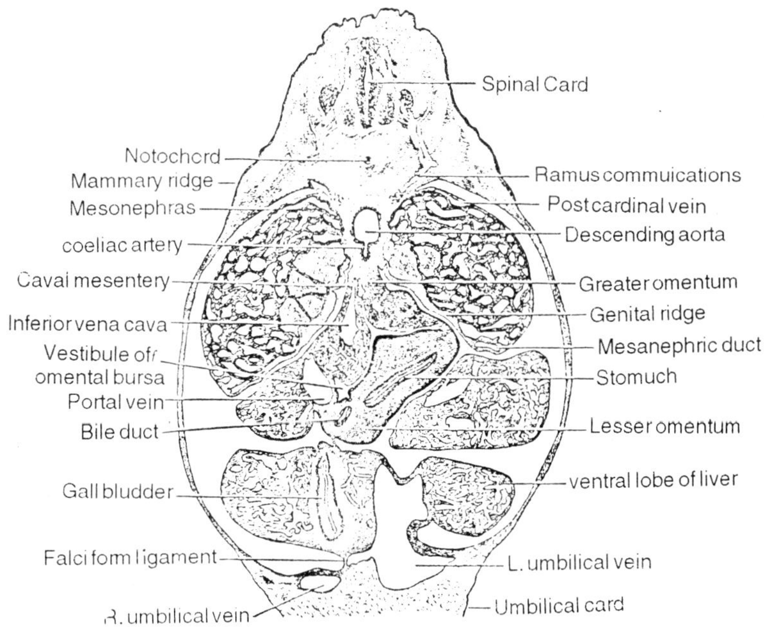

Fig. 3.19.   Transverse section through pyloric stomach and gall bladder of 10 mm. pig embryo.

Section through Stomach and Liver. It is the *stomach* and lobate *liver* that feature this level. The stomach has rotated partially, so that its original dorsal margin is now dropping to a position on the embryo's left whereas the primitive ventral margin is raising correspondingly on the right. These margins are to be the *greater* and *lesser curvatures*, respectively. The stomach is attached dorsally by the *greater omentum*; ventrally the *lesser omentum* passes to the liver. This ventral mesentery splits into halves and is continued as a peritoneal reflection around the liver; the component layers then come together again as the *falciform ligament*, attaching the midventral border of the liver to the body wall. Both the body wall and the abdominal viscera are thus seen to be surfaced with a continuous sheet of mesothelium underlaid by mesenchyme; this serious investment is the *peritoneum,* in which a parietal and a visceral division is recognized.

The liver shows paired dorsal and ventral lobes. The right dorsal lobe is fused dorsally to the greater omentum. The connection forms the *caval mesentery,* in which courses the *inferior vena cava.* Between the attachments of the stomach and liver, and to the right of the stomach, is the *vestibule* of the omental bursa. Midventrally in the liver is the *ductus venosus,*sectioned just at the point where it receives the intraphepatic continuation of the left *umbilical vein* and a branch from the *portal vein.* The liver tissue is a complicated network of trabeculae and sinusoids; the component *liver cords* are composed of hepatic cells, surrounded by the endothelium of the sinusoids; red blood cells are developing here at this stage.

The *mesonephroi* are becoming prominent organs. Along their ventral margins course the *mesonephric ducts;* each shows a connection laterally with the terminal segment of a *collecting tubule.*

Section through Pyloric Stomach and Gall Bladder. The section grazes the third thoracic *ganglia,* passes slantingly through the pyloric end of the *stomach, then* cuts the *con non bile duct,* and finally passes lengthwise through the *gall bladder* superficially embedded in the substance of the *liver.* Within the distance of a few sections it is easy to demonstrate the continuity of these components parts of the digestive system. The *greater omentum* of the stomach is larger and more folded than in the previous illustration, and the *omental bursa* (lying horizontally over the stomach) is correspondingly expansive. It opens into a vertical space which is the *vestibule;* traced caudad a short distance the region of the vestibule, opens into the peritoneal cavity through the *epiploic foramen* of (winslow).

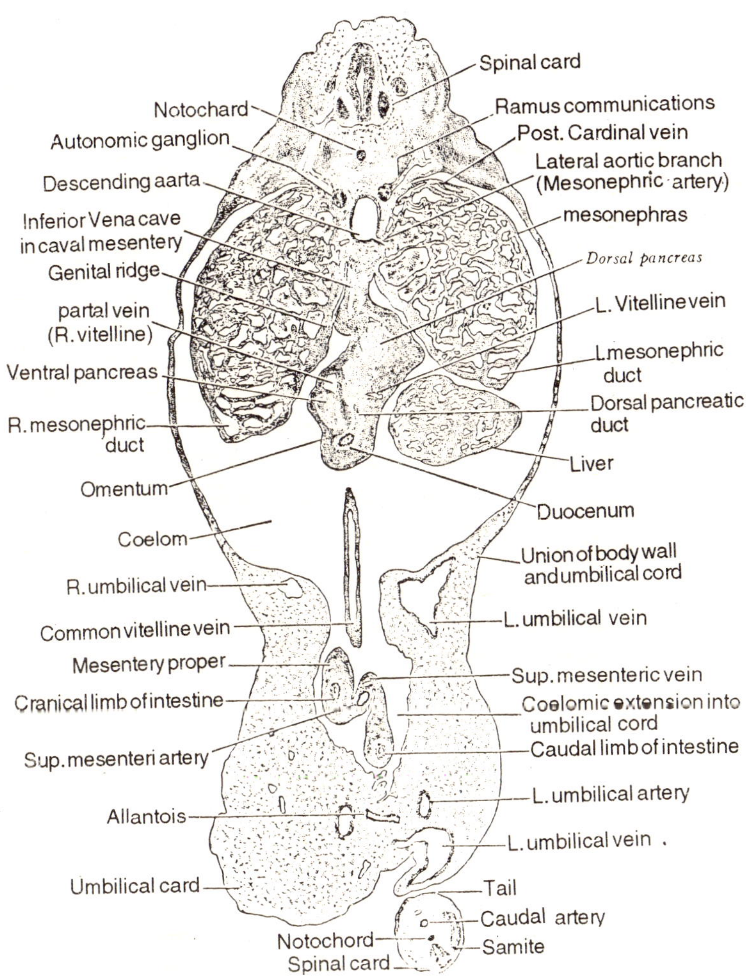

Fig. 3.20. Transverse section through pancreas, intestinal loop and tail of 10 mm. pig embryo.

Beneath the stomach is a caudal portion of the *lesser omentum.* The blood supply of the stomach-pancreas region comes from the *caeliac artery,* which is seen emerging as a ventral branch of the aorta.

The liver has nearly been left behind, and its dorsal and ventral sets of lobes are now separate. Associated with the liver are several veins. Dorsally is the *inferior vena cava,* about to leave the liver within a lip-like fold, the *caval mesentery.* Also in the right upper lobe is the *portal vein.* Ventrally, where the umbilical cord joins the body wall, are the *umbilical veins;* the larger left vessel is entering the left ventral lobe of the liver on its way to the ductus venosus. On each dorsolateral surface of the trunk is a thickened ectodermal ridge, which represents the *mammary ridge.* At the same horizontal level a *ramus communicans* passes from the left nerve trunk to an *autonomic ganglion.* The bottom of the figure includes a little of the insertion of the *umbilical cord* on the abdominal wall.

Section through Pancreas, Intestinal Loop and Tail. The bulging *mesonephroi* are conspicuous. *Mesonephric corpuscles,* with vascular *glomeruli* indenting them, are medial in position. The glomerull receive *mesonephric arteries,* arising as lateral branches from the aorta; one of these is shown. The *mesonephric tubules* are contorted and variously sectioned; they are lined with a cuboidal epithelium and empty into the *mesonephric duct* coursing along the ventral margin of the gland. On the mesial surface of each mesonephros the epithelium is thickened; from these *genital ridges* the sex glands will differentiate. The *posterior cardinal veins* lie on the dorsal surfaces of the mesonephroi, and transient *ventral veins* of the mesonephroi course along the ventral borders. The *inferior vena cava* is a vertical slit in the caval mesentery. Traced caudad a short distance it joins the right subcardinal vein, which continues this important venous channel down the trunk. The *subcardinal veins* are prominent vessels, seen well at a slightly lower level on the ventromesial surfaces of the *mesonephroi.* There they interconnect by a large anastomosis.

The *duodenum* lies within its dorsal mesentery (*mesoduodenum*). The duct of the lobulated *dorsal pancreas* is shown arising directly from the duodenal wall. More to the right is a section of the *ventral pancreas* which traces cephalad to its origin from the stem of the common bile duct. On each side of the dorsal pancreas are portions of the *vitelline veins*, the right at this level being a part of the *portal vein*. A few sections caudad this vessel can be followed across a transverse anastomosis to the left vitelline vein where the *superior mesenteric vein* from the mesentery attaches. Beyond the duodenum the vitelline veins have fused into a common vessel, cut lengthwise in the present section; it can be followed to the yolk sac, where the right and left components again separate.

The ventral body wall is continuous with the *umbilical cord*. This contains as extension of the embryonic *cvaelom*, and a portion of the *intestinal loop* within its *mesentery*. Between the two intestinal limbs are the *superior mesenteric artery* and *vein*. The former is a ventral branch of the aorta; the latter joins the portal vein. The *allantois* is flanked by *umbilical arteries*, while *umbilical veins* are cut in the cord and again as they enter the body wall. The tip of the recurved *tail* shows as a separate section; its structure is simple.

Section through Cloacal Membrane. To maintain the proper relations with sections already studied, this and succeeding sections through the curved, caudal region are shown dorsal side down. The caudal end of the embryo is small. Its laggard differentiation, in comparison to higher levels, is reflected in the less specialized *spinal cord* and *somites.* the slender *tail-gut* is cut across. Between the *notochord* and tail-gut is the continuation of the aorta, known here as the *caudal artery*. On each side of the latter lies the termination of a *posterior cardinal vein*. The ventral half of the section is featured by an epithelial plate that represents the solid *cloacal membrane*. The plane of section is such that the fusion of ectoderm with the entoderm of the plate is shown only at one end.

Section through Subdividing Cloaca. Tracing a short distance down the series from the previous level, the tail-gut joins the *cloaca* and the latter gains a cavity. Still farther, at the present level, the cloaca is separating into a dorsal *rectum* and a ventral *urogenital sinus.* (If should be understood that the recurvation of the tail-end of the embryo makes caudal progress in the sectioned series actually cephalad on this part of the embryo.

Section through Allantois, Urogenital Sinus and Rectum. This illustration (and the four that follow) include only the caudal, recurved part of the embryo. In the ventral body wall is seen the *allantoic stalk,* accompanied by the *umbilical arteries;* slightly cephalad in the series they all enter the umbilical cord. More dorsad are the crescentic *urogenital sinus* and the *rectum.* now separate. The caudalmost portion of the *caelom* tapers to an end between the two.

Section through Stems of Mesonephric Ducts and Allantois. The *colon* is contained within a portion of the mesentery that is specifically named the *mseocolon.* In the body wall are tributaries of the *umbilical veins,* and mesial to them are the *umbilical arteries.* The *allaitcic staik* is sectioned as it opens into the *urogenital sinus.* Dorsal to the sinus is a section of *rectum,* separated from the from the sinus by a crescentic prolongation of the *caelom.* The rectum has no mesentery. The horns of the urogenital sinus receive the *mesonephric ducts.*

Section through Lower Limb Buds and Ureteric Stems. The section cuts through the middle of both lower *limb buds.* Like the upper set already studied, they consist of undifferentiated mesenchyme contained within an epithelial sac whose apex shows marked thickening. Mesial to the limb buds are the *umbilical arteries,* which in turn lie lateral to the *mesonephric ducts.* The left mesonephric duct is cut at just the proper plane to show the *ureter,* or duct of the mesonephric duct is cut at just the proper plane to show the *ureter,* or duct of the metanephros, being given off dorsally. The right ureter is sectioned transversely and appears as a separate tube. The *colon* is cut lengthwise; tracing cephalad in the series, it becomes continuous with the colon and rectum.

Section through Mesonephric Ducts and Ureters. Continuing down the series the ureters can be traced for some distance. The present illustration is a small part of the section close to Fig. 3.37. In it the right *ureter* is seen. Also the right *mesonephric duct* is cut lengthwise (frontally) as it leaves the mesonephros on its way to connect with the urogenital sinus.

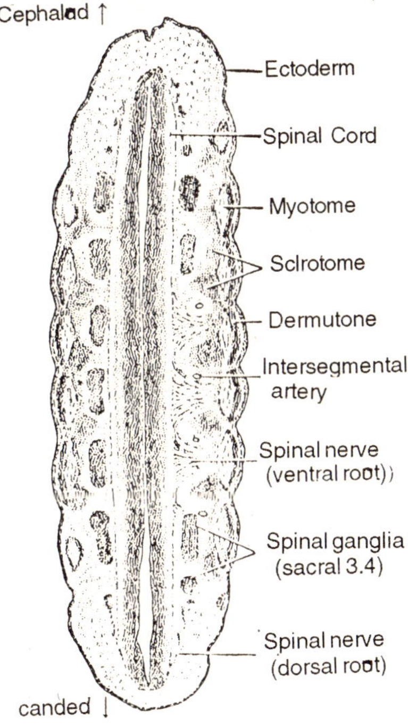

Fig. 3.21.   Transverse section through curved back of 10 mm. pig embryo.

Section through Metanephroi. The ureters are found to terminate in the *metanephroi*, Figured here. Each of these kidney primordia consists of two parts. Internally there is a dilated expansion of the ureter that represents the *renal pelvis;* from it, first the *calyces* and then the system of *collecting tubules* will bud and grow. The periphery of the double primordium is a mass of condensed mesenchyme, derived from nearby nephrotomes; this tissue will differentiate into the *secretory tubules* of the kidney. *Iliac veins,* branches of the posterior cardinals, are passing into the limb buds.

Section through Curved Back. Owing to the lumber curvature, this section is actually frontal. The *spinal cord* is cut lengthwise. It is flanked by *spinal ganglia,* except midway on the left side where a slightly deeper plane includes several *spinal nerves* The ganglia include the third lumbar to the fourth sacral. The sometimes show spindle-shaped *myotomes* and more laterally located '*dermatomes.*' The medial side of each somite is a *selerotome;* this shows subdivision into a caudal denser and a cranial less dense half. Recombination of the dense half of one somite with the sparser half of the somite next caudad will produce a definitive *vertebra. Intersegmental arteries* appear between some of the sometimes on the left side.

ANATOMY OF AN EIGHTEEN MM. PIG EMBRYO

Most of the important organs are laid down in 10 mm. embryos. Older stages are chiefly instructive, therefore, to demonstrate the growth and differentiation of parts already present, rather than the introduction of new ones. Dissections show perfectly the form and relations of organs, their relative rates of growth and changes of position. Since the illustrations indicate better than descriptions the several structures and their states of development, only certain selected features will be mentioned.

External Form. The neck and back are much straighter than before, but the ventral body is still highly convex. The head is relatively larger, the umbilical cord smaller. The sense organs are prominent, and the face, with snout and jaws, is plain. The branchial grooves and cervical sinus have disappeared from the neck. The limbs show indications of proximal and distal divisions, and the hand and foot are paddle-like Several mammary gland primordia occur along the mammary ridges, now locate, more ventrally. The genital tubercle has become a distinct phallus.

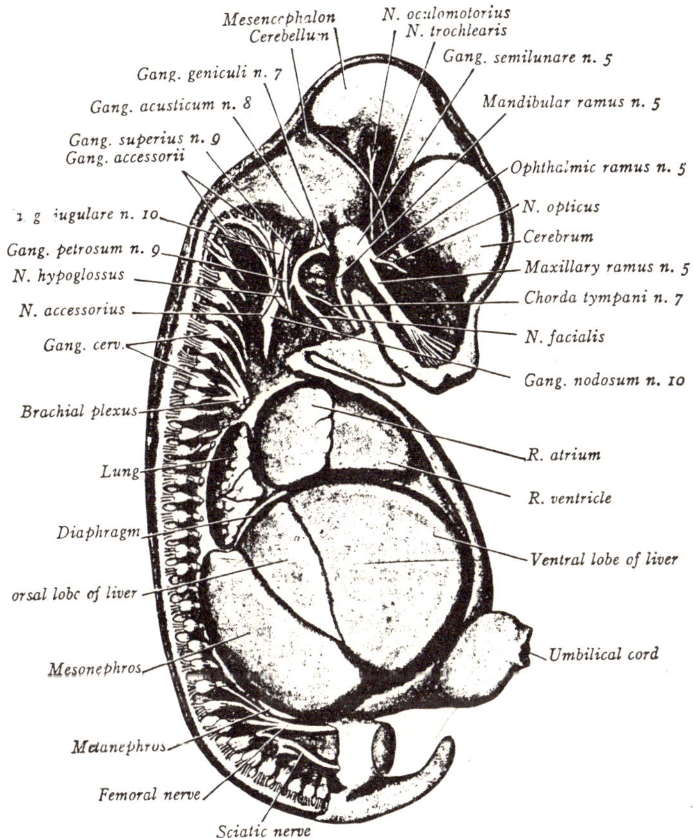

Fig. 3.22.   Lateral dissection of 18 mm. pig embryo, viewed from right side.

Lateral Dissection.   The cerebral hemispheres are larger and the cerebellum is appearing. Beneath the cerebellum is the prominent pontine flexure of the brain, pointing ventrad. Nerves and ganglia show clearly; the brachial and lumbo-sacral plexuses, opposite the limbs, are noteworthy. The liver and lungs are relatively larger and more plainly lobed than before; the heart and mesonephroi are smaller.

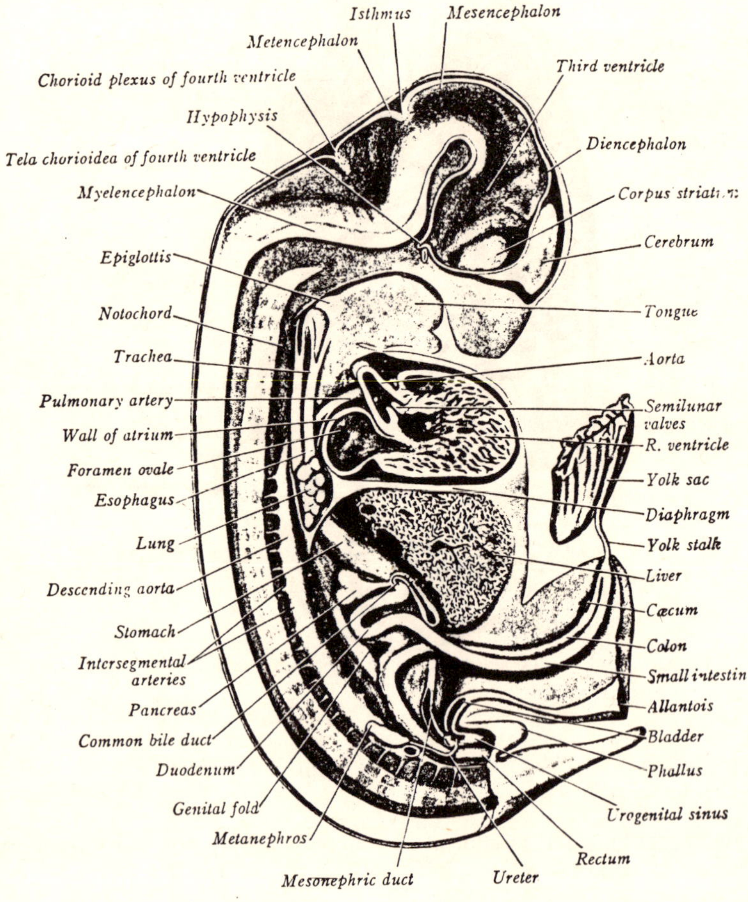

Fig. 3.23. Median dissection of 18 mm. pig embryo, after removal of right half.

Midsagittal Dissection. The corpus strain has developed in the floor of the cerebral hemisphere, a chorioid plexus invades the fourth ventricle, and neural (posterior) lobe of the hypophysis is growing into association with the detached Rathke's pouch. Sclerotomic primordia of vertebrae condense about the notochord. The viscera show only quantitative changes from the 10 mm. stage, but the urogenital sinus and rectum are now separate, as are the aorta and puimonary artery. The intestinal loop has rotated until the cranial and caudal limbs lie right and left, respectively. The caecum is conspicuous and a urinary bladder has developed between the allantois and urogenital sinus.

Ventral Dissections. The lungs, septum transversum, stomach, intestine and mesonephroi are the chief organs seen in the 15 mm. Of special interest are the beginnings of the Mullerian ducts.

ANATOMY OF A THIRTY-FIVE MM. PIG EMBRYO

External Form. The embryo is straighter, slenderer and its ventral surface less protuberant. The head, with its prominent snout, is shaping like that of a lower mammal, and the neck becomes distinct. Digits have appeared on the elongate extremities. The umbilical cord and tail are losing rapidly in relative size.

Lateral Dissection. The spinal cord and brain are relatively smaller, but the latter is becoming highly specialized and folded. The cerebral hemispheres are large, and olfactory lobes extend forward from the rhinencephalon. The body of the embryo elongates faster than the spinal cord, so that the spinal nerves, at first directed at right angles, course obliquely in the lumbo-sacral region. Note especially how the viscera have 'receded' caudad and how the liver dominates the abdomen as the mesonephros loses prominence. The kidney is exceptional in that it shifts cephalad.

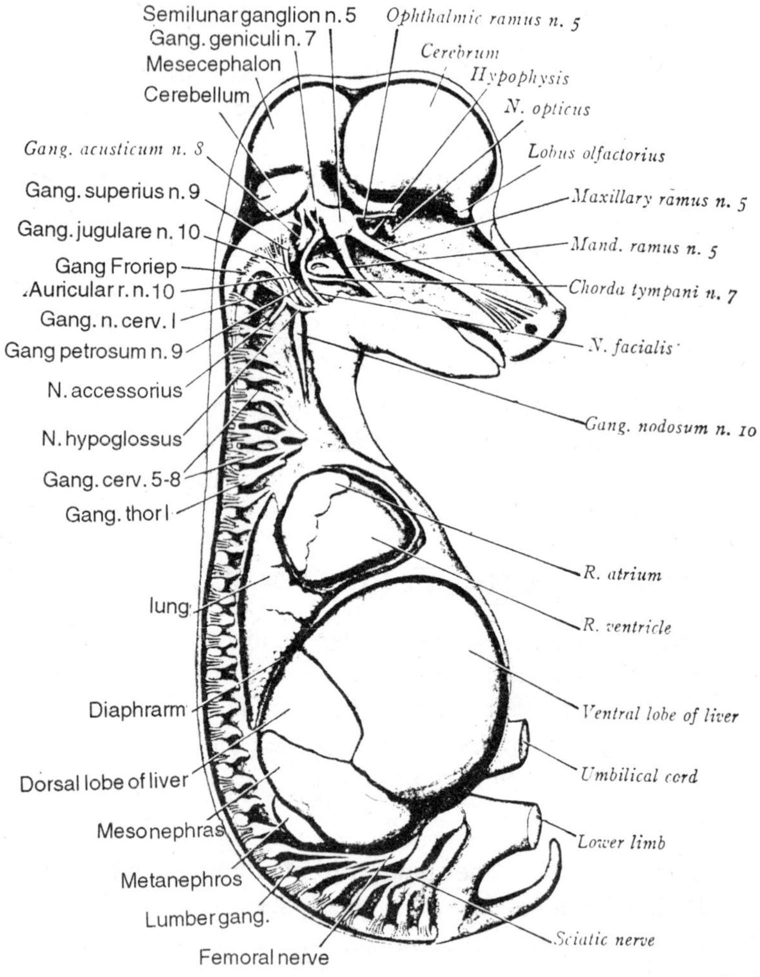

Fig. 3.24. Lateral dissection of 35 mm. pig embryo, viewed from right side.

Midsagittal Section. New features of the brain are the olfactory lobes, the chorioid plexus of the third and lateral ventricles, the thalami, the epiphysis and the consolidated hypophysis. The primitive mouth cavity is now divided by the palatine folds into the upper nasal passages and lower oral cavity. Of the viscera, the distinct genital and suprarenal glands and the enlarged metanephros command attention, as does the coiling of the intestine. The ureters have acquired separate openings at the base of the bladder, and the urethra extends to the tip of the phallus.

Ventral Dissection. The chief new features are the markedly lobate lungs and the longer Mullerian ducts with expanded upper ends. The mesonephros is nearing its maximum absolute size.

4

The Development of Mouse

The laboratory mouse has the longest history of genetic experimentation of any animal. Unlike *Drosophila*, developmental hypotheses and phenotypic tests of gene effects accompanied the studies of inheritance from the beginning of genetic analysis. In consequence, more is known about the developmental genetics of the mouse than that of any other vertebrate. The mouse as a model developmental system has a twofold interest. In the fist place, it may be taken as a representative vertebrate. Although the early development of some vertebrates such as amphibians appears very different from that of mammals, these differences may be less significant than was once thought. Furthermore, the later development of body plans and organ systems in the five vertebrate classes have many clear similarities. In the second place, the mouse is a mammal, and as such is the most convenient model system for human development. Given this fact, it has special interest for us. Indeed, a number of human genetic disorders have their analogues in identified mouse genetic syndromes. The recently devised technique for genetic transformation of the mouse germ line may prove applicable to humans eventually and increases the relevance of mouse biology to human biology.

For mouse genetics, the primary genetic source material consists of nearly 300 highly inbred lines. These strains contain numerous genetic factors of interest. By making the appropriate crosses between them, it is often possible to identify the number and genetic map positions of the genes responsible for particular developmental conditions. To date, over 400 genetic loci in the laboratory mouse have been identified. A large number of these markers are biochemical, the structural genes for particular proteins. The use of such biochemical markers in developmental studies often allows genetic, molecular, and cytological analyses to be combined. An additional feature of the mouse that provides relatively easy experimentation is its relatively short generation time of 90 days.

In this chapter, we will look at the first stages of mouse development, from the beginning of oocyte development through the period of cleavage and independent embryonic growth. By the end of cleavage, when the embryo consists of about 120 cells, it contains two major cell populations, the outer trophectoderm (TE) layer and an enclosed group of cells, the inner cell mass (ICM). This preimplantation embryo is termed a "blastocyst." It is the blastocyst that implants in the uterine wall, beginning the phase of maternal-dependent growth and development. The embryo proper (the fetus) eventually develops from cells of the ICM; the later postimplantation stages of development will be taken up.

The primary focus in this chapter will be on the nature of the morphogenetic system in the early embryo and its origins in the oocyte. The mouse oocyte has a special interest in this respect. In recent times, the mouse embryo has been viewed as a completely plastic system in which all early cellular commitments are impressed on essentially identical blastomeres by environmental forces. In effect, the oocyte has been interpreted as having no built-in morphogenetic instructions (Davidson, 1976). If this interpretation is correct and extends to the oocytes of other mammals, it would represent a major difference in the early developmental biology of mammals from that of possibly all other animal systems.

LIFE CYCLE AND EARLY DEVELOPMENT

Although mouse developmental biology is considerably more complex than that of either the nematode or the fruit fly, its life cycle is simpler, being the place in one continuous phase, from the moment of fertilization until birth, the entire sequence taking place inside the mother's body.

Embryonic and fetal growth in the mouse takes approximately 20 days from fertilization to birth. The first 4.5 days are spent as an unattached, independent embryo, undergoing cleavage. When the embryo reaches the late blastocyst phase, it implants in the uterine wall, and the major phase of embryonic development begins. In contrast to most animals, the pace of the first cleavage divisions is very slow. However, the rates of cell division and developmental change increase following implantation, and accelerate further with the establishment of the placental connections to the mother's circulatory system and nutrient supply. During the first stages of postimplantation development, the three germ layers from within that portion of the embryonic structure that gives rise to the fetus, and by days 8-9, organ formation has begun. By 7.5 days, only 3 days after the 120-cell stage, the embryo proper consists of approximately 10^5 cells.

Because the embryo develops inside and attached to its mother's body, a special issue arises in mammalian development; the role of the mother in the developmental process. Is there a continuing "instructive" maternal role in development? This question has been investigated by culturing embryos in various nonuterine (ectopic) sites, particularly in males. Although such embryos show some abnormalities, the results indicate that the embryonic developmental sequence is governed by the embryo itself; the sole function of the maternal environment is to provide adequate nutrition to the developing individual. Nevertheless, the question of maternal environment is pertinent in assessing putative maternal effect mutants, since early maternal effects on oocyte development must be distinguished from later maternal influences on embryonic development. This distinction can be made by transferring early embryos to surrogate mothers of a different strain, or, if preimplantation embryos are being tested for such effects, by comparison with control embryos in vitro culture.

In general, preimplantation development is most readily analyzed in vitro. To obtain sufficient numbers of embryos for study,females are induced to "superovulate." This action is produced by injection with gonadotrophic hormones prior to mating. Normally, a female will release up to 10-20 eggs per ovulatory cycle; superovulated females can release several times this number of mature ova. The fertilized eggs can be readily removed from the female reproductive tract for further experimental analysis.

GENETICS AND GENOME STRUCTURE

The diploid chromosome number of *Mus Musculus* is 40. Of the 20 chromosome pairs, 19 are autosomal and one is the sex chromosomes. Sex determination in the mouse is by an X-Y chromosome system, with XX individuals being female and XYs male, as in *Drosophila*. However, a fundamental difference in sex determination exists between the fruit fly and mammals. In the latter, sex is determined by the presence or absence of the Y chromosome, which establishes maleness, rather than by the Z: A ratio, as in *Drosophila*.

The genetic nomenclature is similar to that of *Drosophila*, Genes are given italicized symbols that are usually abbreviations of their names. In general gene designations begin with a capital letter unless the gene was first identified by a recessive mutation (e.g., *qk* for quaking).

The development of chromosome banding techniques. G-banding and Q-banding, which distinguish specific homologous chromosome pairs, made possible the assignment of genes to particular chromosomes in the karyotype. This method uses translocations that move known genetic marker from one chromosome to another, in combination with an analysis of the changes in karyotype structure produced by the translocations, as detected by the banding techniques. (Until the advent of the chromosome banding techniques, genes could be assigned only to genetic linkage groups, but the identity of particular linkage groups with particular chromosomes could not be ascertained.)

A map of the mouse genome is given in the Appendix. All chromosomes are acrocentric and numbered from largest to smallest. The sex chromosomes are simply designated X and Y. The Y chromosome is similar in size to the smallest autosome pair (chromosome 19) but appears to be empty of genes, aside from male fertility genes and the locus for the H-Y antigen (a surface male-specific histocompatibility antigen). The nomenclature for chromosomal rearrangements is similar to that of *Drosophila*.

The molecular organization of the mouse genome raises many of the same fundamental questions encountered in connection with the genomes of the nematode and the fruit fly. The major gross difference between these genomes is in size. At 3×10^9 base pairs, the haploid mouse genome is nearly 20 times larger than *Drosophila*. Of this total 8-10% consists of highly repeated satellite DNA, principally a 140-bp, adenine-thymine (AT)-rich repeat, reiterated 10^6 times in the centromeric heterochromatin. The single copy fraction is approximately 60-76% and the midrepetitive fraction is 15-25% of the genome.

The relative dispositions of single copy and repetitive sequences in the mouse genome are the subject of some dispute. The standard interpretation of eukaryotic interspersion patterns has been that there are two major types. In this interpretation, the most common pattern consists of single copy DNA sequences in 1000-to 300-bp stretches, interspersed with short midrepetitive sequences about 300 bp in length. The alternative interspersion pattern is designated the "*Drosophila* pattern," because it was first found in *D. melanogaster*. In this arrangement, long single copy stretches (averaging more than 13,000 bp) are interspersed with long blocks of midrepetitive DNA (5000 bp or more in length).

In the original studies, all mammalian genomes were judged to follow the standard, short interspersion pattern. However, *Moyzis et al.,* (1981a) have reevaluated these findings and concluded that much of the apparent difference between genomes in the interspersion pattern disappears when they are compared by identical renaturation methods. As these authors stress, the categories of "single copy" and "midrepetitive" are operational groupings, and the particular renaturation conditions employed can shift sequences from one category to the other. They show that under stringent reannealing conditions, the rodent genome shows the *Drosophila* pattern, with long single copy stretches (7200 $\pm$ 2000 bp) interspersed with long blocks of midrepetitive sequences. These midrepetitive blocks, in turn, consist predominantly of "scrambled clusters" of shorter (300-bp) blocks. Interestingly, an analysis of the long blocks of midrepetitive sequences in *Drosophila* shows that many of these too consist of similar scrambled clusters.

The functional significance of the sequence interspersion pattern remains unknown. For the reasons discussed in connection with the structure of the *Drosophila* genome, individual midrepetitive sequences are unlikely to have major, direct roles in controlling the expression of individual genes. However, it is not impossible that they have some general function in chromosome folding pattern or replication that indirectly influences gene expression. Conceivably, germ line transformation with engineered genes altered in neighbouring midrepetitive sequences, followed by analysis of their expression patterns may help to answer these questions.

OOGENESIS, FERTILIZATION AND PREIMPLANTATION DEVELOPMENT

Like all animal development, the beginnings of mouse development are to be sought in the structure and properties of its oocyte. The development of the mouse oocyte takes place in two discrete phases. The first phase extends from midembryogenesis until 5 days after the birth of the female mouse. It lasts from the formation of oogonial stem cells by the primordial germ cells to the formation of immature oocytes, which are held in late diplotene of meiosis I. The second phase begins with the attainment of sexual maturity and involves oocyte growth and maturation. This takes place in small batches of oocytes within each estrous cycle.

In its general aspect, the initial development of the ovary in the mouse bears a certain resemblance to that in *Drosophila*. The primordial germ cells, which look identical in the two sexes, are distinguishable from other cells by their round shape and diffuse chromatin, characteristics similar to those of pole cells in *Drosophila*; however, they are most readily detected and tracked by their high alkaline phosphatase content. As in the fruit fly embryo, these primordial germ cells undergo a major migration from their posterior site of origin to the gonad rudiment. This migration, between 8 days postconceptus (pc) and 13.5 days pc. takes place from the caudal end of the embryo proper (the primitive streak) through the hindgut to the genital ridges (the somatic rudiment of the gonad). In female embryos, the arrival of these primordial germ cells stimulates the proliferation of the epithelial cells of the genital ridges; the primordial germ cells themselves also commence multiple rounds of cell division, to form oogonia. The oogonia subsequently undergo two waves of inward migration. Most of those in the first wave degenerate, but the cells of the second wave come to rest in the peripheral (cortical region of the developing ovary. These cortical oogonia continue to proliferate, although remaining connected by many intercellular bridges. As we have seen, this syncytial arrangement is common in early germ line cells, and presumably serves to facilitate intercellular molecular transfer and synchrony of nuclear divisions.

During the migration to the ovarian cortex, groups of oogonia become surrounded by prefollicle cells of mesodermal origin. Eventually, each pro-oocyte is covered by a monolayer of these somatic cells to form a "unilaminar follicle." Pro-oocytes enter the first meiotic division but arrest at late diplotene until just prior to ovulation. Development of the unilaminar follicle is accompanied by the loss of syncytial connections between pro-oocytes. The proliferation of follicle cells and the secretion of extracellular material serve to isolate the follicles further from one another.

Oocytes remain quiescent until sexual maturation, when they are hormonally stimulated to commence growth and development in small groups. Although the estrous cycle takes 5 days, each preovulatory growth periodlasts for 20 days, with most growth taking place between 20 and 6 days prior to ovulation. During this period, the growing oocyte increases in diameter from 20 to 80 μm and the surrounding follicle cells multiply to reach 50,000 per follicle. During these final stages, these final stages, the oocyte comes to occupy fluid-filled chamber, the "antrum," within the follicle. The follicle cells act as feeder cells for the oocytes; 85% or more of the metabolites required by the oocyte are passed to it through the follicle cells.

The postnatal phase of oocyte development involves both growth and differentiation of the cell. A characteristic sequence of differentiative changes in the major cytoplasmic organelles—ribosomes, endoplasmic reticulum, nucleoli, mitochondria—occurs, and marked protein synthesis takes place in both oocyte and helper cells. As in many other animal oocytes, synthesis and accumulation of histone proteins and of tubulins occurs. These proteins are subsequently utilized in cleavage. The tubulin subunits may comprise as much as 1% of the total protein in the mature oocyte, thus constituting one of the most abundant protein fractions in the egg (as they are in the fruit-fly oocyte).In common with other cellular differentiations, the oocyte growth phase features stage-specific alterations in the pattern of polypeptide synthesis, as assayed in two-dimensional separations.

This period is also one of active transcription, of both ribosomal and informational RNAs. The synthesis of poly A⁺ mRNA is part of this transcriptional program. When [³H]uridine is used as the radioactive label, 9.6% of the total counts in the mature oocyte are found in poly A⁺ mRNA Bachvarova and De Leon, 1980). Indeed, the fraction f stored RNA that is in the form of mRNA is substantially higher in the mouse ovum than in the sea urchin or in the frog *Xenopus,* where the proportion is closer to 1-2% of total stored RNA.

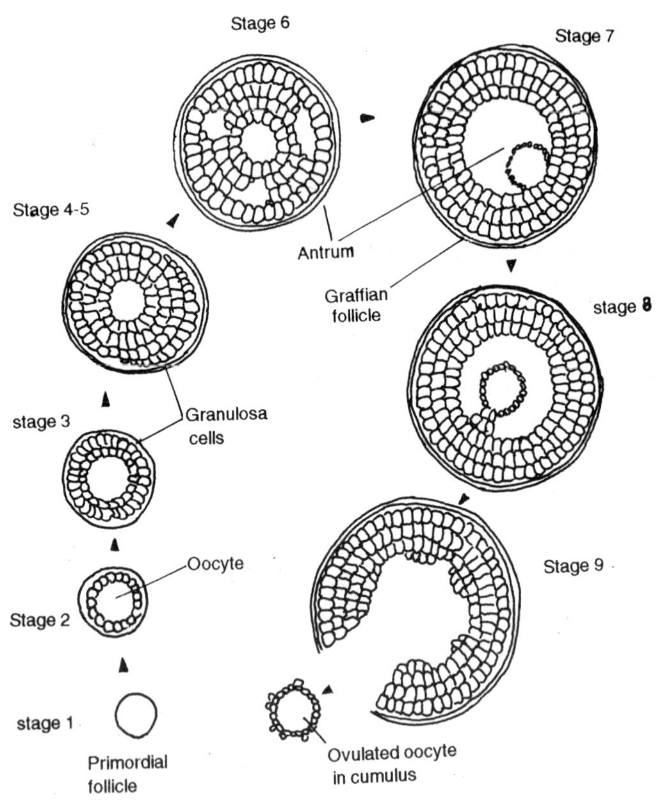

Fig. 4.1.   Stages of mouse oocyte and follicle growth.

The period of oocyte growth and differentiation ends about 6 days prior to ovulation and is succeeded by a 5-day phase of follicle cell proliferation and antrum development. In the last 9-10 hr before ovulation, the oocyte reenters meiosis and completes the first meiotic division, giving rise to the first polar body. The oocyte nucleus then proceeds directly to meiosis II without re-formation of the nuclear membrane but halts at metaphase until fertilization. Ovulation involves release of the ovum from the follicle into the oviduct. It is surrounded by a polysaccharide-rich envelope, termed the "zonapellucida," and a group of follicle cells, the "cumulus oophorus." Because receptivity to mating closely correlates with ovulation, the time of fertilization can be gauged fairly accurately under conditions of controlled mating. Fertilization involves penetration by the sperm through the cumulur oophorus, which is shed during passage down the oviduct, and the zona pellucida. Following entry, the sperm head is converted to the male pronucleus and loses its complement of chromosome-coating protamines, which are replaced by histones and acidic proteins from the maternal cytoplasm. Fertilization triggers resumption of meiotic progression in the oocyte nucleus; completion of meiosis is quickly followed by extrusion of the second polar body. The two pronuclei then move slowly into apposition. (DNA replication takes place in the pronuclei and is completed during this lengthy migratory phase.)

During the first 3 days following fertilization, the zygote progresses through four cleavage divisions to the 16-cell stage, the morula. In the following day and a half, the solid ball-like morula develops into the blastocyst. The blastocyst is a hollow, nearly spherical fluid-filled ball of cells, and, as mentioned earlier, it is both the implanting structure and the carrier of the precursor cells of the true embryo.

Cleavage in mammals is a slow process. In the mouse, the first cleavage does not take place until 24 hr postfertilization, after the prolonged migration and fusion of the pronuclei. Successive divisions take place at approximately 12-hr intervals. The eight-cell stage may be a developmentally critical juncture. It is at this point that the heretofore loosely knit blastomeres undergo the process of compaction. During compaction, the cells transform from spherical to columnar and develop a variety of intercellular junctions. The net result is the drawing together of the ball cells. Compaction is almost certainly important in creating certain properties of polarity within the blastomeres of the eight-cell state; the significance of such polarity will be discussed later.

From the eight-cell stage on, cleavage results in the formation of distinct inner and outer cell populations. By the 16-to 32-cell stage, the inner cells have begun to secrete fluid into the developing central cavity, the blastocoele,. By the 60-cell stage, approximately 3.5 days after fertilization, the embryo is at the early blastocyst stage and consists of two cell populations. The outer layer of approximately 45 cells constitutes the trophectodermal (TE)) layer and the internal 15 cells comprise the ICM. Formation of the early blastocyst is followed by reexpansion of the embryo to fill the zona pellucida.

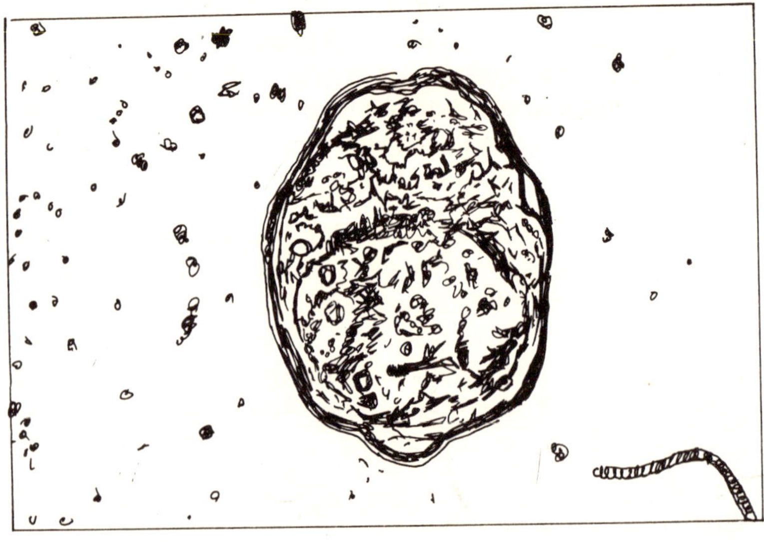

Fig. 4.2.   An expanded 3.5-days blastocyst. diameter 70-80 μm.

The final phase of preimplantation development involves

the production of two further cells groups. The ICM gives rise to group of endodermal cells that eventually cover the entire inside of the blastocoele. And the TE diverges into a cap of "polar TE" cells, which comprise the sides of the blastocyst. By the late blastocyst stage, some of these mural TE cells have begun to become polyploid; eventually, these large polyploid cells become the "primary giant cells" of the implanting blastocyst wall.

EXPRESSION OF THE ZYGOTIC GENOME DURING PREIMPLANTATION DEVELOPMENT

The changeover from maternal to zygotic genome control in the early mouse embryo has been tracked by several means. One of these is analysis of the stores of maternal mRNA and their depletion during preimplantation development. As noted above, the poly A$^+$ mRNA pool is unusually large, amounting to approximately 10% of the total RNA in the zygote. This maternal RNA pool is found to decline in parallel with the prelabeled maternal RNA. The measurements show an initial depletion of prelabeled mRNA to 60% of the prefertilization content during the first 24 hr of development (to the two-cell stage), followed by a more gradual decline between the cell-cell and early blastocyst stages to a final value of 30%. The actual decline may in fact be considerably greater; inhibitor experiments and tests of in vitro translation of extracted maternal mRNA indicate a rather rapid inactivation or depletion of these stores by the mid-two-cell stage.

Despite the large starting maternal mRNA pool, the embryo becomes dependent on transcription from its own genome early in development. Blockage of transcription by either actinomycin D or α-amanitin stops development at any point after the first cleavage and reduces amino acid incorporation from the two-cell stage on. This process is significantly different from the early developmental programs of sea urchin and amphibian embryos. In these animals, actinomycin D treatment during cleavage has little immediate effect on amino acid incorporation or cell division. However, mRNA synthesis is shut off and development arrests at the postcleavage stage.

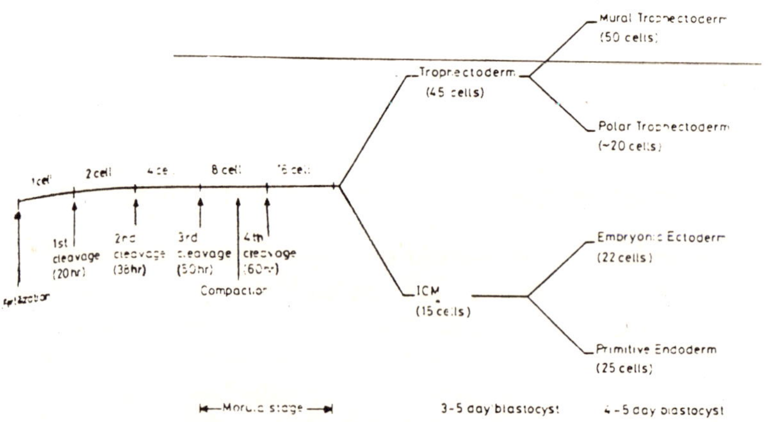

Fig. 4.3.   Time course and stages of preimplantation development.

In accordance with the inhibitor study results, labeling of new RNA synthesis by radioisotope incorporation shows that zygotic genome transcription begins early. Using [³H]uridine incorporation, *Knowland* and *Graham* (1972) placed the beginning of transcription in the embryo at the middle of the two-cell stage. With the same label. Levey et al. (1978) similarly located the beginning of nonribosomal nuclear RNA and of poly A⁺ mRNA at this point. However, when [³H]adenosine, which is taken up more efficiently than uridine, is used, the synthesis of new species of nuclear RNA and poly A⁺ mRNA is found to begin in the middle of the one-cell stage. Nevertheless, formation of labeled poly A⁺ messengers at the one-cell stage involves primarily addition of poly A tails to preexisting maternal mRNAs.

Given the early expression of the zygotic genome, when does the cytoplasmic composition of the embryo begin to change over from maternally encoded products to those characteristic of the embryonic genome? Bio-chemical experiments suggest that the transition begins no later than the two-cell stage. However, results of both inhibitor and labeling experiments are subject to various interpretations. Independent estimates for ascertaining when the zygotic genome begins to affect embryo cytoplasmic composition can be obtained by genetic means. Such tests involve detection and discrimination of paternal genome-encoded products from maternal ones, while maternal products may derive either from the oocyte or from the maternally derived chromosomes of the zygote, paternally encoded products in the embryo can come only from the zygotic genome.

In *Caenorhabditis* and *Drosophila*, it is possible to test for paternal gene activity by determining whether rescue from lathal maternal effects takes place in embryos with wild-type paternal alleles. In the mouse, where few maternal effect mutants are known (see the following section), one must resort to biochemical methods for detecting such paternal gene expression. One such method is to score for a quantitative increase in gene expression programmed by a paternal gene. An example is provided by a test for B-galactosidase expression in early embryos, an enzyme encoded by a gene — on chromosome 9. During development to the blastocyst stage, this enzyme, like that of two other lysosomal hydrolases, undergoes a 50-to 100-fold increase in activity. This activity increase in mediated by a closely linked, *cis*-dominant regulator gene site, designated *Bgl-s*. One allele of this *cis*-dominant regulator, *Bgl-s^h*, produces a twofold higher level of activity than the other known allele, *Bgl-s^d*. To establish the time of paternal gene expression for B-galactosidase in embryonic development, *Esworthy* and *Chapman* (1981) mated female mice homozygous for the *Bgl-s^d* (low activity) allele to males with the *Bgl-s^h* (high activity) allele and collected embryos at successive times after fertilization for measurement of enzyme activity. The strains employed were nearly identical in genotype ("congenic"), except for the region on chromosome 9 containing the structural gene and its linked regulator.

Figure 4.4 shows the plotted increase in enzyme activity in hybrids as a function of developmental stage when normalized to the activity increases seen in parallel *Bgl-s^d x Bgl-s^d* crosses. The result show a paternal allele-mediated increase in activity that extrapolates back to 36-45 hr, or shortly after the second cleavage division. That the increase mirrors a true turn-on of gene expression rather than an effect of heterosis is shown by the reciprocal cross (with a high maternal background) results, which did not give a comparable increase. The findings indicate that the cytoplasmic composition of the embryo begins to reflect zygotic genome expression by the two-cell stage. Comparable tests with the chromosome 5 structural gene for B-glu-curonidase give a similar result.

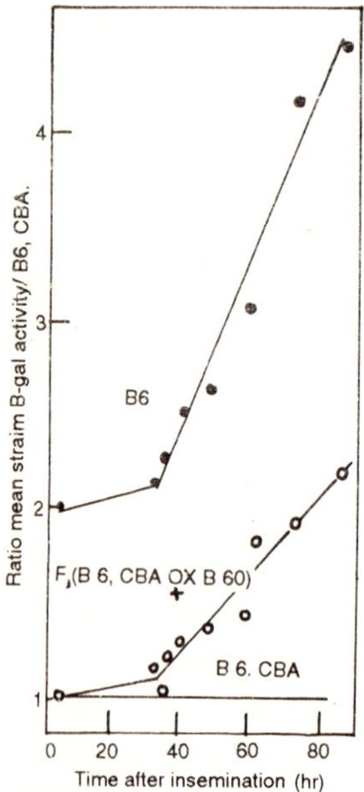

Fig. 4.4.   Time course of paternal expression of β-galactosidase in mouse embryonic development.

From the two-cell stage on, zygotic genome expression accelerates, its products increasingly dominating the cytoplasm Ribosomal RNA synthesis cannot be detected in the one-cell embryo, but by the mid-two-cell stage it accounts for a significant fraction of the new transcripts. General transcription rapidly accelerates in the eight-cell embryo, the point at which compaction occurs. By the early morula state, rRNA synthesis is approximately 16-fold more rapid than in the two-cell stage.

In contrast, protein synthesis is active throughout early development and even in the unfertilized oocyte. Much of this early synthesis is carried out on maternal templates. However, detailed analysis, employing two-dimensional gel electrophoresis, shows that zygotic genome-encoded protein products are detectable by the early two cell stage.

One of the most interesting findings to come from such protein synthesis studies concerns the occurrence of postranscriptional regulation in the maternal mRNA pool. These regulated messengers code for a group of proteins of 35 kd, which first appear during the late one-cell stage. The scheduled synthesis of these proteins continues unabated in the presence of the transcription inhibitor α-amanitin, at a drug concentration sufficient to inhibit all in vivo RNA ploymerase II activity. Furthermore, when maternal mRNAs are extracted from unfertilized eggs and placed in an in vitro protein-synthesizing system, these proteins are made in quantities comparable to those of two-cell embryos. If follows that the responsible mRNAs must be held in some state that is unfavourable for translation, until the early two-cell stage, and are then released by some signal to the translation apparatus. One possibility is that this signal is a timed polyadenylation of these messengers, permitting transport out of the nucleus, because the onset of translation correlates with the approximate period in which long poly. A tails are added to preexisting maternal mRNAs.

The cellular functions of these posttranscriptionally controlled proteins and the significance of their controlled translation are unknown. We noted earlier the existence of a parallel phenomenon in *Drosophila*. A third example of such posttranscriptional control has been reported in the sea urchin, involving delayed translation of a histone H3 maternal mRNA. One possible explanation for the phenomenon is that such posttranscriptionally regulated mRNA s might code for proteins required in abundance by the early embryo, and therefore must be supplied by maternal templates, but that their "premature" translation could cause some kind of damage to the embryo. This suggestion might be tested by examination of the effects on development of early injection of the purified proteins.

The question of mechanism is also unresolved but may involve different processes in different organisms or even for different mRNA species within the same animal embryo. *Drosophila* involves sequestration in cytoplasmic RNP particles, but the sea urchin histone mRNA seems to be retained by some selective mechanism in the nucleus. Delayed polyadenylation might, in principle, be a general mechanism of posttranscriptional control in the mouse, but the particular MRNAs that undergo delayed translation are stored as polyadenylated species.

Although changes in electrophoretic patterns of labeled polypeptides can be used as an index of changes in gene expression, they are not an infallible guide to changes in protein synthesis. About 50% of the changes occurring in early preimplantation development, as registered on two-dimensional separations, result from reproducible patterns of posttranslational modification—the addition of nonprotein moieties, such as glucosamine and phosphate, to preexisting polypeptides that cause a change in migration behaviour. Some of these posttraslational modifications are set in motion by oogenesis itself and occur in unfertilized eggs (Pratt et al., 1983). From the eight-cell stage, there is relatively less change in the overall distribution pattern of labeled polypeptides in two-dimensional gels from whole embryos. Thus, during the period of blastocyst formation, the overall protein synthetic pattern appears relatively stable.

Despite this relative constancy in protein synthesis between the eight-cell/compaction stage and the early blastocyst, a few significant changes in gene expression take place. The first is a key regulatory event in the late morula state, as the embryo goes from a 16-cell structure to one of 32 cells, which is necessary for the secretion of fluid that takes place during blastocyst formation. This fluid initially accumulates between cells and then collects in the developing blastocoele, the entire process being referred to as "cavitation."

Smith and *McLaren* (1977) have shown that the onset of cavitation requires five DNA replication cycles. Presumably, the signal to begin cavitation involves either a direct "counting" of these replication cycles or the attainment of a characteristic nucleocytoplasmic ratio during cleavage. In investigating the molecular biology of this developmental event. *Braude* (1979) found that the onset of cavitation requires-transcription and concluded that this transcriptional event occurs during or as a result of the fifth replication cycle. The nature of the link between the fifth replication cycle and transcription has not been elucidated. Since zygotic genome transcription has commenced well before this point, the control event cannot be a simple titration of transcription inhibitor by DNA as occurs in *Xenopus*.

The second set of events concerns the divergence of the first two distinct cell populations, the ICM and the TE cells. *Van Blerkom et al.* (1976) were the first to detect differences in patterns of polypeptide labeling between these cell groups. They dissected ICM and TE cells from 3.5-day (expanded) blastocysts, incubated the two fractions separately with [35S]methionine, and then analyzed the patterns of labeled polypeptides with two-dimensional separations. Although the majority of spots were shared between the two cell groups, as expected from the general constancy of labeled polypeptides in whole embryos between the 8-and 64-cell stages, a few distinctive ICM and TE spots were detected. Thus, by the late blastocyst stage, the two cell populations, which are recognizably different in position and beginning to be distinct cytologically, can be distinguished by molecular markers.

Handyside and *Johnson* (1978) subsequently showed that the "ICM-Specific" and "TE-specific" cells are first detectable between the early morula (12-to 25-cell) and late morula 25-to 30-cell) stages. When the inner cells from embryos of these stages were isolated by the technique of immuno-surgery these inside cells were found to label just the shared and ICM-specific polypeptides and not the TE-specific ones. Some of these differences may reflect posttranslational modification rather than new gene expression per se, but the significant point is that the molecular events of differentiation are detectable before there are cytologically distinct ICM and TE cell populations.

The sequence of changes in gene expression and macromolecule composition that take place in the preimplantation embryo. The entire set of changes constitutes a complex picture. However, the salient feature is that zygotic genome expression begins early in this embryo, and zygotic genome products increasingly dominate the molecular composition of the embryo from the two-cell stage on.

DETERMINATION EVENTS IN THE PREIMPLANTATION EMBRYO: GENETIC ANALYSIS

In *Drosophila,* the analysis of early embryonic determination has involved two principal kinds of genetic technique. The first is the creation of genetic chimeras by removal of genetically marked cells from the blastoderm stage and subsequent culture in recipient embryos or larvae. The fates of the donor cells in these chimeras are used to establish the timing of the first determinative events. The second genetic technique is the isolation of maternal effect mutants that produce aberrations in the morphogenetic system of the egg. These experiments have permitted a genetic characterization of the mechanisms of determination.

Investigation of determination mechanisms in the early mouse embryo has relied principally on the first method, chimera construction. Extensive mutant hunts are impossible because of the limit number of progeny that can be screened; furthermore, the collection of existing mutants includes extremely few with maternal effects. However, the chimera experiments alone have proven instructive about the timing of determination events in the preimplantation embryo. This information, in turn, provides some clues to the mechanisms of determination in this embryo. We will first look at the methods of chimera analysis used in the mouse and the application of these techniques to analysis of early determination. In the next section, we will examine some of the hypotheses about determination in the mouse embryo that have been proposed.

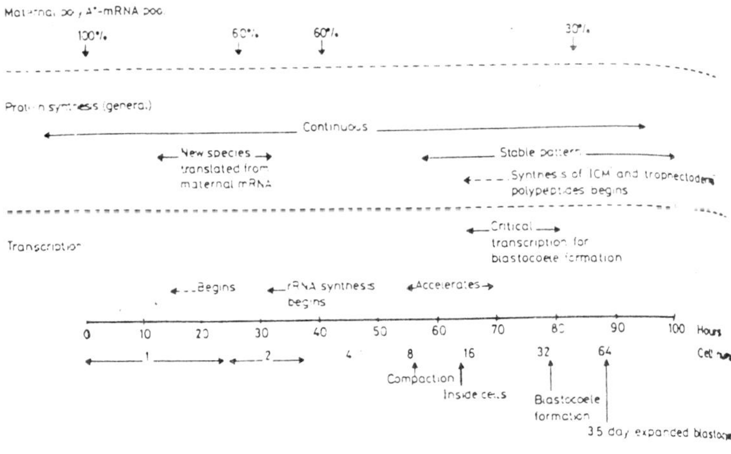

Fig. 4.5.   Summary of zygotic genome expression and depletion of maternal mRNA in mouse preimplantation development.

Chimera Construction and Analysis; Methods

There are several methods of constructing chimeric mouse embryos. The first early embryo aggregation was invented by *Tarkowski* (1961) and *Mintz* (1962). In this technique, eight-cell embryos from two genetically distinguishable strains are stripped of their zonae (mechanically, or now usually by digestion with pronase) and then brought into contact; fusion normally occurs within a period of 2-4 hr at 37°C. The composite morulae are then transferred to the uterus of a pseudo-pregnant foster mother, where implantation and subsequent development take place. (Such foster mothers are typically prepared by prior mating to vasectomized mice.) A substantial proportion of such morulae develop and give rise to normal newborn mice. In experiments in which the two input genotypes differ in their coat color phenotype, the coat color pattern of the chimeras can have a striking appearance. Note: the term "allophenic mice" is also found in the literature to denote chimeric mice.)

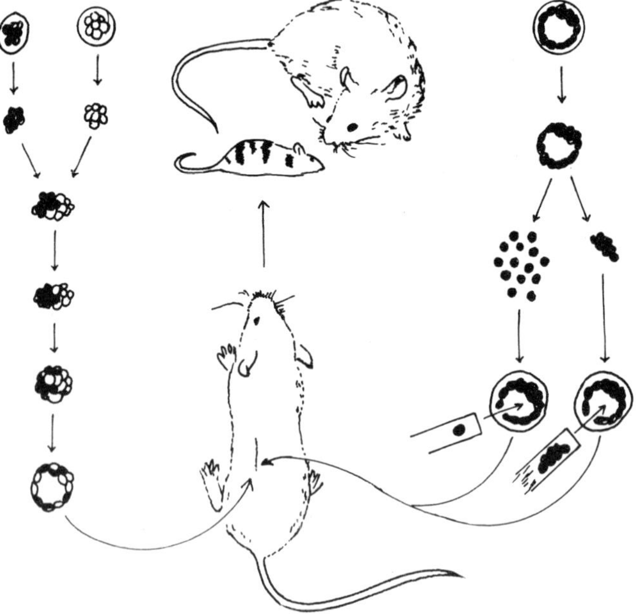

Fig. 4.6.   Principal methods of chimera production in the mouse. Morula aggregation shown on the left, blastocyst injection on the right.

The second method, introduced by *Gardner* (1968), permits the assay of the developmental capabilities of cells of different stages, either singly or in small groups, without the major disruption of preimplantation development that embryo aggregation entails. In this procedure, single or multiple cells of the desired stage and genotype are injected into host blastocysts through a narrow incision. The recipient blastocyst is then placed in a pseudopregnant foster mother; those that have not been irreparably damaged by the injection procedure give rise to chimeric fetuses. The method allows the testing of developmental capabilities of any of the cells in the preimplantation embryo. The method can also be used to look at developmental potentialities of cells in postimplantation stages.

A potential drawback of the cell injection method is that the blastocoels environment may affect the developmental paths taken by the injected cells. To avoid this complication, a third method has been invented. In this scheme, the cells to be tested are aggregated with eight-cell morulae to create chimeric embryos.

The informativeness of all chimera experiments depends on the sensitivity or resolution with which the cells of different genotypes can be resolved. When chimeric embryos are allowed to develop to birth, the use of coat color markers is adequate, at least for detecting respective contributions to the epidermal tissues. However, when internal tissues are to be traced, or when intermediate stages of development are to be examined, histochemically distinguishable genetic markers must be used. These are usually enzymatic or antigenic markers.

Biochemical markers are broadly classified as either *direct* or *indirect*. With direct cell markers, individual cells can be scored in tissue sections, permitting fine-scale tracing in situ of the location and relative contribution of the two genotypes. Plus or minus activity enzyme activity differences and surface antigenic markers are in this category. Indirect cell markers are those that can be scored only in bulk tissue preparations. The most usual indirect genetic marker is an electrophoretic difference in an enzyme, that is electromorphic alleles. With indirect cell markers, the chimeric embryos are dissected into their component tissues, which are ten assayed separately for the two genotypic contributions. The two most commonly used indirect markers are electromorphic differences for the enzymes isocitrate dehydrogenase (IDH) and glucose phosphate isomerase (GPI). For the latter, minor electrophoretic differences can be detected in proportions as small as 1-3% of total activity.

In general, direct markers give more information with less chance of error. With indirect markers, there is always the problem of cross-contamination of tissue samples by cells of the other genotype. However, it is not always possible to find strains differing in alleles for a good direct marker. The recent development of techniques for discriminating cells of differing genotype at the major histocompatibility locus (H2) of the mouse may solve this problem. Because of the lack of suitable direct markers, most experiments have relied on an indirect cell marker, principally the GPI electromorphic difference.

Chimeras and the Analysis of Early Determination Events

The analysis of determination events in preimplantation development by chimera construction is similar to that used in the analysis of *Drosophila* determination in the blastoderm. One takes cells of genotype A, from a defined stage or location, and places them in host embryos of genotype B. Subsequent development is followed, and the relative contributions of A and B to the various tissues of the later embryo nor adult are scored. If all tissue types have some donor cell (genotype A) contribution, then these donor cells must have been totipotent. If, on the contrary, they are found in only certain tissues, the original donor cells are presumed to have experienced some previous determinative events.

The chimera technique has been used to explore three fundamental questions about determination in the preimplantation embryo: (1) does the divergence of ICM from TE entail a restriction in developmental potency of either cell type? (2) If such restrictions do occur, when do they take place? (3) Does the formation of endoderm from ICM or the divergence of polar and mural TE involve determinative events?

The topic that has received most attention is the nature of the divergence between ICM and TE cells. It has long been known that these two cell populations differ distinctly in their embryological fates. The ICM of the late blastocyst contributes directly to the fetus and to certain of the extra-embryonic membranes, while the TE contributes directly to tissues involved solely in the implantation process. By 3.5 days pc, when the blastocyst consists of 60-64 cells, these two cell populations differ in distinctive cytological features. These differences include the kinds of cellular junctions each displays, the presence of microvilli on ICM and their absence from TE, and recognizable differences in mitochondrial structure. Given the difference in fates between ICM and TE, is it based on an irreversible difference in developmental capability, and if so, when does this difference become established?

The experiments of *Rossant* (1975a) demonstrate that the cells of the ICM are already determined by the 3.5-day, expanded blastocyst stage. In these experiments, the donor ICMs were aggregated with recipient eight-cell morulae and transplanted to the uterine horns of pseudopregnant foster mothers, then allowed to develop for periods ranging from 8.5 to 18.5 days. The genetic marker used for tracing the ICM-derived cells was GPI, with the donor cells specifying the b electromorph and the recipient embryos specifying the a form. Following termination of development, the fetuses and associated embryonic structures were isolated and dissected into the ICM-derived embryo plus membranes fraction and the TE-derived fraction.

The analysis of the chimeric embryos showed a substantial and highly preferential contribution of donor GPI enzyme to the ICM-derived tissues, with little or no contribution to TE-derived fraction. Of the 15 embryos showing a donor cell contribution, 12 showed an exclusive contribution to the embryo plus membranes fraction. The three apparent exceptions showed only a minor contribution, which probably reflected contamination with ICM-derived tissues. The results provide firm evidence that ICM cells from 3.5-day expanded blastocysts are already restricted in their development capacity. A similar result has been reported using a direct cell marker (an antigenic difference) in rat-mouse chimeras, in which rat ICM cells were injected into mouse blastocysts.

TE cells from 3.5-day blastocysts are comparably restricted in their developmental potential. When TE vesicles are emptied of their ICM cells before fostering, they give rise to trophoblast cells (the TE cells that mediate implantation) and undergo implantation. However, they cannot give rise to ICM or to those structures normally derived from ICM. Furthermore, injection of TE cells into 3.5-day blastocysts never yields embryos with a donor cell type contribution. The results show that by the 60-to 64-cell stage, the cells of the ICM and TE are committed to distinct and restricted pathways of development.

When does this restriction in developmental potential arise? The experiment of *Kelly* (1975) provides a lower bond. She dissociated four-cell embryos in vitro and then allowed each blastomere to divide once, to give four "octet pairs. "The members of each pair were then separately aggregated with individual eight-cell carrier morulae and fostered. Donor cells carried the *Gpi-1* allelic form, while host morulae carried *Gpi-1*. In addition, the donor strain was albino, so that chimeras that developed to term could be distinguished by coat color. Fourteen chimeras, derived from seven octet pairs, had donor cell contributions in the fetuses. In five of the pairs, donor cells had participated in the formation of both ICM- and TE-

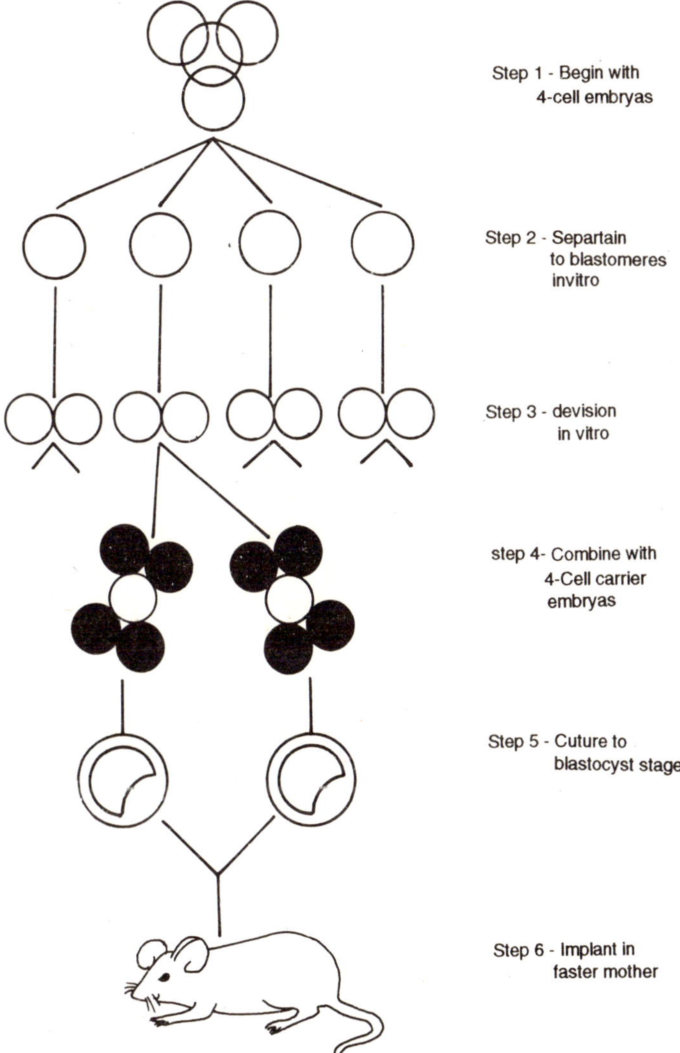

Fig. 4.7.   Test for totipotency of blastomeres at the eight-cell stage in the mouse embryo. See text for details.

derived fractions. A substantial donor cell contribution was also seen in all chimeras allowed to develop to term. Evidently, the blastomeres of the eight-cell embryo do not show segregation of developmental potential for either ICM or TE. This work demonstrates that blastomeres at the eight-cell stage are totipotent. Therefore, the first determinative events occur between the 8- and 64-cell stages.

To pinpoint the time of developmental restriction more precisely, the technique of *"immunosurgery"* was used by *Handyside* (1978) to isolate and the test the capabilities of inner cells at successive intermediate stages between compaction and the expanded blastocyst. Immunosurgery on mouse embryos involves the selective lysis of outside cells by exposure of the embryos to rabbit anti-mouse serum, followed by washing and treatment with complement (a substance that lyses antibody-reacted cells). When the treatment is applied to embryos of successively more advanced stages and the isolated inner cells are tested in vitro for their differentiative capabilities, the inner cells of later morulae and early cavitating blastocysts are found capable of generating TE derivatives. However, the ICMs of 3.5-day expanded blastocysts do not possess this capacity and produce only endodermal layers.

A somewhat different result, obtained with the same procedure but with embryos from a different strain, was obtained by *Hogan* and *Tilley* (1978), who found some 3.5-4.0 ICMs capable of generating TE-derived structures. These authors suggested that the timing of the restriction event differs somewhat between different strains and that it is a progressive, gradual process, with some potential for reversal up to the 4.0-day stage. Such gradual loss of developmental potential could reflect either a decreasing plasticity of all cells or a progressive shift in the composition of the inner cell population from those with TE developmental capacity to those that do not have it.

Do these restrictions in cellular developmental capacity have nuclear state correlates? Two sets of published results suggest that they do. *Illmensee* and *Hoppe* (1981) compared the abilities of single nuclei from ICM and TE cells of expanded blastocysts to support development in eggs when substituted for the zygotic nuclear inheritance. They reported that ICM nuclei had a significantly higher capacity than TE nuclei to support development, even obtaining three mice that developed to term with substituted ICM nuclei. *Modlinski* (1981) performed a similar comparison without removing male and female pronuclei. He compared the ability of nuclei from morulae (12- to 18-cell embryos) to that of ICM and TE nuclei to support embryonic development by transplanting nuclei directly into newly fertilized eggs without pronuclear removal. Under these conditions, the transplanted nucleus fuses with the zygote nucleus to form a tetraploid hybrid nucleus. When nuclei from either morulae or ICM cells are tested in this manner, approximately half of the surviving transplants develop to the blastocyst stage; in contrast, TE nuclear transplants invariably stop at earlier stages. The results of these nuclear addition experiments are thus similar to those obtained with nuclear substitution.

Less is known about the divergence of cell types within the ICM and TE cell populations in the final 24 hr of preimplantation development than about the initial separation of ICM from TE. However, the delamination of the internal layer of endoderm from the ICM probably also involves a determinative restriction. The endoderm cells of the late blastocyst differ from the ectodermal cells of the remaining ICM in three respects: they show more intense histochemical staining, they possess a more extensive endoplasmic reticulum, and their surface membranes have a "rough" character that contrasts to the "smoother" surface of the ectodermal ICM on dissociation in vitro under the appropriate conditions.

Gardner and *Rossant* (1979) purified rough (endodermal) and smooth (ectodermal) inner cells and injected them singly or in small groups into 3.5-day blastocysts, the latter marked with a different GPI electromorph. Early endodermal cells contribute solely to the endodermal layer of the yolk sac, while ectodermal cells contribute solely to the mesodermal

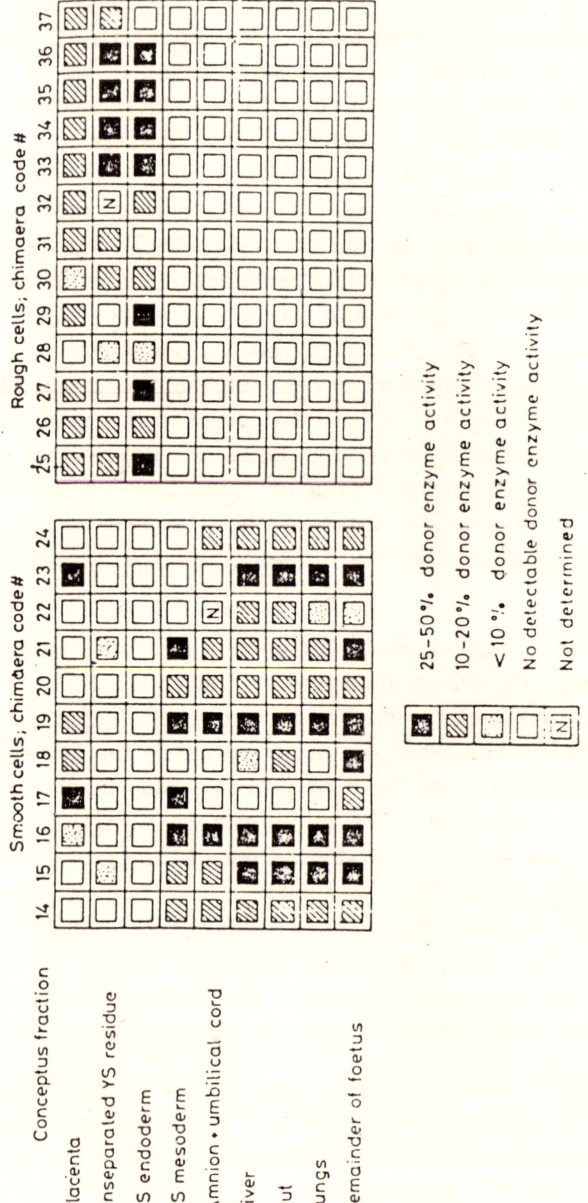

Fig. 4.8.   Developmental fates of Smooth (eclodermal) and Rough (endodermal) cells from 4.5-day *p.c.*

layer of the yolk sac and the fetus proper, but never to the endoderm of the yolk sac. (The structure of the early postimplantation embryo will be described). Evidently, by the time the two ICM-derived cell populations of the late blastocyst have separated, they have diverged both in fate and in their respective developmental potency.

The case of the two TE cell populations may be subtly different. A principal difference between polar and mural TE cells is in their nuclear character Polar TE cells remain diploid and retain their division capacity, while mural TE cells become polyploid and lose this capacity. The retention of division capacity by polar TE cells seems to be a consequence of their association with the ICM in some manner, this contact serves to maintain the polar TE as diploid cells. However, while mural TE cells cannot become polar TE cells, the reverse transformation can and does occur, as polar cells are pushed to the sides and assume positions away from the ICM. Determination in the TE cell populations is this a one-way street, not the mutually exclusive pair of possibilities exhibited by the other divergences.

Mutant Analysis of Preimplantation Development

A genetic analysis of the mechanism of determination in the early embryo requires maternal effect mutants—specifically, those affected in the morphogenetic system of the egg. To date, no such mutants have been identified among the very few maternal effect mutants known in the mouse. However, among the few known and suspected maternal effect mutants of the mouse, none has yet proven informative about the morphogenetic system of the egg or about distinctive control events in the early embryo.

The dearth of mouse maternal effect mutants at first seems puzzling Maternal effect mutants are abundant classes in *C. elegans* and *Drosophila*. As we have seen, minimal estimates of the number of loci in these organisms that can mutate to this phenotype are in the range of 10-20% of the total gene set; the real proportion of gene functions expressed in oogenesis its probably considerably higher in both of these animals. Oogenesis in the mouse involves as complex a cellular differentiation as any in this animal. Even if the proportion of genes expressed in oogenesis is smaller in mammals than in invertebrate systems (because mammals have more genetic functions), the absolute number of oogenesis genes and hence potential maternal effect genes should still be large. Indeed, there are very few early developmental mutants in the mouse at all, although zygotic genome expression begins early and is essential for preimplantation development from the early cleavage divisions.

The explanation is probably connected to the particular features of gene expression in early development, and specifically to the *relative demand* for certain gene products in oogenesis early development in comparison to later stages. Many of the partial maternal effect genes in *Drosophila*, those that function in other stages of development, are first detected as maternal effect mutants precisely because the demand for their gene products is accentuated during early embryogenesis relative to later stages. In the fruit fly, cleavage and early development occur very rapidly and make large metabolic demands on the stored materials in the embryo. In the embryo of the mouse, the situation could easily be the reverse. Cleavage and early development are leisurely processes compared to subsequent development. Under these conditions, a hypomorphic mutant of a gene expressed during both early and later developmental stages might show the more severe deficiency at the later stage and might therefore be classified as a later lethal.

It should eventually be possible to estimate the number of genes that are expressed in oogenesis and early development relative to later stages by molecular methods. As more cloned sequences of genes expressed in later stages become avilable, it will be possible to test for the presence of their transcripts in the later oocyte or early embryo by the technique of in situ hybridization. Conversely, if and when cloned sequences are made by copying mRNAs of early embryos, the same procedure can be used to estimate the proportion of early expressed genes that are also expressed at later times. Such studies may help to illuminate some of the general features of the patterns of change in gene expression from early development to later stages, a prerequisite for framing a theory of gene expression control in development.

Nevertheless, genetic methods must remain an adjunct to the molecular studies for ascertaining which functions are truly essential in early development and in which cells they are required. As methods of mutagenesis in the mouse improve, more early zygotic mutants should become available; their characterization can be expected to expand greatly our understanding of early development in the mouse.

An illustration of the genetic analysis of an early zygotic mutant is provided by a chimera analysis of the *yellow* (A^y) mutant, the classic mutant of mouse genetics. Early work established its time of action as within the period of preimplantation development. The experiments also showed that the effects on blastocyst development are delayed if the homozygous mutant embryos are allowed to develop in wild-type mothers, suggesting that part of the lethal effect involves the maternal environment of the developing embryos. However, the controversy over A^y has concerned the principal site of action of the gene. The mutant allele could be a general cell lethal, or it might act initially in either the ICM or the TE, with secondary effects on the other cellular component. Direct observation of developing homozygous mutant embryos has only produced conflicting interpretations and failed to settle the issue.

The controversy is potentially resolvable through chimera construction. The principle of such experiments is the same as that of the reciprocal pole cell transplantations in *Drosophila*. If one chimeric combination produces a mutant phenotype but the reciprocal combination does not, the locus of gene action is directly identified in the phenotypically mutant chimera. In practice, the main difficulty is similar to that in the *Drosophila* pole cell transplantations—knowing which embryos from the donor strain are the homozygous mutant ones. For any recessive lethal, the homozygous mutant individuals can be obtained only by intercrossing the heterozygotes (m + X m/+) which comprise only one-quarter of the progeny.

In the particular case of yellow, the homozygous mutants are obtained by intercrosses such as A^y/a^e X A^y/a^e. where a^e is the viable allele (it produces a black coat color). In 3.5-day donor embryos, before the lethal effect sets in, the three genotypes (homozygous yellow, homozygous viable, and heterozygotes) are indistinguishable.

Papaioannou and *Gardner* (1979) employed a statistical approach to avoid this problem. They used the progeny from two crosses as their sources of ICM and TE components : A^y/a^e X A^y/a^e (the experimental cross) and A^y/a^e X a^e/a^e (backcross). The intercross produces one-quarter homozygous mutant embryos, while the backcross produces none. A difference in embryonic yields from ICM or TE components between the two crosses can signify the cell group in which the lethal mutation is creating its effect. In experiment 1, the two sets of ICMs were tested and compared by being placed into surgically emptied blastocysts of a control albino strain (CFLP). In experiment 2, the two sets of TE cells were tested by emptying the blastocysts of their ICMs and placing control (CFLP) ICMs inside the shells. The two pairs of reciprocal chimeric embryo groups were then transplanted to hormonally prepared foster mothers and analyzed for blastocyst development, measured as the ability of embryos to produce decidual swellings (*implantation*).

The detailed predictions for the various possible outcomes are given in Figure If A^y/A^y are defective in either ICM *or* TE, then the two experiments should yield nonequivalent results: the chimeras with a defective mutant component should show a 25% deficit of chimeras from the intercross relative to the

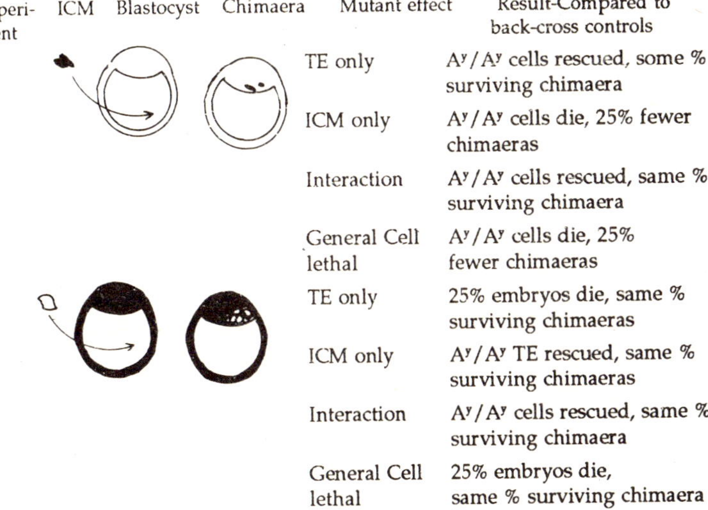

Experiment	ICM	Blastocyst	Chimaera	Mutant effect	Result-Compared to back-cross controls
				TE only	A^y/A^y cells rescued, some % surviving chimaera
				ICM only	A^y/A^y cells die, 25% fewer chimaeras
				Interaction	A^y/A^y cells rescued, same % surviving chimaera
				General Cell lethal	A^y/A^y cells die, 25% fewer chimaeras
				TE only	25% embryos die, same % surviving chimaeras
				ICM only	A^y/A^y TE rescued, same % surviving chimaeras
				Interaction	A^y/A^y cells rescued, same % surviving chimaera
				General Cell lethal	25% embryos die, same % surviving chimaera

Fig. 4.9.   Predictions of survival outcomes of chimeras involving A^y/A^y embryos from intercrossed $A^y/+$ parents.

TABLE 4.1

Reciprocal ICM/TE Transplantations to Test for the Cellular Site of the A^y Locus of Action

Experiment 1: $A^y//a^e$ strain ICMs injected into control blastocysts

	Intercross	Backcross
Implantation Number of decidual swellings/number of transferred embryos	54/62 = 87%	70/80= 88%s
Proportion of chimeras Number of chimeras/total number of conceptuses	44/53 = 83%	62/66 = 94%

$Chi^2 = 4.02$ $P < 0.05$

Experiment 2: Control ICMs injected into blastocysts of the A^y/a^e strain

	Host blastocyst	
	Intercross	Backcross
Implantation Number of decidual swellings/number of transferred embryos	73/99 = 74%	89/98 = 91%
Proportion of Chimeras Number of Chimeras/total number of conceptuses	47/55 = 85%	68/82 = 83%

$Chi^2 = 1.4; P > 0.20$

backcross embryos. As an additional control, the overall rates of chimera production (using donor GPI as a marker) within the surviving conceptuses were assayed.

DETERMINATION IN THE PREIMPLANTATION EMBRYO: PROPERTIES AND MECHANISMS

A key question about the events that separate ICM from TE is whether they are initiated by some form of preexisting organization in the mouse oocyte. A specific hypothesis of this kind was initially formulated by *Dalcq* (1957) on the basis of fixed preparations of mature mouse oocytes. He reported that these specimens exhibited differentiated "dorsal" and "ventral" cytoplasm and proposed that these cytoplasms were differentially segregated to blastomeres by the eight-cell stage. In the Dalcq hypothesis, the dorsal cytoplasm eventually gave rise to the ICM and the ventral cytoplasm to the TE.

It now appears, in fact, that the regional differentiation of fixed eggs reported by Dalcq was an artifact of the fixation process. Furthermore, the idea of a strict separation of ICM- and TE -forming potential was disproved by the experiments of *Kelly* (1975), described above, which demonstrated that blastomeres at the eight-cell stage are equally capable of giving rise to ICM and TE cells. Although the observations of totipotency at the eight-cell stage do not eliminate the possibility that a segregation of ICM and TE "determinants" occurs at a later stage, this too appears unlikely. The molecular and embryological evidence suggests that the developmental restrictions that accompany separation of ICM from TE take place relatively gradually and with a certain degree of reversibility or lability.

If the determinative process is not rigidly fixed by the cytoplasm that each blastomere inherits from the egg, then the divergence of ICM from TE must involve some form of differential cell-cell or cell-environment interaction. The obvious physical differences between these two initial cell groups is in their relative positions: the ICM cells derive from cells that are inside the blastocyst, while the developing TE cells are at all times outside and exposed to the environment. As first noted by *Mintz* (1964), this difference in physical position constitutes a difference in "microenvironment" that might shape the initial divergence in fate and developmental capacity between the two cell types.

The first experimental findings to support this "inside-outside" hypothesis of determination were presented by *Tartowski* and *Wroblewska* (1967). They examined the products of in vitro development of isolated blastomeres from both four-cell and eight-cell embryos and concluded that all cells at these stages had the capacity to give rise to vesicular (blastocyst-like) forms. Thus, there was no suggestion of a segregated ability to form ICM specifically from any of the balstomeres of these two stages. Significantly, however, the isolated one-and two-cell products of eight-cell embryos were found to form a higher percentage of purely trophoblastic vesicles (lacking ICM) than did blastomeres from four-cell embryos. This difference is explicable if fluid secretion from the outside cells, and hence blastocyst formation, tends to occur only after a particular number of cleavage divisions—for instance, after the fifth cleavage division. Under these circumstances, cell clusters derived from one out of eight blastomeres (i.e., single blastomeres from eight-cell embryos) would necessarily be smaller at the time of fluid secretion than mini-embryos derived from one out of four blastomeres and would be less likely to have inside (proto-ICM) cells. In this view, the capacity to form ICM is solely a function of being inside at the time fluid secretion begins.

The strongest experimental support for the inside-outside hypothesis comes from chimera experiments in which cells from four-or eight-cell embryos were dissociated from one another and placed singly or in groups, either internally or externally, on carrier embryos. The fate of the donor cells, marked by prior labelling of their nuclear DNA with

[³H]thymidine or with GPI-isoenzyme difference, was then followed in subsequent development. The results showed a clear biasing of dell fate toward either TE or ICM development as a function of donor cell positioning on the recipient embryos. Cells placed externally contributed over 90% of their descendants to the outside region of the blastocyst, while internal placement produced a less dramatic but still substantial (40%) contribution of cellular descendants to the ICM. Furthermore, when 8-to 16-cell embryos were enclosed by six unlabeled embryos, aggregates were formed that developed into larger blastocysts; of these several showed labeling only within the inner cells and developed into normal appearing embryos by 13 days. In these embryos, cells that would have developed into TE had evidently been channeled by their position into forming ICM. Finally, chimeras with externally placed cells that were allowed to develop to birth showed reduced contributions from the donor cells to the development of the coat, as scored by coat colour markers, regardless of their specific genotype. This result is the expected one if external cells are preferentially channeled toward TE formation and away from ICM formation. In sum, the results indicate that position rather than intrinsic blastomere cytoplasmic inheritance is a primary factor in influencing blastomere fate.

How might cellular position translate into a molecular mechanism for the differential restriction of cell fate? The first explanation proposed was that of *Ducibella* (1977), who suggested that inside cells find themselves within a qualitatively distinct microenvironment, created by the formation of apical zonular tight junctions between the TE cells that seal off the ICM cells completely from the external medium. The unique microenvironment of the internal cells then leads, by a series of steps, to the final differentiative and determinative steps.

However, two facts tell against this explanation. First, zonular tight junctions only develop at about the 32-cell stage *after* the initial divergence of the two cell types at the molecular level. Therefore, the first events in divergence take place without

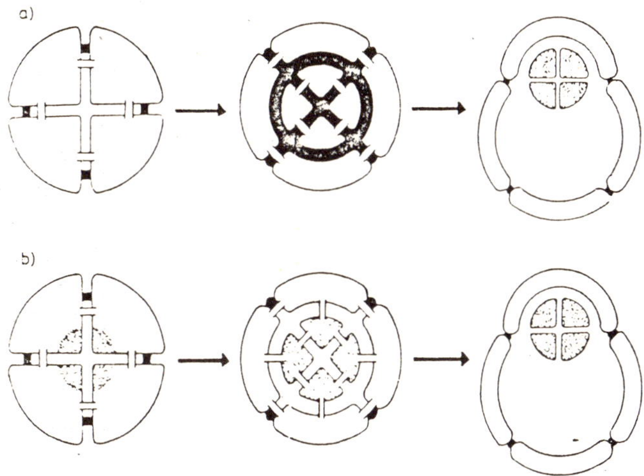

Fig. 4.10. Two models of positional fate setting in the mouse embryo. (*a*) The permeability block microenvironmental hypothesis; (*b*) the polarization hypothesis.

a communications barrier between the two cell types. Second, it has so far provent impossible to create by artificial means the postulated microenvironment. One observation, in particular, is fairly striking: injection of an entire eight-cell embryo into a fully expanded blastocyst failed to prevent the outer cells of this enclosed embryo from turning into TE despite the tight permeability block to the external environment provided by the host blastocyst.

The microenvironment hypothesis of Ducibella might be termed an extreme "instructionist" one : the environmental difference is a complete, qualitative difference that automatically produces a corresponding qualitative cellular difference. However, if the initial environments of outside and inside cells are not different in kind but only in degree, the initial cellular differences may also be subtle ones of degree. In slightly different terms, the initial difference may be quantitative, involving a graded property of some kind.

Johnson et al. (1981) proposed a hypothesis along these lines. Their "polarization hypothesis" stated that inside-outside differences stem from a radial asymmetry within the embryo in which individual cells are polarized with respect to molecular and cellular properties at their (outer) apices and (inner) bases. In this view, cleavage in the compacted eight-cell embryo, occurring perpendicularly to the radial axis, separates cells with "inner" (presumptive ICM) and "outer" (presumptive TE) values.

This hypothesis makes compaction a key event in creating or amplifying a radially-graded property. Compaction also draws the eight cells together into an organized unit in which communication is maintained through extensive intercellular contact and small intercellular channels at gap junctions. The importance of compaction in the separation of ICM and TE has been demonstrated experimentally using treatments that reversibly inhibit compaction during development. With one treatment that inhibits compaction without affecting cleavage, a delay in compaction beyond a certain time produces blastocysts that lack a genuine ICM component, yet the cells of these aberrant blastocysts synthesize the full spectrum of proteins characteristic of normal blastocysts. Under these conditions, the basic gene expression program is activated, but in the absence of cell compaction, none of the cells evidently perceive themselves as ICM.

An important second set of observations that support the hypothesis concerns the existence of individual cell polarization at the eight-cell stage. Whether in the intact embryo or in isolated cell pairs, cells at this stage become columnar and axially polarized, showing an outer cap of microvilous extensions that act as binding sites for a fluorescent ligand (fluorescein isothiocynate-concanavalin A). When the cells of the eight-cell embryo divide in vitro, most divisions generate a polarized cell (possessing the ligand binding sites) and a non-polarized cell (lacking these sites): the former correspond to outer cells, the latter to inner cells. Some 2/16 cell couplets (pairs of cells at the 16-cell stage) of this kind. In the normal embryo, these two types of cells tend to form two discrete lineages, polar cells dividing to give pairs of polar cells and apolar cells dividing to give apolar cells. The beginnings of the ICM-TE lineage divergences are foreshadowed in these cytologically defined lineages.

Does the suggestion of a radial gradient conflict with the discovery that cell fate can be biased in one direction or the other by the appropriate changes of position within the early embryo? There is no essential conflict if the postulated gradient, with its high point at the center and decreasing along all axes out to the periphery, can "entrain" cells by means of cell-cell communication to assume the characteristic gradient value at any position. *Hillman et al.* (1972) point out that their results are consistent with a gradient hypothesis if the gradient source exists at the center of the embryo; transplanted blastomeres would tend to acquire the radial gradient value of their new position.

However, whatever the origins of polarity characteristics, whether they arise in the embryo or preexist in the oocyte, the maintenance of the two cellular lineages requires continuing interactions between polar and apolar cells. While isolated 2/6 (polar :apolar) couplets give rise predominantly to four-cell (4/32) clusters containing two polar and two apolar cells, synthetic couplets consisting of two polar (TE-like) cells produce some ICM-like cells following division. The results indicate that it is the continuing interactions between polar and apolar cells that tend to maintain the two cell type lineages.

The results indicate a complex origin of the "morphogenetic instructions" in the early mouse embryo involving both preexisting organization and cell-cell interaction. Position *is* important in fate setting, as shown by the experiments of *Hillman et al.* (1972), but it is position with respect to the site of origin within the oocyte mass and to the other cells of the early embryo rather than to the external environment that is a primary importance.

There are two other features of the ICM-TE divergence that deserve note. The first is that there may be certain regularities in cleavage that affect cell fates. From a study of cleavage rates and spatial patterns in partially flattened embryos in vitro, C.F. Graham and his colleagues conclude that at the two-cell stage, one blastomere always cleaves first and its progeny tend to occupy internal positions preferentially. If the observations pertain to normal embryos, they suggest that a topogenetic bias leads to a biasing of cell fates. The other observation is that the first inner cells show more rapid cycles of nuclear replication than the outer cells. In this system, as perhaps in others, it appears that cells with a greater range of developmental potential show more rapid rates of nuclear DNA replication.

The origins of the second pair of determinative events in the preimplantation embryo are more obscure than the ICM-TE divergence. However, they presumably also involve the intrinsic properties of the progenitor cells (established by the earlier events) and the interaction of these cells with each other and with the environment. The divergence of the two kinds of TE cells—polar and mural—seems to depend, as mentioned earlier, on the special interaction of polar cells with the ICM mass. In the absence of that interaction. TE cell invariably become mural TE. The divergence of endoderm from ectoderm in the ICM may, however, depend primarily on different relative exposures to the blastocoelic fluid. When ICM cells are placed into empty zonae, they invariably develop into endodermal cells. In intact embryos, the embryonic ectoderm, enclosed between endoderm and polar TE. remains totipotent and is the ultimate source of the fetus.

In both the TE-ICM and the endoderm—ectoderm divergence, it is the relatively enclosed inner cell group that retains the undifferentiated appearance of early morula cells and the grater set of prospective potencies. The high point of developmental capacity in the preimplantation embryo is thus at or near the center; cells formed from the periphery differentiate first and exhibit a smaller range of fates.

POSTIMPLANTATION DEVELOPMENT IN MOUSE

In appearance, a chicken egg and a mouse egg are very different. Most obviously, they differ vastly in size, roughly 5-6 cm average diameter in the former compared to 80 μm in the latter. Despite this difference, the basic embryology of the chick and the mouse have much in common. The difference in egg size reflects the different ways of provision for the nutrition of the developing embryo. The chick embryo has all of the nutrients that it needs prepackaged into the egg, in the form of yolk and albumin, while the mouse embryo must derive virtually all of the material and energy it needs for its growth and development directly from its mother. The mammalian arrangement requires that the embryo construct a much more elaborate set of extraembryonic membranes than the chick, membranes that will house it and connect it to its maternal parent.

Nevertheless, the relative simplicity of chick embryogenesis provides a suitable introduction to the mouse and will be briefly described here. The unfertilized chicken egg consists of a large mass of yolk capped by a small disc of cytoplasm containing the egg nucleus. Fertilization of the membrane-enclosed egg occurs as the egg passes down the oviduct, before the hard outer shell is formed, and it succeeded by rapid cleavage divisions, which generate a disc of approximately 80,000 cells by the time of laying. This cell layer or blastoderm consists of a relatively thin central area (the *"area pellucida"*) and a thicker outer ring (the *"area opaca"*). At one side of the area pellucida, a secondary thickening about five cells deep is apparent. This thickening is termed the "primitive streak" and constitutes the direct embryonic precursor region. By a series

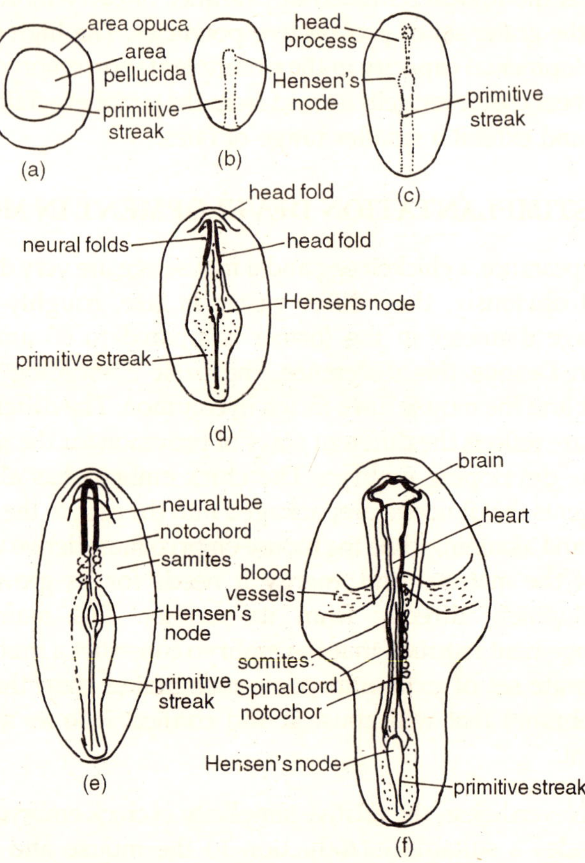

Fig. 4.11. Stages of development in the chick embryo. See text for description.

of cell movements, the primitive streak soon comes to be composed of two cell layers. The upper, ectoderm-like layer is termed the *"epiblast"* and the lower, more endodermlike layer the *"hypoblast,"* The embryo itself develops exclusively from the epiblast, while the hypoblast gives rise to the major extraembryonic membranes of the yolk sac. These membranes separate the embryo from the yolk and serve as a conduit for nutrients from the yolk to the developing chick. The cell movements that generate and separate the epiblast from the hypoblast are complex and constitute the process of gastrulation; they involve migration of cells toward and through the primitive streak area.

The *primitive streak* defines the future a-p axis of the embryo. The next major phase of development involves the creation of the major longitudinal (axial) sequence of structures in the embryo. The primitive streak lengthens, and this lengthening if followed by the formation of a condensation at the anterior end, to form the structure known as "Hensen's node." Hensen's node is mesodermal and gives rise to the main mesodermal elements of the early embryo. A portion elongates in the anterior direction, underneath the outer ectodermal cells, and thereby lays down the precursor cells of the notochord, a transitory central element in vertebrates. The most anterior point of this migration marks the position of the *"head process,"* the feature cephalic region. The remaining portion of Hensen's node migrates posteriorly, under that portion of the ectoderm that gives rise to the neural plate, the precursor of the spinal cord. This migration is accompanied by the delimitation of the somites, which are paired mesodermal, segmental units. The somites are the ultimate source of the axial skeleton and ribs, the body musculature, and the major part of the dermis of the adult bird. As somite formation progresses, the neural plate closes to form the neural tube. The ensuing developmental events in the chick embryo are so closely correlated with the stages of somitogenesis that chick embryos can be reliably staged by reference to the number of somite pairs that have formed at any given time point. By the end of somite formation, the body plan of all of the major organ systems in the chick is discernible.

The mouse embryo also develops from the primitive streak, but the preamble to its formation is considerably more complicated. At the time of implantation, which occurs in the mouse at about 4.5 days postoconceptus *(p.c.),* the balstocyst consists of just four cell types: the polar and mural TE cells and the primary ectoderm and primary endoderm. A full 2 days elapse during implantation before a detectable primitive streak arises. The complete sequence of events from implantation through the first stages of organ formation has been described by Snell and Stevens (1966), and is summarized here.

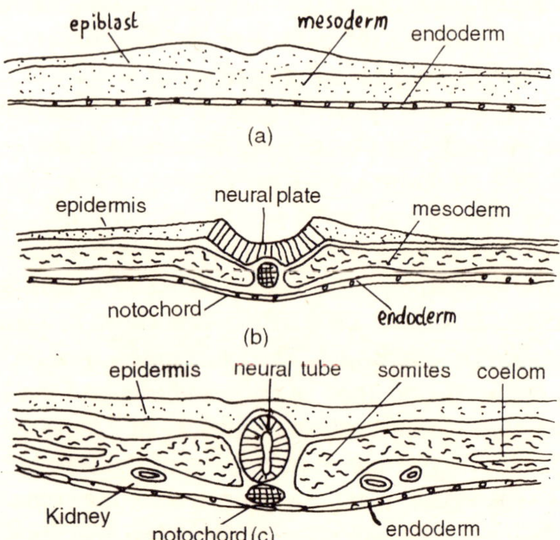

Fig. 4.12. Development of the chick neural tube. (*a*) Primitive streak stage; (*b*) neural plate development; (*c*) neural tube stage.

At the time of implantation, the internal endoderm covers the ectodermal ICM cells but has not spread around the inside of the blastocoele. During the next 1.5 days, while the embryo is embedding itself in the uterine wall, the primary endodermal layer comes to cover both the inside of the blastocoele and the inner ectodermal cell mass. The latter, covered with a stocking of endodermal cells, projects progressively farther into the blastocoelic cavity, and the whole structure is now referred to as the *"egg cylinder"*; the blastocoele at this point is designated the *"yolk cavity"* (which in the avian embryo is filled with yolk).

(This early invagination of embryo precursor material into the yolk cavity is peculiar to mice and rats, and does not take place in other placental mammals, even in more primitive rodents. The typical mammalian pattern is initially similar to the avian one; the ectoderm is a disc sitting atop the endodermal layer and invaginates only at a much later stage. The murine arrangement, in which ectoderm comes to be enclosed by endoderm, is referred to in classical embryological texts as the "inversion of the germ layers.")

The primary endoderm now consists of two distinct-cells layers. The layer that covers the inside of the yolk cavity is referred to as the *"distal"* (or *"parietal"*) endoderm., while that covering the egg cylinder is termed the *"proximal"* (or *"visceral"*) endoderm. The proximal endoderm blankets both the embryonic ectoderm—so called because it is the precursor of the embryo proper—and the extraembryonic cells; the latter are derived from the polar TE cells, which retain their capacity for cell division. Capping the extraembryonic ectoderm in the 5.0-day embryo is the ectoplacental cone, also derived from the polar TE and a major precursor of the placenta. At this stage, a small cavity within the embryonic ectoderm can be discerned, the proamniotic cavity, which is the precursor of the amniotic cavity, the fluid-filled sac containing the fetus.

The entire structure enclosed by the visceral endoderm is called the *"yolk sac,"* and from this point on, the visceral endoderm can be equated with the hypoblast of the chick embryo. The yolk sac enlarges within the yolk cavity, and by 5.5 days the embryonic ectoderm has begun to delaminate the second of the germ layers of the embryo, the mesoderm. Concurrently, the cavities within the embryonic and extraembryonic portions of the egg cylinder have enlarged and temporarily joined. Cavities have also begun to appear within the mesoderm; these cavities unite to form the beginnings of the exocoelom, which by 7.25 days occupies the space between the ectoplacental and amniotic cavities. The exocoelom is separated from the amniotic cavity by the amnion, a membrane composed of a layer of mesoderm (on the exocoelomic side) and layer of ectoderm. By 8.5 days, the exocoelom has surrounded the amniotic except for the region occupied by the fetus. By this stage, the ectoplacental cone has fused with the chorion (the membrane bounding the exocoelom opposite the amnion) and the allantois, a structure derived from the mesoderm that serves as a connection between the maternal blood supply and the fetus. The chorioallantoic membrane is connected to the fetus through the richly vascularized membrane that bounds the exocoelom, the yolk sac splanchnopleure.

The fetus originates at 6.5 days, with the appearance of the primitive streak as a thickened region on one side of the embryonic ectoderm, near the developing allantois. Between 6.5 and 10 days, all of the major organ systems take shape within the primitive streak through a complex series of morphogenetic movements accompanied by extensive growth. The process beings with the production and proliferation of mesodermal progenitor cells at the proximal (allantoic) end of the primitive streak; this position marks the future caudal end of the fetus. As ectodermal cells migrate through the primitive streak, they move both laterally and distally toward the future cranial end of the embryo. This migration of cells into and through the primitive streak constitutes gastrulation in the mouse embryo. The a-p orientation of the embryo is thus established or revealed by the laying down of the mesoderm in a caudocranial direction. As this process continues, mesodermal cells progressively occupy the space between the visceral endoderm and the embryonic ectoderm, except for the narrow central section that forms the spinal region of the fetus; this early primitive streak stage is very similar to that of the chick.

At 7.0 days p.c., the head process has taken shape at the most distal end of the egg cylinder. As in the chick, the head process gives rise to the notochord; it also contributes to part of the endodermal lining of the gut. The head process grows both proximally, on the side of the egg cylinder opposite the primitive streak, and laterally. By 7.5 days, the growing head process and the mesodermal sheets have met; throughout most of the embryo, ectoderm is separated from primary endoderm by the mesodermal layer, except for the midsaggital strip of head process directly in contact with the ectoderm.

The early primitive streak stage is a period of exceptionally rapid growth and transformation. During this period, the longitudinal elements, namely, the digestive tract and the axial system, begin to form. The gut begins as two invaginations of the yolk, sac on opposite sides, near the junctures of the embryonic and extraembryonic ectoderm. The foregut precursor pushes in at the cranial end, beginning at 7.0 days, and 7.75 days has become a pronounced inpocketing: the hindgut has just begun to form at this point. The blind ends of the foregut and hindgut invaginations eventually break through to the outer surface of the embryo, giving rise to the mouth and anus, respectively. The invagination of the foregut also lays the foundation of two other key organs, the heart and the brain. The most dorsal region of ectoderm that is pushed in is the head fold structure, which develops into the brain. The heart arises from the mesoderm lying between the head fold and the outer endoderm.

The axial system comes into being from the lateral sheets of mesoderm that border the notochord, the so-called paraxial mesoderm. As in the chick embryo, the axial system arises in a strict, linearized sequence of somite formation that begins at the cranial end of the embryo, just posterior to the embryonic brain, and progresses to the caudal end. Somitogenesis therefore proceeds in the opposite direction to that of mesoderm formation. The first pair of somites arise at about 8 days. Initially, pairs of somites are laid down at the rate of one per hour; subsequently, the process slows to one pair every 2-3 hr. Somitogenesis is complete by about 13 days of development, when 65 pairs of somites have been formed; of these, 30 pairs will form of the skeleton, musculature, and dermis of the body (neck, thorax, and lower back) and 35 pairs will form these elements in the tail. Each newly formed somite shows a trace of a boundary between the anterior and posterior halves, and each vertebra arises from cells of the posterior half of one somite and the anterior half of the next. Differentiation of the somites is accompanied by dispersal of the mesenchymal cells that compose them, leading to an obliteration of the intersomitic boundaries.

The stages of neural development, the precursor of the spinal cord and the other major component of the axial system, are closely correlated with the stages of somitogenesis. The neural tube develops from the closure of the neural plate area just above the notochord and between the sheets of paraxial mesoderm. The strict correlation of the somite stage with other events in normal development permits one to stage embryos according to their somite number, as in avian embryos.

By 10 days p.c., the major elements of the circulatory system, also formed from the paraxial mesoderm, have developed. The turning of the embryo also completes the formation of the gut by generating the midgut and joining the fore-and hindguts. During the succeeding 10 days of development in utero, the major and minor organ systems undergo progressive differentiation. A photograph of a 26-somite embryo and a diagram of the 13.5 day embryo, showing the major organ rudiments, are presented in Figure 4.12. A timetable of some of the major events in postimplantation development is given in Table.

Groth Size and Morphogenesis
One of the most general and remarkable features of embryogenesis is the co-ordination of growth with the events of development. This co-ordination ensures that the newborn or newly hatched progeny of a given species or strain are of a relatively uniform size and that the internal and external structures are correctly proportioned.

The existence of mechanisms for co-ordinating development with growth becomes obvious when embryo size is altered experimentally. The most dramatic of such regulatory responses are those seen when isolated blastomeres from early embryos give rise to whole animals. For instance, if one separates the blastomeres of the four-cell sea urchin embryo, each cell develops into a perfectly shaped and normally functioning pluteus larva, although only one-quarter the normal size and containing one-quarter the normal number of cells. Similar regulative capacities are displayed by the first blastomeres of frogs and starfish; in mammals, normal newborn rabbits have been obtained from isolated blastomeres of eight-cell embryos.

TABLE 4.2

Developmental Landmarks in Postimplantation Development[*]

Event	Days p.c.
Implantation	4.5
Proamniotic cavity appears	5.0-5.5
Primitive streak detectable	6.5
Foregut appears	7.0
Neural plate forms	7.75
Somite formation begins	8.0
Chorioallantoic placenta forms	9.0-10.0
Primordial germ cells appear at the base of the allantois	7.75-8.0
Primordial germ cells arrive in genital ridges	10.5-11.5
Melanoblasts leave the neural crest	8.5-9.5
Neural tube development completed	10.0
Forelimb buds appear	9.2
Hindlimb buds appear	10.0
Somite formation completed	13.0
Hematopoiesis begins in the fetal liver	9.5-10.0
Hematopoiesis begins in the fetal bone marrow	16-17
Birth	19-20

[*]Times cited are approximate; inbred mouse stains frequently show a delayed timetable.

Major adjustments can also occur much later in development. Individual duck blastoderms, for instance, if operated on prior to the emergence of the primitive streak, can give rise to four independent embryos. In *Drosophila*, severe limitations of the food supply during the period of larval growth can lead to the production of miniature, although otherwise fully normal, fruit flies.

In the mouse, growth regulation can occur at both early and intermediate points of development. Chimeric blastocysts made from two or more embryos are invariably larger than normal blastocysts, yet develop into newborn mice of standard size. At some point, there must be a "downward" regulation of size to yield a normal-sized embryo. The point has been localized to between 5.5 and 6.0 days days p.c.; it occurs, therefore, after implantation but before primitive streak formation. This response probably reflects a limitation of the nutrient supply to the larger blastocystic embryos, with a consequent slowing down of cell division rate.

"Upward" regulation of the growth of abnormally small embryos also occurs, although after primitive streak formation. Unlike the sea urchin embryo, viable mammalian embryos formed from single blastomeres of the two and four-cell stage do not give rise to one-half or one quarter size animals, but to newborns of normal size. The size adjustment apparently occurs between days 10.0 and 11.0 p.c., and takes place shortly after the maternal circulation is made fully available to the embryo, with complete development of the placenta. The nutrient supply may thus make possible the growth spurt, but the signal to do so must come from the embryo itself, which somehow "knows" that it is small.

The most striking of upward regulation of growth occurs following treatment with mitomycin C (MMC). This drug, when injected intraperitoneally into mothers at the beginning of primitive streak development, produces extensive random cell death in the embryo. At 7.5 days p.c., the epiblast of MMC-exposed embryos contains only 10-15% as many cells as that of control embryos, the dead cells being sloughed off.

Despite this extraordinarily large reduction in cell number, the majority of embryos manage to catch up to control embryo size by 12.0-13.5 days, and many are born alive. The primary defect in the survivors is a partial or complete sterility, reflecting a permanent reduction in germ cell number effected by the MMC treatment.

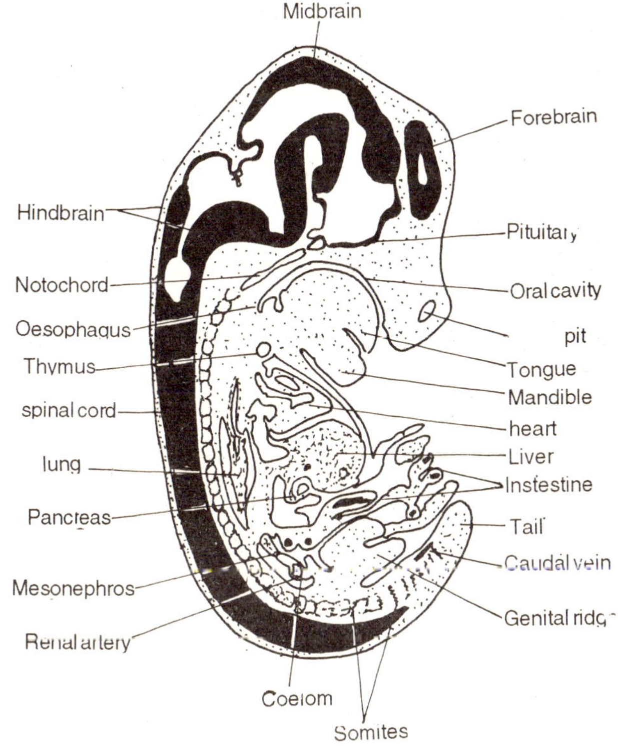

Fig. 4.13. Schematic longitudianl section of a 13.5-day embryo.

Examination of MMC-treated embryos shows that the developmental time course of both ectodermal and mesodermal components is affected, with a visible retardation of neural tube development and of somitogenesis. However, the effects on the neural tube and the somites are partially uncoupled, with the neural structures experiencing the more severe delay.

It follows that the stages of development of the two major components of the axial system are neither obligatorily coupled to, nor dependent on, one another during the period of recovery (nor, by inference, during normal development).

An intriguing aspect of somite development in MMC-treated embryos is that while the normal number of somites is produced in the catchup embryos, the individual somites are smaller and have fewer cells (Tam, 1981). The conservation of somite number at the expense of individual somite cell content suggests that there exists some form of global partitioning mechanism within the presomitic mesoderm. This mechanism operates prior to the first segmentation event and ensures that the total mass is divided into the correct number of somite pairs. The existence of such a pattern-setting mechanism is, at least superficially, reminiscent of the partitioning of the *Drosophila* blastoderm surface into segments, although somitogenesis appears to have a much greater regulative capacity (and operates with greater cell numbers).

Gene Expression in Postimplantation Development

The events of implantation mark a point between the early but limited cell diversification of preimplantation development and a much more extensive set of cellular differentiations. Is this acceleration in cell diversification accompanied or driven by a corresponding set of qualitative changes in the gene expression pattern? The extensive evidence for *Drosophila* and the less complete data for *Caenorhabditis* indicate that successive developmental stages and different cell types exhibit a large set of conjointly expressed genes, with comparatively few qualitative differences between these gene sets. Is the mouse similar or different in this respect?

Answers to this question have been sought primarily by protein labeling and electrophoresis techniques and by RNA-DNA hybridization techniques. The results indicate that, as in the fruit fly and the nematode, there is a broad base of shared gene expression between different cell types and stages during mouse development. At the same time, they show that large-scale quantitative shifts in gene expression or gene product accumulation occur during development.

Labeled Polypeptide Patterns

The first developmental events in the postimplantation embryo involve the elaboration of the principal extraembryonic membranes and the development of the primitive streak. By 7.5 days of development, the polar TE has given rise to distinct ectoplacental cone (EpC) region, a layer of giant cells (GCs) that surround the diploid core cells of the EpC, and a layer of extraembryonic ectoderm (EE), which includes the chorion. The future embryonic region is separated from the exocoetom by the amnion and consists of the region of the primitive streak and, opposite it, a region primarily composed of embryonic ectoderm (EmE).

Johnson and *Rossant* (1981) assayed the molecular correlates of this first stage of postimplantation change by comparing the polypeptide labeling patterns of the three polar TE lineages with each other and with embryonic ectoderm. The different regions were dissected, labeled in vitro, and the polypeptides then extracted and separated on two-dimensional gels. In a large number of such comparisons, 50 spots, from a total set of more than 500 detected, were found that distinguished the four cell types from one another. Of these, six were unique to EmE, one to EpC, and three to GCs. All of the remaining polypeptides in this set of 50 were shared between different subsets of the four tissues. Although the comparisons themselves provide no direct information about developmental relationships between the tissues, an in vitro culture experiment showed that cultured EE cells first develop an EpC-type profile that is then succeeded by a GC-type profile. By this criterion, the EE cells may be equivalent to the original polar TE cells and serve as a reservoir of EpC and GCs. The principal finding is that within the four ectodermal cell groups compared, which differ in cytology, function, and fate, there is a large degree of similarity in the detectable polypeptide synthesis pattern.

Later periods of development show much the same similarity of proteins synthesis. A detailed study of polypeptide synthetic patterns in embryonic mouse organs, employing [S] methionine pulse labeling in vitro, was reported by *van Blerkom et al.* (1982). The organ systems tested included liver, brain, kidney, forelimb, hindlimb, yolk sac, lung, and fetal tail (essentially somites) from 10- to 13-day embryos. For all stages and all organs, a total of 850-1000 polypeptide spots were detected. For the various organs surveyed, an average of three organ-specific protein synthetic differences were found, or less than 1% of the total. In comparison, quantitative shifts specific to organs or stages were relatively more common. In the lung, for example, 5 qualitative changes in spot pattern and 15 quantitative ones were measured in the period from 10 days p.c., to early postnatal development. For the brain, from day 10 p.c. to day 10 postnatally, 1 qualitative change and 20 quantitative changes were detected. The very small number of observed changes in this survey was not a reflection of limitations in the labeling or electrophoretic procedures; labeling with ^{14}C amino acids or changing the pH range in the separation increased the total number of detected spots without altering the percentages of qualitative change in pattern.

The results of this and other studies not cited here are all in general agreement: there is little difference in polypepeptide labeling patterns between cells that are markedly different in phenotype. The two-dimensional separation procedure, however, may only resolve the various housekeeping proteins, all of which should be present in comparable amounts between different cell types. However, not all housekeeping proteins are necessarily in the abundant or semiabundant classes. *Galau et al.* (1976) have presented arguments that physiologically significant metabolic functions can be carried out by proteins produced from the rare mRNA class. Conversely, some of the detectable polypeptide spots may be produced from relatively rare mRNAs if translation efficiencies differ between certain species. Nevertheless, whatever the cellular functions of the detected proteins or the relative abundances of their mRNAs, they must represent only a portion of the expressed protein-coding genes. In contrast, RNA-DNA hybridization methods provide a more inclusive measure of the complete set of the number of expressed genes.

mRNA Pol Studies

Several measurements of mRNA sequence diversity in mouse cells have been made. Sequence complexity was estimated in these studies from the kinetics of hybridization of poly A$^+$ mRNA to cDNA made from such RNAs; the advantages of this approach were described earlier. The chief disadvantage is that only minimal estimates of sequence diversity are obtained. The true sequence diversity is underestimated in two ways. Firstly, any mRNA species that are transported to the cytoplasm without poly A tails will not be purified; hence, these sequences will be missing from the cDNA preparation. In *Drosophila*, at least, these molecular species apparently comprise a substantial portion of the total mRNA pool. Secondly, rare mRNAs that are present in some but not all cell classes will not be detected because of dilution in the total mRNA pool. Nevertheless, if the presence or absence of polyadenylation is characteristic of a particular mRNA sequence, the measurement of sequence complexities in different cellular poly A$^+$ mRNA pools will provide meaningful comparisons between different cell types. The three studies described below show that most mouse tissues have approximately 10,000 poly A$^+$ mRNA species, of which the great majority are shared between cells of very different types.

In a comparison of such mRNAs from adult liver and brain and from whole 14-day embryos. *Young et al.* (1976) found that the three mRNA pools are complementary to 0.7, 1.7, and 0.7% of the sequence of the mouse genome, respectively. If one takes 20000 b.p. as a typical mRNA length, these percentages correspond to approximately 12,000 different mRNA species for the liver, 25,000 for the brain, and 10,000 for the 14-day embryo. (Although the brain is well-developed in 14-day embryos, the lower sequence complexity for whole embryos relative to the brain presumably reflects the dilution of rare brain-specific sequences in the former.) By performing cross-hybridizations to near saturation in order to detect sequence overlap, Young et al. found a large measure of qualitative similarity between the mRNA pools of adult liver and brain. However, the rates of cross-hybridization, which primarily reflect the more abundant species, were lower than those in the homologous hybridizations. The measurements therefore indicate that the most abundant species for each of the two tissues were less prevalent in the other.

A similar analysis by *Hastie* and *Bishop* (1976) of mRNA populations of adult liver, brain, and kidney produced estimates of 13,000, 11,600 and 11,500 different structural gene sequences in the three messenger pools, respectively. (The distinctly higher sequence complexity of the brain mRNA pool reported in the previous study was not seen in this one; the reason for the discrepancy is unclear). For all three tissues, three discrete abundance classes could be distinguished, consisting of the very abundant species (4-9 types), the intermediate (480-750), and the rare (11,000-12,200). Of the rare 11,000 poly A$^+$ mRNA sequences of the kidney, 9000-10,000 were estimated as shared by liver and brain. As in the study of Young et al., the abundant mRNA species for each tissue were also detected in the mRNA pools of the other two but were in the intermediate or rare class. Altogether, a total of 15,000-16,000 mRNA species are found within the three tissues. Each tissue sampled in the analysis represents one of the major germ layers of the fetus— the liver, endoderm; the kidney, mesoderm; the brain, embryonic ectoderm—yet the vast bulk of the structural genes expressed as poly A$^+$ messengers are shared between them. The main qualitative differences are in the rare mRNA classes, although quantitative differences of the abundant and intermediate classes are a marked feature.

Ideally, one would like to trace the entire pattern of transcript of diversification that occurs from the beginning of fetal development, in the primitive streak stage, on. Unfortunately, this has not yet proven possible because of the difficulty in obtaining enough early embryonic material. An alternative approach is to use neoplastic cells whose characteristics mimic those of early embryonic cells. These cells are termed "embryonal carcinoma (EC) cells" and have been widely used as an analogue of the early embryonic cellular condition.

EC lines come in at least four grades of developmental capacity, ranging from those that give rise to all cell types (totipotent lines) to those with wide but incomplete capacities *(pluripotent lines)* to those incapable of giving rise to any differentiated cells (nullipotent lines). *(Nullipotency* is associated with long-term culture and accumulated chromosomal abnormalities). A few lines have specialized differentiative capacities, such as a particular capacity to give rise to muscle cells under a differentiative stimulus. The different grades of developmental capacity may provide analogues of the various stages of embryonic cell specialization.

If toti- or pluripotent EC cells are truly similar to early embryonic cells, then a measurement of transcript diversity in EC cell lines should give a hint of the gene expression pattern in ICM or primitive ectoderm cells. *Affara et al.* (1977) measured EC mRNA transcript diversity in two pluripotent cell lines and two specialized, EC-derived myoblast (muscle precursor) cell lines. For the pluripotent EC lines, a transcript diversity of 7700 different poly A$^+$ mRNAs was estimated, and for the myoblast lines a transcript complexity equivalent to 13,200 poly A$^+$ mRNA species. By cross-hybridization tests, all of the EC mRNA sequences were found in the transcript pools of the myoblast cell lines, but the latter cells contained a substantial number of rare mRNA species not found in the EC cell line. As before, the abundant and intermediate abundant sequence classes that characterize one cell type are less abundant in other cell types.

Taking EC cells as an analogue of early embryonic cells,

the comparison, suggests that divergence of cells from the early embryonic cell state is accompanied by an expansion of gene expression, principally in the class of mRNAs present in just a few copies per cell. However, other factors, such as the different origins of the various cell lines, may have contributed to the differences between the EC and myoblast cell lines.

Another investigation showed that terminal differentiation is accompanied by relatively little change in overall transcript diversity. The cells of a Friend erytholeukemic line—a virally transformed cell line with the potentiality to differentiate into red blood cells—can be induced to differentiate into mature red blood cells by exposure to the permeabilizing agent dimethyl sulfoxide (DMSO). When these cells, whose sequences are all common to EC cells apart from a few in the most abundant class are induced to differentiate, the principal changes occur in the most abundant species. These changes include a decline in the frequency of several sequence types and the appearance of new ones, including, most characteristically, the hemoglobin mRNAs. Thus, neither commitment to a given pathway of differentiation nor the process of overt differentiation itself necessarily entails large-scale changes in overall gene transcript diversity relative to the (presumptive) early embryonic cell state. Whether the biologically significant changes are primarily those in the abundant classes or whether changes in all classes are equally important remains an open question. However, in all specialized cells whose physiological function depends on a certain protein or group of proteins, one change will always be an increase in the concentration of the abundant mRNA(s) that code for those protein(s).

Changes in cellular mRNA concentrations may occur either through regulation of transcription or through changing efficiencies of transcript processing. There is some evidence that control of processing plays some role in regulating mRNA abundances. When the nuclear RNA pools of different mouse cell types whose cytoplasmic mRNA compositions differ are examined, they are found to contain sequences for the same rare mRNA types. Thus, the nuclei of brain cells have copies of all of the sequences present as mRNA in the cytoplasm of kidney cells and vice versa. However, the nuclear sequences that do not become converted to mRNAs are rare, being present at only one or a few copies per nucleus. These results indicate that there may be some transcription of all potential mRNA sequences in all cells.

On the other hand, the regulation of the more abundant species may be subject to specific transcriptional control. *Derman et al.* (1981) cloned several sequences from cDNA made from liver poly A^+ mRNA, and identified 11 cDNA clones corresponding to abundant or semiabundant mRNAs that are much rarer in two other cell types that were examined-hepatoma cells and mouse L cells. For all 11 of these "liver-specific" sequences, the rate of pulse labeling of these sequences in nuclei in vitro—which serves as a measure of the instantaneous transcription rare—was found to be significantly higher in liver nuclei than in brain nuclei. In contrast, the in vitro pulse rate labeling of mRNA complementary to five cDNA sequences shared by several cell types was found to be equivalent in liver and brain cell nuclei. The experiment indicates that the more abundant mRNA sequences in cells are more concentrated precisely because the transcription rate of their homologous gene sequences is specifically enhanced. Transcriptional regulation in eukaryotes may therefore involve primarily modulations in transcription rate that shift sequences between the more and less abundant categories, while the presence or absence of certain rare sequences as mRNAs may involve some degree of sequence-specific posttranscriptional processing.

POSTIMPLANTATION DEVELOPMENT : GENETIC ANALYSIS

Postimplantation development is a continuous sequence of events but may be divided arbitrarily into two phases. The first extends from implantation at 4.5 days to approximately 6.5 days p.c., during which the extraembryonic membranes and the primitive streak are formed. The second and longer phase, from 6.5 days to birth, is the period of organ rudiment formation, organogenesis, and fetal growth. Everything that precedes the primitive streak stage involves the elaboration of structures necessary for the housing and nutrition of the embryo proper that arise from the primitive streak.

A large part of contemporary murine developmental genetics has centered on the delineation of the various component cell lineages of the tissues and organs of the postimplantation embryo by means of genetic techniques. Because mitotic recombination is either rare or nonexistent in mammalian embryos, it cannot be used to reconstruct the detailed history of these lineages. Instead, principal reliance has had to be placed on the construction and analysis of chimeras. As in the study of preimplantation lineages, early embryo chimeras are first made either by morula aggregation or by injection of cells into blastocysts. The chimeric embryos are then implanted in pseudopregnant foster mothers and, following the requisite period development, the animals are dissected and the tissue or structure of interest is analyzed in terms of the input marker phenotypes. From the pattern of distribution of the marked cells, deductions can then be formulated about the nature and/or number of the founding cells. This form of analysis has proven quite successful for preimplantation development, as we have seen, and for early postimplantation development. For post-primitive streak development, the method has been applied principally to estimating the number of founder cells for different organs or tissues; in these studies, the interpretations, while interesting, have been consistently controversial. A second form of clonal analysis is that based on the inactivation of single X chromosomes that occurs during the early development of two-X (female) embryos; it has also been employed to estimate founder cell numbers. However, these analyses also suffer from ambiguities that affect the interpretation of the results. The findings from these approaches, and the difficulties, are reviewed below.

Origins of the Extraembryonic Membranes and Fetal Germ Layers

At implantation, the TE lineage of the early blastocyst has split into two distinct cell groups. The mural TE cells, which occupy the sides of the blastocyst, have begun the transformation to primary GCs. These cells eventually achieve ploidy levels of several hundred; their function is to achieve the initial implantation of the blastocyst in the uterine wall. The other TE cell group, the polar TE, produces the Epc and EE cells. The presumptive cell derivation relationships in the polar TE lineage wee described earlier; the EE cells may be the primary type and give rise to EpC cells, which in turn generate secondary GCs at the most proximal end of the embryo.

Although there has been little controversy about the TE lineages, the origins of at least two fetal tissues have been the subject of debate in the recent past. One of these tissues is the fetal endoderm. From the classic embryological description of the formation of the gut, it had been assumed that the proximal endodermal layer gave rise to the lining of the gut; without question, endodermal cells from this layer are carried along in the invaginations that give rise to the fore- and hindguts. However, this source is difficult to reconcile with the development of the avian embryo: in the latter, the entire embryo, including the gut, arises from the epiblast. The avian hypoblast, which appears similar in function and location to the proximal endoderm, does not contribute to the embryo proper.

The origins of the germ line, the gamete-producing cells, have been the second contentious issue. Direct histological tracking of the primordial germ cells (PCGs) had placed their source, like that of the gut, in the proximal endoderm. A number of observations have forced a revision of this view; it is now apparent that the germ line, like all the other components of the fetus, comes from the EmE.

A feature of the PGCs that has made it possible to trace

them in early development is their high level of alkaline phosphatase (AP) activity. By staining for AP activity, was able to detect presumptive PGCs in the 8.0-day p.c. embryo near the caudal end of the primitive streak, within the base of the allantois. Twelve hours later, the main cluster of high AP cells was observed in the proximal endoderm and, 12-24 hr later, within the invaginated hindgut rudiment. From this position, the PGCs then move by an active, chemotactic process, along the dorsal mesentery within the developing body cavity, to the genital ridges, which they then colonize. (This pattern of movement occurs in embryos of both sexes; all of the differences in germ cell maturation between the sexes begin to be apparent after colonization of genital ridges). The process of migration is complete by about 11 days p.c. From the point at which they first become distinguishable to the end of migration, the number of histochemically detectable PGCs increases from about 100 to over 1000.

Two genetic approaches establish that the early epiblast or primitive steak is the source of the PGCs. The first involves the phenomenon of X chromosome inactivation in female embryos. Although female mammals, like female fruit flies, have two X chromosomes to the male's single X, mammals do not use the fruit fly's system of transcriptional dosage compensation. Rather, in each somatic cell of the female embryo, one X chromosome is totally inactivated through chromosome condensation. The inactive state is then inherited by all of that somatic cell's progeny. This transcriptional inactivation happens after the blastocyst stage in all cells of the embryo. In the endodermal layer, inactivation occurs early, and it is always the paternal X chromosome that is inactivated. In the epiblast (the EmE and all its derivatives), inactivation is slightly later and is random, with half of the cells experiencing inactivation of their paternal chromosome and the other half that of their maternal X. If the germ cells were of endodermal origin, they should all show an active maternal X only. In fact, the PGCs within the genital ridges always show the epiblast pattern, a random inactivation of the maternal and paternal Xs.

Chimera experiments supply an even more direct proof that the germ line comes from the primitive ectoderm. These experiments not only pinpoint the source of the germ line but also establish that it does not arise from a defined germinal plasm. This fact marks a significant difference from the eggs of many insects, nematodes, and anuran amphibians, among other animals. The first experiments indicating that the germ line arises from non-specialized blastomeres were those of *Kelly* (1975). She found that in each of two four-cell embryos dissociated to individual blastomeres, at least three of donor cells gave rise to germ line chimeras. In all cases, the germ line chimeras also displayed mosaicism in their somatic tissues, showing the absence of segregation for germ line-forming ability by as late as the second cleavage division.

Even more conclusive are the experiments of *Gardner* (1977), who injected single ICM cells from 3.5 day p.c. blastocysts or single embryonic ectodermal cells from 4.5 day blastocysts and scored for subsequent somatic and germ line mosaicism. For the 3.5- and 4.5-day donors, there were, respectively, four out of six and two out of seven somatic chimeras that also showed germ line chimerism (as evidenced by progeny possessing the donor germ line marker). Although the number of chimeras in these experiments was not large, the frequencies were typical for single-cell injections. The experiments show that there is no specialized germ line determinant in the mammalian oocyte and that the primary ectoderm is the source of the germ line. In mammals, therefore, the continuity of the germ line is maintained through the soma rather than independently of it. The same is true of the avian embryo; reciprocal interspecies chimeric hypoblast-epiblast combinations of quail and chic reveal an epiblast origin for the germ line.

The complete sequence of major tissue derivations in the mouse embryo, reconstructed primarily from chimera studies, is summarized. The chief conclusions are that all the cells of the fetus are derived from a portion of the EmE and that the placenta is composed of several discrete lineages of both the TE and the ICM.

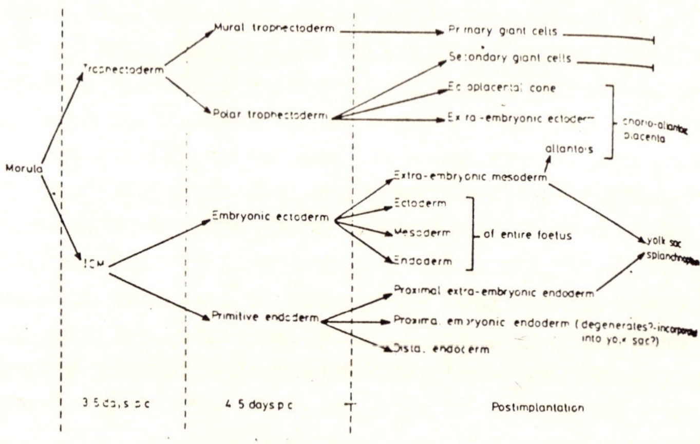

Fig. 4.14.   Principal cell lineages in the pre- and early postimplantation mouse embryo: a summary.

Genetic Analysis of Tissue and Organ Foundation

The period of major tissue foundation in the mouse embryo occurs within 24-hr of the formation of the primitive streak. However, many of the details of histogenesis and organogenesis are obscure. The two aspects that have received most attention are the first cellular commitments in organogenesis—whether they constitute restrictions in developmental capacity (determination) or depend on cell movement—and the number of founder cells for each organ. One uncertainty that affects the analysis concerns the precise timing of the initial developmental assignments. Presumably, the first restrictions of developmental potential within the primitive streak occur between 6.5 days p.c. and the beginning of somitogenesis. By the late somite stages in avian embryos, distinct determinative states have been imposed; somites from the thoracic regions of the embryo generate vertebrae with ribs when transplanted to other regions of the embryo, but lumber somites, those from the lower back region, never form ribbed vertebrae. If one assumes that cells of the early primitive streak mouse embryo are unrestricted, as they are in the avain embryo, a considerable degree of narrowing in developmental potential occurs between the early and late primitive streak stage.

Experimental embryology offers one set of approaches to establishing the nature and timing of the first developmental commitments in primitive streak embryos. One method is to cut out and culture pieces of these embryos at specific stages and to see what they form. The range of structures that can be produced by a given piece from a given stage serves as a measure of the autonomous developmental capacities of that piece; if specific embryo fragments always produce specific characteristic structures, the fragments may be presumed to have experienced some assignment of developmental role. *Snow* (1981) has reported the results of such a study. Isolated pieces of early (7-day p.c.) to late (8.5 day p.c) primitive streak embryos and the complementary deficiency embryos were cultured in isolation for 24-hr in vitro and then examined for the structures formed. In these tests. examination of the deficiency embryo in conjunction with the fragment allows one to determine whether duplication or regeneration of structural occurs. A high frequency of differentiation (> 70%) of the cut pieces and an even higher frequency for the deficiency embryos (>90%) was observed.

These results can be related to the prospective fates of the different regions assayed. In all cases, the autonomous development of those tested pieces that produced recognizable structures corresponded to the prospective fate of that region in the unoperated primitive streak embryo. Thus, the most anterior third of the 7.5 day embryo, which is destined to give rise to the head fold, brain, and foregut, produced those structures in vitro. The middle third, which gives rise in vivo to the major mesodermal elements, was found to produce sometimes, neural tube, and some heart structures in vitro. Finally, the most posterior section, whose fate is to produce tailbud, hindgut, and PGCs, produced those structures plus allantois and no others. A further trisection of the posterior region permitted a mapping of the allantois to the most posterior section of the primitive streak, and the PGC-forming capacity to the section immediately anterior to this one.

What developmental property is mapped in this experiment? *Snow* (1981) has described the result as being that of an "allocation map," to indicate a preliminary regional localization of capabilities (without necessarily being fixed, determinative states). On the other hand, *Beddington* (1982) has argued that such an inference is invalid because the tested pieces are large and consist of more than one tissue type. In principle, the presence of a combination of cell types might establish or stabilize a range of specific inductions, each of which triggers a differentiative pathway whose end product is scored at the termination of the culture period. On the basis of the ability of late primitive streak ectodermal cells, labeled with [³H]thymidine, the colonize different regions and germ layer derivatives, *Beddington* has suggested that the ectodermal cells at least are undetermined in late primitive streak embryos. A difficulty with this view is that colonization and replication of cells in a heterotopic site do not necessarily signify that the cells have taken on the developmental state of the cells with which they cohabit.

To date, therefore, the techniques of classical embryology have left unsolved the problem of when determination in the primitive streak embryo takes place. It may be, however, the before states of determination can be assessed, a fuller description of cell allocation in the early embryo is necessary. Allocation presumably precedes or accompanies determination and characterization of the timing and number of founder cells allocated to various tissues or organ primordia should provide a basis for the subsequent analysis of determination.

In principle, clonal analysis should be able to furnish information on the number of founder cells present at allocation from the estimates of numbers of descendant clones produced within the organ or tissue at the time of allocation. In the fruit fly, one first determines the approximate time of allocation by finding that point at which mitotically generated recombinant clones cease to mark neighbouring primordia (whose proximity can be judged from the gynandromorph fate map). Mitotic recombination is then induced at that state, and from the fraction of tissue or organ occupied by the marked cells, the number of founder cells can be calculated.

The mouse embryo demands a different strategy because marked cells cannot be induced by mitotic recombination in mammals; because one cannot generate marked clones at will during development, one cannot directly estimate the time of allocation in mammalian embryos. Rather, one must introduce the marked cell(s) at arbitrarily early times, either by direct physical placement to create chimeras or by the use of X chromosome inactivation in female embryos. If a female fetus is heterogygous for an X-linked gene—for instance, for a gene encoding electrophoretically distinct forms of an enzyme—half of its cells will express one form, the other half the other form. X inactivation therefore provides a natural form of genetic mosaicism within heterozygous female embryos, and thereby furnishes a second potential means of clonal analysis in the mouse embryo. By measuring the proportion of cells within in tissue that express one allele or the other, the experimenter can arrive at estimates of founder cell number. Unfortunately, as in chimera construction, the genetic heterogeneity is not under experimental control and occurs before primitive streak formation.

With either chimera construction or X inactivation, therefore, the genetic marketing of cells occurs before the presumptive times of fetal tissue and organ allocation. As discussed earlier, the consequence is that marked structures tend to consist of disproportionately excessive amounts of marked tissue, producing systematic underestimation of founder cell number.

There is a further aspect of mammalian cell development, typical vertebrate development in general, that can complicate the analysis. In *Drosophila*, the descendants of single cells tend to remain contiguous; hence, most visible coherent clones are true individual descendant clones. In the mouse, this situation is much less true. Extensive cell mingling and cell migrations take place, particularly in later development. The result is that a given early clone will usually fragment into separate daughter subclones. If one mistakenly equates patches of marked cells with individual descendant clones, the estimate of primordial cell numbers will be inflated to a degree corresponding to the fragmentation of descendant clones. Furthermore, the estimate can be affected depending on whether cell migration occurs before or after tissue allocation, some of the effects of different patterns of migration and coherent clonal growth on final patch sizes and discreteness. When mingling is very extensive, individual patches may consist of two or more subclones that happen to be contiguous.

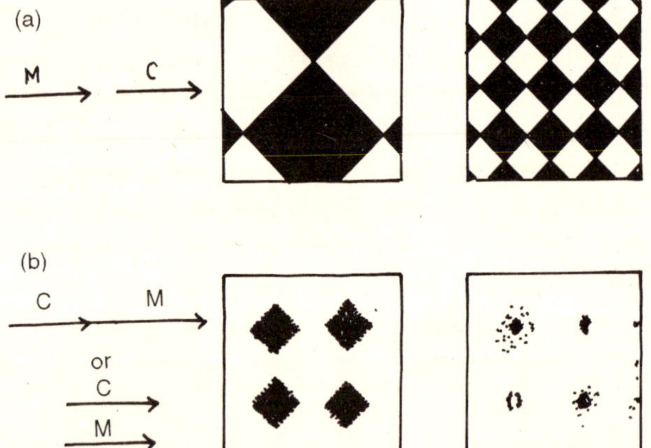

Fig. 4.15. Patch pattern as a function of the timing of coherent clonal growth and cell migration (*a*) Change in a pattern of large patches (left) produced by a period of migration (M) followed by coherent daughter clone growth (C) without subsequent migration; (*b*) patterns produced by clonal growth followed by migration or by a balance of clonal growth and migration.

The consequences and complications of all of these factors in estimating founder cell numbers in the mouse have been reviewed by *McLaren* (1976a) and *West* (1978). The ultimate result is that estimates from clonal analysis in the mouse are considerably less certain than the corresponding estimates in the fruit fly. Nevertheless, the mammalian studies have yielded minimum estimates of founder cell number and have illuminated the detailed patterns of cell growth in mammalian development.

Chimeras have been analyzed in two distinct ways to estimate founder cell numbers. One approach employs indirect cell markers—such as enzyme electrophoretic variants, which can be scored only in bulk preparations— and involve statistical estimates of founder cell number from the proportions of the component genotypes in the chimeras. The second set of methods utilizes direct cell markers, which can be scored in situ by direct inspection or histochemical means. In these methods, the physical disposition of the marked cells within the total cell population provides clues to the number of founding cells or clones and to their growth pattern within the tissue.

The earliest approach to estimating founder numbers with the chimera technique was statistical. It was based on the assumption that formation of a tissue or organ primordium consisting of a fixed number of cells, can be likened to a lottery drawing in which cells are picked at random from a common precursor pool until the correct number has been drawn. In any large set of two embryo-aggregation chimeras, this method of primordium formation should generate a range of binomially distributed genotype compositions; the shape of the distribution is function of the number of founding cells.

Instances involving very small numbers of founder cells will illustrate the point. If a given tissue primordium is always started by one cell, then none of the samples of that tissue from a group of chimeras will be mixed composition; all will be genetically pure. If a primordium is always composed of two founder cells, then the fraction of chimeras that will have tissue of pure genotype a will be one-fourth (= 1/2 x 1/2) that of pure genotype b will also be $^1/_4$, and the fraction of chimeras that will have tissue will be one-half (= 2 x $^1/_2$ x $^1/_2$). By the same reasoning, if the cell number is three, then one-(fourth of the chimeras will be genetically pure (2 x 1/8 for each genotype) and three-fourths will be mixed. Since, in practice, it is often easiest to score the fraction of embryos with non-chimeric tissues, some of the early estimates of founder cells number were based solely one the fraction of mixed embryos that were judged non-chimeric for the structure in question.

A variant of this approach is to use relative similarities in composition between different tissues as an indicator of possible common tissue origins. An example is that of the analysis by *Wegman* and *Gilman* (1970) of two blood cell tissues in comparison to the skin of the mouse. Mixing two input genotypes that differed for a white blood cell marker (an immunoglobulin), a red blood cell marker (hemoglobin), and a coat colour marker, they observed a correlation in nine chimeras between the composition of the immunoglobulin and hemoglobin markers, with a much smaller degree of correlation between these blood cell markers and that of coat colour. The correlation between white and red blood cell genotypes was inferred to indicate a common cellular origin of the white and red blood cell lineages. From the observed frequency of chimerism in each of the markers, primordial cell numbers of both hemopoietic and surface ectoderm were calculated as being no fewer than five cells each.

There are two principal difficulties with all such statistical estimates. The first is the most important; the calculations are based on the implicit assumption that there is *complete and random cell mixing until the moment of tissue foundation.* The consequences of this assumption were first described by *McLaren* (1972). If there is a period or regional coherent clonal growth preceding the moment of tissue allocation, then the estimate of founder cell number will in reality be an estimate of the number of marked coherent clones from which the primordium originates. The greater the extent of coherent clonal growth preceding tissue allocation, the lower the numerical estimate of cells (= clones) will be. Indeed, all such estimates for fetal tissues probably reduce to the number of *original fetal progenitor cells present of the introduction of genetic heterogeneity.* For chimeras made by morula aggregation, the estimates of presumptive founder cell number for a given fetal structure therefore, in all probability, refer to the number of cells present at aggregation that contribute descendants to the structure in question, or about five to eight cells.

The estimate of five to eight cells derives from the following considerations. From single cell injections into blastocysts, it is apparent the single ICM cells can contribute to most or all fetal tissues and hence organs. Therefore, most or all ICM cells in the late blastocyst probably contribute to all fetal tissues. (Such widespread decendancy of single ICM cells presumably reflects a thorough mixing of cells prior to or during primitive streak formation.) Tracking prelabeled cells in morula aggregates, however, shows little cell mixing until blastocyst formation with the consequence that the ICM derives primarily from the five to eight cells that are present internally in the aggregate. Indeed, most binomially derived estimates of fetal tissues founder cell number are in the range of two to five or slightly more cells.

There is a second problem with the statistical approach: the influence of selection during cell growth within chimeras. When the cells of the two input strains differ in background genotype, they often contribute differentially to different tissues or organs. A consistent underrepresentation of one genotype in a tissue relative to its frequency in other tissues is a certain indication of relative tissue-specific differences in growth or competitive ability between cells of the two genotypes. Such selective differences can, of course, influence the relative contributions to several tissues, undermining the presumptive significance of correlative frequencies for inferences about similar or disjoint cellular origins.

The analysis of chimeric tissue in situ, using direct cell

markers, eliminates some of the ambiguities inherent in the statistical approaches. By permitting the direct visualization of marked cells within the analyzed tissues, the *in situ* screening techniques directly yield information on clonal growth patterns. This information, in turn, can help of furnish some minimal estimates of founder cell numbers for the structures or tissues in question. When the screening is carried out only on adult or newborn mice, the inferences remain indirect and are affected by the uncertainty as to whether individual founder cells or clones or adjacent founder cells are being estimated. However, if the marker or markers of choice are expressed throughout development, and if screening is carried out throughout this period, the method can yield clear results. In practice, few such studies have yet been carried out. But the use of cell autonomous antigenic markers may make this approach feasible in the near future.

The first example of an *in situ* analysis of chimeric patterns concerned the origins of the melanocytes, the cells responsible for pigmentation of the mouse coal. The melanocytes are the daughter cells of the migratory mesen-chymal cells termed "melanoblasts" that originate in the neural crest, the region of epithelium that marks the site of dorsal closure of the neural tube. The neural crest gives rise to two major populations of migratory cells, those that migrate deep into the body between the neural tube and somites to give rise to neurouns of the autonomic nervous system, and a second population that migrates just underneath the surface of the ectoderm. It is the latter group that includes the melanoblasts.

By making chimeras between strains that differ in melanocyte pigment-forming capacities, it is possible to determine something about the clonal histories of the melanoblast cell populations. Such experiments were first performed by *Mintz* (1967). Chimeras were made between albino (*c/c*) and wild-type (*C/C*) strains and between strains differing in melanocyte genotype for colour (e.g., black *B/B* and brown *b/b*). In all of the experiments, the striking outcome was the production of striped chimeric mice. Initially, *Mintz* reported that the patterns showed a regular alternation of the different colours, but larger samples have shows that there can be considerable variability in the width of the stripes.

Despite this variability, an underlying or "archetypal" striping pattern can be descerned when all the chimeras are compared. The wider bands observed in the animals simply consist of multiple unit widths, where the unit width corresponds to the narrowest single stripe seen. The striping pattern itself is explicable if colouration in the coat is produced by small numbers of melanoblast clones arranged in an initial linear series along the longitudinal axis in which the clones that neighbour any particular clone may either be of the same genotype (giving a broader band) or of the alternative genotype (giving a central band of unit width). Comparison of the banding patterns in many chimeric mice suggests that the head region consists of three unit stripe widths, the body six, and the tail eight. Because the left and right sides at any position along the longitudinal axis can be of different colour, the clonal origins of the left and right sides must be independent. Equating stripe widths with founder clones, one obtains 17 founder clones per side, or a total of 34 melanoblast founder clones. The independence of the left and right sides suggests that the cells that give rise to melanocytes are committed to do so before dorsal closure of the neural tube is complete and also that the founder cells/clones begin their lateral migration before closure takes place. Were it otherwise, cross-contamination would produce a high frequency of mixed unit stripes. With dorsal closure beginning at 8.5-9.0 days p.c., the migration of melanoblasts must take place before this period.

Does the estimate of 34 melanoblast founder clones mean that there are only 34 founder cells of the melanoblast lineages? Not necessarily. From the fact that the clarity of the striping is sharper in chimeras composed of cells from more distantly related strains than from more closely related ones, it appears that there is a partial sorting out according to genotype. If a certain amount of self-segregation is occurring early in the melanoblast lineage, the founder clones might consist of two or more cells of like genotype. However, the number is unlikely to be more than two or three, since the higher the number the more likely would be the occurrence of "split" unit stripes, and these are not seen.

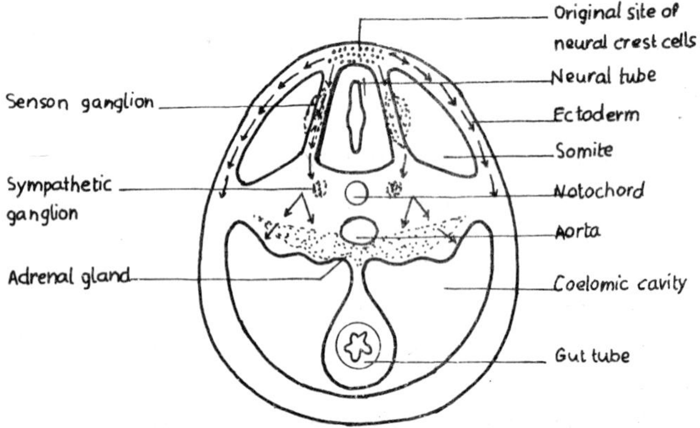

Fig. 4.16.   Two pathways of neural crest migration. The outer arrows indicate cells that give rise to melanocytes; the internal arrows denote neural crest cells that give rise to parts of the neural system and the adrenal gland; stippled areas indicate neural crest—derived areas.

The technique of chimera analysis has also been used to explore the origins of mesodermal structures derived from the somites. It has been possible to analyze the clonal origins of vertebrae, for instance, because certain common laboratory strains differ in their details of vertebral morphology Utilizing these morphological differences as indices of genotype and making chimeric mice from such strains, *Moore* and *Mintz* (1972) concluded that both the left and right halves and the anterior and posterior halves of each vertebra have separate and distinct clonal origins; these findings indicate that each vertebra is derived from a minimum of four clones. This result fits the known facts of somite development well; each vertebra is formed by the fusion of the caudal (posterior) and cranial (anterior) halves of neighboring somites and by those of the left and right somite pairs. The estimate of four clones is a minimum estimate of the number of founder cells because of the possibilities of cell type self-assortment and differential selection within founding cell groups.

A different approach, using the indirect cell marker Glucose Phosphate Isomerase (GPI), was used to examine the origins of skeletal muscle; do these muscles arise by cell fusion or by repeated nuclear division without cytokinesis? The approach relies on the fact that many enzymes are made up of two or more polypeptide subunits. For such polymeric enzymes, the existence of two alleles specifying different electrophoretic variants within the same cell results in the production of "hybrid" enzyme molecules; these molecules are distinguishable from the parental enzyme forms by their intermediate position on gels after electrophoretic separation. If skeletal muscles arise by fusion of cells, then chimeric muscles from strains that differ in GPI alloenzymes should make hybrid GPI. On the other hand, if each syncytial muscle arises by repeated nuclear divisions from a single cell, then the muscles of chimeric mice should make only the two pure forms of the enzyme. The analysis of skeletal muscle from such chimeras revealed the presence of substantial amounts of hybrid enzyme in addition to the two parental alloenzyme forms. Muscle syncytia must therefore arise through fusion of cells rather than as clones of individual nuclei.

In a later study, *Gearhart* and *Mintz* (1972) analyzed the origins of both somites and muscles, using GPI as the marker. They found that the great majority of individual somites from 8-to 9-day embryos contained both forms of GPI. Individual somites are therefore note clones, but originate from at least two mesodermal, cells. In the analysis of certain individual eye muscles, each of which was formed from a single myotome of a single somite, 22 out of 33 muscles assayed were found to contain both GPI alloenzymes and hybrid enzyme. The results show that each somite-derived myotome is also formed from two or more myoblasts. It was also observed that neighbouring somites may share some cellular ancestry.

All of the results discussed above have involved restrospective analyses of events: inferences about the earliest events are made from the tissue composition of the adults. Relatively few studies have attempted to determine the history of a tissue by tracing the development of marked clones in chimeric tissue. One example of such an analysis is that of *West* (1976) on the growth patterns of cells in the retinal pigment epithelium (RPE) of the eye. This monolyer of epithelial cells lies between the neural retina and the enclosing outer structure of the eyeball, the choroid. By making wild-type/albino chimeras and examining cross sections of the RPE, West showed that cell mixing in the RPE is extensive until about 12 days p.c. At that point, each clone is little more than one cell on average. This period of extensive cell mingling is then succeeded by a period of more coherent cell growth, with some cell migration taking place during continued growth of the monolayer. Final clone sizes of marked tissue seen in section consist of five to six cells.

Such one-dimensional sections through a monolayer give a final-gained picture of clonal growth. Examination of the total area of the RPE in chimeras provides a more complete picture of its history. When the two-dimensional surface of the PRE is analyzed in wild-type/albino chimeras in which the pigmented cell population in the RPE is a minority component, distinct radial patterns of growth can be discerned. From the size of the sectors, and on the assumption that each sector corresponds to a founding clone, the results indicated that the RPE is initiated by a relatively small number of founding clones, probably more than 10 but fewer than 250. Synthesizing the cross-sectional and whole surface observations, it appears that the RPE is founded by a relatively small group of founder cells that generate descendants in a radial pattern of outward growth, initially with much local cell mixing the subsequently with greater coherent clonal growth.

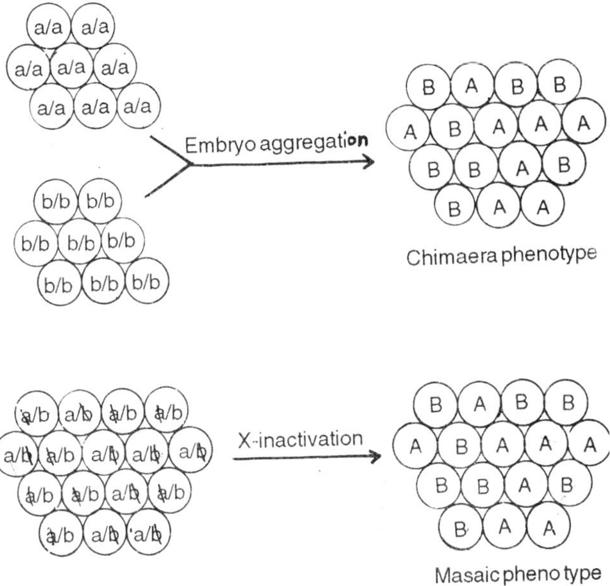

Fig. 4.17. **Two** ways to produce mice with a dual-composition cell population. Top, chimera construction bottom. X chromosome inactivation in an X chromosome heterozygote.

X Chromosome Inactivation Analyses

The naturally occurring phenomenon of X chromosome inactivation in female mammalian embryos furnishes the second major form of clonal analysis in the mouse. Although X inactivation, like chimera production produces two phenotypically different populations of cells, the difference is generated intracellularly within individual heterozygous embryos rather than by the forced mixture of cells from two genegically distinct embryos at the outset. An additional difference between the methods is that X inactivation analyses usually require the use of X chromosome markers, while chimera analysis can employ any chromosomal markers, whether autosomal or sex-linked. The few autosomal genes that can be utilized in X chromosome mosaicism studies are those located on pieces of autosomes that have been translocated to the X.

These translocated fragments experience the X-inactivation process, and if the translocated piece and the normal homologue differ for a scorable marker, a mosaic pattern of expression is produced. The most useful of these translocations is Cattanach's translocation, which is a transposition of part of chromosome 7, containing the wild-type allele for the albino locus (C), into the X. When the standard seventh chromosome homologue contains the recessive (C) allele, inactivation of the translocated chromosome produces an albino clone in the descendants of that cell.

As drawn schematically, the phenotypic outcomes of chimera construction and X inactivation look identical. In reality, the patterns of spatial heterogeneity in chimeras and X inactivation mosaics often look somewhat different even when the same markers are employed. In general, the chimeras show more variable patchiness and larger patches. Indeed, the two properties are directly related. Variable tissue composition between cells of two genotypes tends to produce larger patches, with any shift from a 1:1 ratio increasing the probability that cells of like genotype will find themselves as neighbours. However, when cell mixing is extensive, as in the early period of development of the RPE, the differences between chimeras and X chromosome mosaics are reduced. Only when there are large differences in background genotype between the component strains of chimeras does the chimeric pattern begin to look different from the X mosaicism pattern because of selective self-assortment.

The principal methods for estimating founder cell number in X chromosome mosaics are statistical. The central premise of these methods is that if sampling for tissue foundation is from a randomly mixed cell pool, the variance in composition for a given tissue between different mosaics in the same series will be an inverse function of founder cell number. When the founder cell number (n) is very small, a large variance between like samples will be observed; when n is large, the variance should be small. If tissues or organs show similar compositions and similar variances, they may stem from a common precursor pool; the residual variance between different samples would then reflect differences in the founder cell numbers of the individual primordia.

The first study of this kind was reported by *Nesbitt* in 1971. The cell marker she employed was Cattanach's translocation, whose frequency of inactivation can be scored in chromosome spreads from mitotic cells labeled with [³H]thymidine early in S phase and scored autoradiographically. (Because the inactive, heterochromatinized X replicates late in S phase. The portion of the cell cycle devoted to chromosome replication, it appears as the nonlabeled X in these conditions). Samples were prepared from several different tissues—spleen, lung, liver, and coat melanocytes—of newborn mice, cultured for 4-5 days to amplify the fraction of dividing cells, and then analyzed by the mitotic spread procedure. The analysis indicated that there are about 20 epiblast cells that act as the precursor of the fetus at the time of X chromosome inactivation and that 20-50 founder cells are allocated within the primitive streak for each of the organ systems analyzed.

A more recent approach to the same problem has been reported by *McMahon et al.* (1983). They used an indirect cell marker, the X-linked enzyme Phospho Glycerate Kinase (PGK), which exists in two electrophoretically distinguishable forms, and a sensitive assay, suitable for small amounts of electrophoresed enzyme from tissue samples were prepared from 12.5-day female embryos, heterozygous for the two alleles of PGK, and analyzed for PGK electromorph composition. The tissues examined were representatives of the three primary germ layers of the fetus—neural ectoderm, heart mesoderm, and liver endoderm—and the germ cells. The analysis indicated that there are 47 fetal precursor, cells and about 190 precursor cells for each of the four tissues examined (at the time of germ layer delimitation within the primitive streak). Although the last number is an average, a high correlation of alloenzyme composition between different tissues from each embryo was observed suggesting that all three primary germ layers and the germ cells arise from similar-sized precursor cell pools. The higher estimates of founder cell number by *McMahon et al.* (1983), compared to those of *Nesbitt,* (1971), reflect the lower variances observed in their study.

The estimate of 190 germs line precursor cells (PGCs) is of special interest for two reasons. Firstly, this number is considerably greater than the number of PGCs detectable by AP staining at 8.5 days, the first point at which the cells can be observed directly. If the statistical estimate is correct, then high AP activity may identify only a fraction of the true number of PGCs. Secondly, the similarity of this number to that for the other three germ layers raises the possibility that PGCs are not actively instructed to be such but may simply be the residuum of unassigned cells. In this view, the retention of totipotency by PGCs is a consequence of their failure to be restricted in fate. It may even be that the caudal end of the primitive streak, from which the PGCs arise, is a region that protects against restriction of totipotency. As discussed earlier in connection with germ line determination in *Drosophila,* the posterior polar plasm from which the germ line of the fruit fly originates may exert its determinant function by providing comparable protection.

As in the chimera studies, the estimates of founder cell number from X mosaicism studies are probably estimates of the number of coherent clones contributing to the tissue rather than the number of founder cells per se. When more reliable information from other techniques on the times of tissue allocation becomes available, both forms of clonal analysis should contribute more precise estimates of founder cell number than are presently obtainable.

DEFECTS IN ORGANOGENESIS: GENETIC ANALYSIS

A very large number of mouse mutants are affected in

organogenesis Many of these mutants were first detected on the basis of their dominant heterozyogous phenotypes—for instance, a degree of runting or a change in coat colour from the parental stock. In homozygotes, the mutant defects often result in embryonic or perinatal mortality. Another class of mutants affected in the development of organ systems are those initially identified on the basis of their behavioural phenotypes; such mutants are usually defective in the nervous system. The *shaker (sh)* and *waltzer (w)* mutants, whose phenotypes are conveyed by their names, are examples of neural defectives.

Until the advent of chimera and mosaic studies, the developmental genetics of the mouse consisted almost entirely of the study of the developmental defects associated with particular mutants. The procedure is the classical one of developmental genetics. One first identifies the set of mutant phenes and then works backward in development to the first observable aberrancy. From these observations, one then attempts to reconstruct the cellular site, and if possible the biochemical basis, of the initial developmental derangement. In a number of instances, however, the etiology is similar to that produced by known specific enzymatic deficiencies in humans and the mouse syndrome proves to have an identical basis (Bulfield 1981).

For the most part, however, the mutant defects that produce severe aberrations of early or middevelopment have not been characterized at the biochemical level. Nevertheless, these mutants represent a valuable store of genetic source material for future investigations. As the techniques for biochemical and cytological characterization advance, the precise biochemical basis of some of these mutant defects will become clear and the characterization of the genes themselves, isolated via recombinant DNA procedures, will elucidate the genetic basis of individual defects in an increasing number of instances.

Perhaps the single most valuable tool in characterizing

the cellular basis of mutant defects has been the electron microscope. With electron micrographs, it is often feasible to determine which cells are defective and to reduce the range of possible explanations of the mutant phenotype (although ambiguities usually still remain). An example of such an analysis will illustrate the usefulness of the approach.

One of the first morphological mutants of the mouse to be characterized was the dominant short-tailed *T* or *Brachyury* mutant, reported by *N. Dobrovolskaia-Zavadskai* in 1927; When short-tailed *(T/+)* mice are crossed. one obtains a 2: 1 ratio of short tailed to normal mice instead of the expected 3:1 dominant to recessive ratio (1*T/T* :2*T/*:1 +/+). This observation suggested that, like A'. *T* is lethal when homozygous. This suggestion was confirmed by *Chesley* (1935), who performed the first detailed examination of the embryos. Using the standard light microscopic techniques available at the time, *Chesley* observed that the first abnormalities in *T/T* embryos appeared in early somite, 8.5 day p.c. embryos. In the posterior third of the embryo, the notochord is absent, the somites are disorganized, and the neural tube shows aberrant morphology. These early defects are succeeded by progressive disorganization of the entire notochordal and neural tube regions and a failure of hindlimb development. Heterozygous *(T/+)* litter mates, incontrast, do not show any defects until about 11 days p.c. At this point, they exhibit a diminished neural tube and notochord in the tail region and minor irregularities of the notochord anteriority. The heterozygous syndrome is thus a milder version, with a later onset, of the homozygous developmental defect. The cause-effect relationships between the various abnormalities could not be determined, but Chesley speculated that the notochordal defect was primary and the neural tube defects secondary.

In a detailed electron microscopic study, *Spiegelman* (1976) confirmed and extended the earlier findings. By 8 days p.c., the *T/T* embryo has a reduced trunk and a correspondingly enlarged primitive streak. In these embryos, the conversion of embryonic ectoderm to mesoderm is blocked; in consequence, instead of forming notochord and somites in the posterior region, the blocked mesodermal cells form excess neural tube material. In effect, the block in mesodermal differentiation from ectoderm leads to a temporary hyperplasia or overgrowth of neural tissue: these epithelial tubes subsequently undergo degeneration. concomitantly with the disintegration of the anterior mesodermal axial structures. (This syndrome of neural overgrowth.

The specific cellar defect producing the *T*-lethal syndrome is revealed by the electron microscope. Unlike the wild-type embryo, in which the neural tube and notochord are separated by a discrete basal lamina, these two cell types meet in the lethal *T* embryos at the ventral margin of the neural tube. There are nobasal laminae in the region of contact, and the neural tube and notochordal cells establish junctions with one another. This failure of normal cells recognition, with subsequent failure to establish a normal barrier between the notochord and the neural tube, must play a role in the ensuing multiple failures of mesodermal organization and differentiation.

The difference in notochordal cell surface behaviour in the mutant could be a direct reflection of a cell surface property that directly impedes normal primitive streak differentiation. It is equally possible that the cell surface change is a secondary event and perhaps only one element of the failure of mesodermal differentiation. The pictures themselves do not settle this point, but only raise it.

To characterize the role of a wild-type gene in development, one requires a complete description of both the cellular and developmental phenotype and the genetic character of the mutation itself. With *Drosophila*, one can usually perform the Mullerian dosage test in order to classify the expression property of the mutation. In the absence of the necessary duplications and deficiencies, one can resort to isolating additional mutant alleles of the same gene. The various mutants can then be compared, and the phenotype produced by a complete or nearly complete inactivation of the gene can be inferred.

In the mouse, these genetic stratagems are usually impossible, for the most part, the necessary duplications and deficiencies are not available for the dosage test, and large-scale searches for non-complementing lethal mutants are impractical.

Only when a mutation creates a distinctive visible phenotype without attendant lethality can one hope to isolate more mutants of the same gene by direct screening, permitting comparative studies. To isolate new mutants, one screens for non-complementer chromosomes, as in *Drosophila*. Heterozygotes carrying the initial mutation ($+/m$) are mated to mutagenized wild-type animals, and the progeny are examined. Any individual in the F_1 showing the mutant phenotype must carry the original mutation m, and a new mutant allele, m^*. These individuals are then outcrossed to isolate the new mutant, which can be distinguished from the original if the original mutation is linked to scorable markers.

Because this method is limited to genes in the mouse that give a visible but fully viable phenotype, it is inapplicable to many genes of developmental interest, Where it can be applied, it has been informative. The procedure is illustrated by the search for mutants of the *albino (c)* locus on chromosome 7. This gene is probably the structural gene for tyrosinase, the enzyme that synthesizes the major melanic pigments. To collect new albino mutants, female heterozygotes *(c/C)*. The C allele being the dominant, wild-type allele) were mated with irradiated wild-type males, and all albino progeny were outcrosed to isolate the new c alleles. A number of these new mutants are lethal when homozygous. The reason for the lethality is not the absence of c activity but the delection of the mutant chromosomes for adjacent essential loci.

Gluecksohn-Waelsch (1979) has described a set of six of these lethal deficiency *albino* chromosomes. One mutant

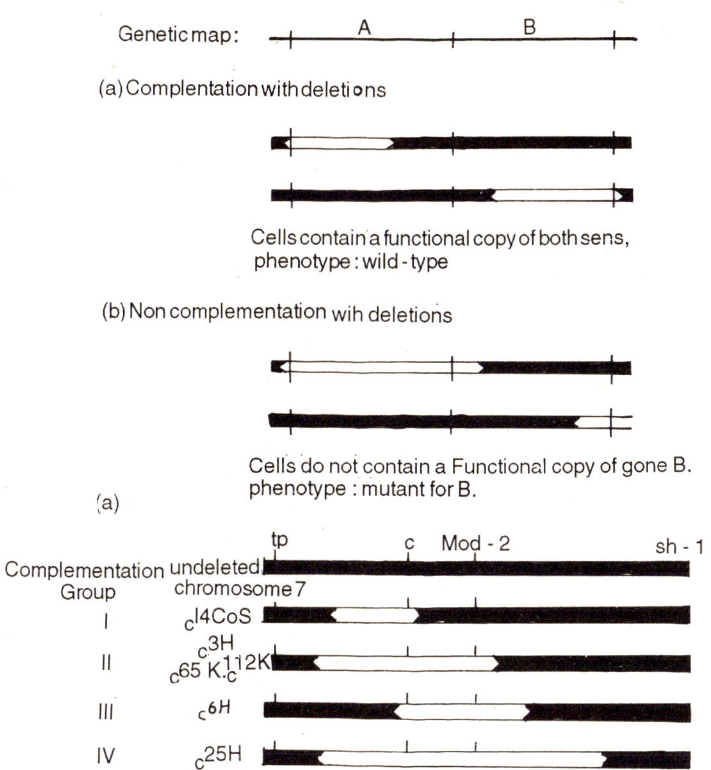

Fig. 4.18. Delection mapping of functions (a) principle of deletion mapping; (b) deletion map of c region.

homozygote dies in early cleavage, one dies in the early egg cylinder stage, and the remaining four exhibit perinatal mortality; these last four exhibit a specific deficiency for six particular liver enzyme activities (while having normal activities for several other liver-specific enzymes). By making heterozygotes of pairs of the different lethal deficiency chromosomes and scoring for the presence or absence of particular mutant phenotypes, one can position the wild-type genes whose activity prevents particular mutant phenes. Such a gene localization procedure is termed "complementation mapping." The liver enzyme defect is between *c* and *Mod-2* (the gene for mitochondrial malic enzyme-2), and the essential early embryonic function(s) lie between *Mod-2* and *sh-1* (one of the genes that gives the shaker phenotype) the results show that isolating new mutants of a locus can allow one, if some of the mutations are deletions, to work "outward" along the chromosome from that locus, identifying new essential gene functions and their linear sequence. In effect, the locus provides a bridgehead for exploring the genetic organization of a chromosomal region. (An analogous molecular approach to the analysis of chromosome genetic organization.

The defect in the four perinatal mutants may be of special interest with respect to genetic regulation. In these mutants, the cause of death appears to be a combination of defects in the liver, kidney, and thymus, with the liver defects being the most serious. The enzymatic deficiencies of the mutants may reflect the loss of a specific regulatory locus essential for their activity. When mouse/rat hybrid cells are constructed by fusing mouse cells homozygous for one of the deletions with rat cells, the hybrid cells are found to express the mouse G6PD activity (one of the deficient activities of the mouse cell); evidently, the rat genome can supply the missing regulatory factor. The immediate cause of the inability to synthesize the affected enzyme activities may be a reduction in receptors for glucocorticoid hormones found in the mutant cells. However, the presence of some receptors in the cells of the deletion homozygotes suggests that the structural gene for the receptors is not missing.

The mortality effects associated with the late-acting mutants may involve certain aberrancies of cellular organelle development. Electron microscopic inspection of the affected organs in the four perinatal lethal homozygotes shows that the membranes of the rough endoplasmic reticulum, the Golgi apparatus, and the nucleus are all abnormal in the liver and the kidney. However, these membrane organelles are normal in appearance in the thymus of the affected embryos, although the thymus itself shows morphological abnormalities. These organelle effects reveal tissue-specific differentiative aspects of certain commonly shared organelles. The existence of such subtle differences had not been anticipated on the basis of conventional cytological studies and adds a further dimension to the known complexity of tissue differentiation. However, some caution is required in accepting this interpretation: the liver organelle defects may be particular secondary consequences of the enzyme or other deficiencies that do not pertain to the thymus.

THE *T*-COMPLEX: MAMMALIAN SWITCH GENES OR DEVELOPMENTAL ANOMALY?

One of the fundamental questions in animal developmental genetics is whether development is governed by major regulatory "switch" genes. We have seen instances of apparent switch genes in the two invertebrate systems discussed earlier. In the nematode, replacement of various cellular lineages can be produced by mutation in a few specific genes, with consequent major alterations in the developmental pathway. And in *Drosophila*, the homeotic mutants, particularly those of the *bithorax* and *Antennapedia* gene complexes, show that certain genes play key roles in the assignment of segmental phenotype. Are there comparable loci in mammalian development?

During the 1970s, several investigators proposed that the so-called *t*-complex on chromosome 17 of the mouse played just such a key regulatory role in early mouse development. Several results have rendered this hypothesis increasingly unlikely, but the developmental genetics of the *t*-complex have produced much information of interest.

The first *t* mutations bearing chromosomes were picked up fortuitously because of an unusual reaction they display with *T*. As described earlier, *T/+* heterozygotes are short-tailed mice and *T/T* homozygotes are embryonic; lethals. When certain mice from wild populations were crossed with *T,* a new class of viable progeny appeared: completely trailless mice. Surprisingly, the condition of taillessness was found to breed true in brother-sister matings from each cross; that is, only tailless progeny were produced in each ensuing generation. Outcrossing revealed that each tailless mouse carried a new recessive lethal, a *t* mutant, in addition to *T* Crude mapping placed the new mutants near *T* on the chromosome now designated number 17. Each tailless strain therefore carries two heterozygous lethals. *t* and *T*. Such strains, in which each member of a homologous chromosome pair carries a lethal mutation and in which only the heterozygous combination is viable, is said to be a balanced lethal strain. The explanation for the perpetuation of the tailless phenotype is that the *T/T* and *t/t* embryos are lethal and only tailless heterozygotes *(T/t)*survive.

The genetic complexity of the *T/t* locus, as it was first designated, emerged only when the different tailless strains were crossed with one another. If all of the *t* mutations are in the same functional genetic unit, then the intercrosses should yield only tailless mice. Thus, if two different *t* mutants are designated t^a and t^b, the intercross (T/t^b) should give only parental genotype, tailless progeny. In fact, some crosses gave normal-tailed mice in as many as one-third of the surviving progeny. These mice were subsequently shown to be of genotype t^a/t^b. In other words, some of the *t* mutants were found to complement each other, permitting normal development. As more *t* mutants were isolated, it became apparent that the different lethal alleles could be placed into six different complementation groups.

In addition to *t* mutants from wild populations, new *t* mutants were isolated in the laboratory from the true-breeding tailless *T/t)* lines, at a rate of about 1 in every 500 or 1000 progeny. These new mutants were isolated as exceptional, normal tailed mice within the stocks, and were shown by the complementation test to carry new *t* alleles, derived by some form of unequal crossing over within the *T/t* complex. Many of these laboratory-derived strains are semi- and fully viable *t* mutants, each of the new mutants being identified as a *t* mutant by the *T* interaction test.

However, direct genetic mapping of the mutant sites could not be carried out because all of the *t* mutants exhibited a region of depressed recombination within the area of the *t*-complex itself. Furthermore, many lethal *t* mutants show unusual inheritance patterns, in particular, a tendency for the mutant allele to be preferentially transmitted from wild-type heterozygotes (*t*/+), but only in the male line. In contrast to this behaviour of the lethals, some of the viable *t*/'s are present in less than 50% of the sperm of heterozygotes. An additional complexity is that combinations of high- and low-transmission complementing *t* alleles do not always behave predictably, but can show partially reversed relative transmissibility in each other's presence.

The causes of the these distortion in transmission ratio are obscure but reflect both early and late developmental effects on spermatogenesis. The perferential transmission effects from *t*/+ heterozygotes apparently reflect direct alterations in the fertilizing ability of the haploid mature sperm. These changes in sperm behaviour are correlated with specific antigenic changes in the sperm head. Each lethal complementation *t* group is associated with a particular set of sperm antigenic determinants; some antigenic determinants are shared between different groups, while other are unique to a particular complementation group.

The potential developmental significance of the *t*-complex became apparent only when the specific lethal syndromes of homozygous inviable embryos were closely examined. It was found that members of each complementation group show arrest of embryonic development at characteristic stages, ranging from early morula to early somitogenesis. The developmental blocks associated with each lethal complementation group, different mutants often show some differences in severity or terminal stage, but the members of each group resemble each other more closely than do members of different groups.

In a review of the genetics and biology of the *t* mutants. *Bennett* (1975) proposed that the developmental effects reflect a major governing role of the presumed wild-type *t*-complex in early mouse development. This hypothesis amplified an earlier suggestion made by *Glucksohn- Waelsch* and *Erickson* (1970). Based on the known effects of the *t* mutants on sperm antigens *Bennett* proposed that normal development involves a fixed sequence of changing cell surface properties, each essential for the next developmental transition. The role of the *t*-complex would be to determine this programmed sequence of changing cell surface properties. Because the earlier embryological studies pinpointed ectodermal cells and derivatives as the chief foci fort mutant effects, the action of the *t*-complex was provisionally characterized as the mediation of a series of switches within the ectodermal lineages. In terms of this hypothesis, the developmental effects of *T* were interpreted as stemming from the cell surface defect observed in the electron micrographs.

TABLE 4.3

Developmental Blocks in Lethal *t* Mutants

*t*Complementation Group	Stage of First Abnormality	Stage of Developmental Arrest	Cell types Affected
t^{12}	Morula	Morula	All blastomeres
t^{w73}	Late blastocyst	Early egg Cylinder	TE
t^{o}	Early egg cylinder	Late egg cylinder	TE. EmE, endoderm
t^{w5}	Late egg cylinder	Preprimitive streak	Mainly EmE
t^{9}	Early primitive streak	Late primitive streak	Mesoderm and EmE
t^{w1}	Late Primitive streak	Early organo-genesis	Ventral neural tube, brain

The most telling argument against the developmental switch hypothesis comes from an analysis of the expression of *t* antigens during development. If the developmental transitions in the early embryo are mediated by the appearance of specific cell surface molecules on particular cell types at specific developmental stages, then one would predict that the sequence of *t* mutant syndromes, from morula to late neural tube stages, would be reflected in a sequence of *t* antigen appearances on the cell surface(s). *Kemler et al.* (1976) tested this hypothesis by preparing antisera specific to the various *t* complementation groups and monitoring the time course of appearance of the *t* antigens on embryos. (The antisera were prepared by immunizing mice with spermatozoa from heterozygotes of the various *t* lethals and then absorbing the antisera first with wild-type sperm and then with lymphocytes, the latter step to remove other non-specific antigenic determinants.) The time course of appearance of the different antigens on heterozygous (*t*/+) embryos was then monitored by treating the embryos first with the specific sera and then with a fluorescent anti-mouse Ig antibody. Under these conditions, only embryos, with the characteristic *t* antigen on their surface fluoresce.

For four mutants representing four different *t* lethal complementation groups (the t^{12}, t^0, t^{w5} and t^9 groups), *t*-specific antigens were detected on embryos as early as the eight-cell stage in all cases. By this test, the *t* mutant cell surface changes all appear early and without temporal relationship to the time of onset of the various developmental aberrancies.

The developmental switch hypothesis also requires the existence of wild-type alleles of the *t* "*mutations*." Much evidence now indicates that the *t* mutants differ from the wild-type in large blocks of DNA within the proximal region of chromosome 17, rather than in single-site mutations. The absence of recombination between wild-type chromosomes 17 and *t* mutant chromosomes in this region reflects their absence of fine-structure homology. Instead of mutant alleles, therefore, particular *t* mutations are said to be "*t haplotypes*." Furthermore, the different developmental aspects of the *t* mutants phenotype—interaction with the *T* locus, embryonic lethality effects, enhanced transmission ratio in *T*/+ heterozygotes, male sterility in many double heterozygotes—are caused by different loci within the block of *t*-chromatin. It has been possible to dissect these different loci genetically because of the occurrence of rare crossovers in the promimal region between *t* haplotypes and marked wild-type chromosomes; the recombinant chromosomes bearing portions of the *t*- complex experience recombination with other *t* chromosomes at normal rates, because of their shared omology, and can be further analyzed. A genetic map of the *t* region, and its wild-type analogue has been constructed by *Siler* (1981). The results explain why no *t* mutants have ever been isolated from non-*t* carrying populations in the laboratory. In contrast, several x-ray—induced mutants of *T*, which was once thought to be part of the *t*-complex, are known.

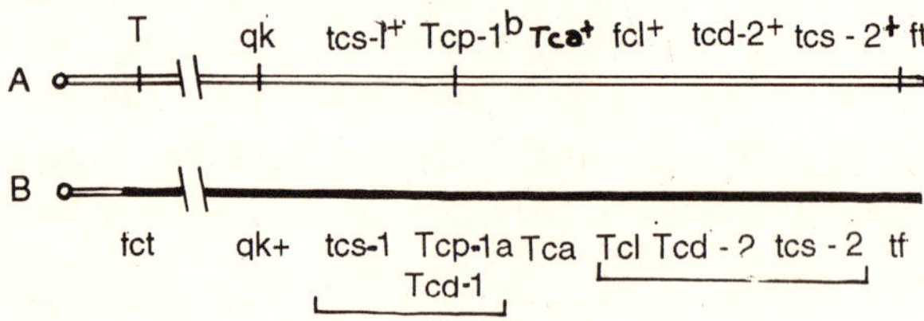

Fig. 4.19.   Map of the *t*-complex region in wild-type and complete *t* haplotype. The solid line represents *t* chromatin. *T, shorttailed : tct*, tail interaction factor, *qk*, quaking; *tes-1* and *tsc-2* proximal and distal sterility factors, respectively ; *Top-1*. testicular *t* complex protein: *Tcd* and *Tcd-2*, proximal and distal segregation distortion factors respectively; *tcl.*, *t*-lethal factor(s) ; *tf*. tufted.

The developmental differences between *t* haplotypes reflect differences in the amount and sequence composition of *t* chromatin they contain. Indeed, the observations suggest an unusual origin of *t* chromosomal material, originally proposed by *Dunn* and *Bennett* in 1971, namely, that all *t* mutants share a common ancestral chromosome introduced from a distinct and partially diverged mouse population. Because *t* mutants are found in all wild *Mus musculus* populations, the hypothesis demands that the progenitor chromosome has spread throughout the world via these populations. Such diffusion is explicable in terms of the genetic properties of the complex itself; recombination suppression would tend to keep the transmission-enhancing factors together with the lethality and sterility factors, with the former acting to keep the *t* chromosomes at high frequency within each population. Only the deleterious effects associated with the *t* chromosome out of the genetic pool.

The direct evidence that all *t* chromosomes derive from a single source is a major testicular protein encoded by a gene (*tcp-1*) in the region of the *t* complex. All non-*t*-bearing mouse strains. including *T/+* strains, produce a slightly basic form of the protein, while all *t* mutants possessing the proximal region of the complex produce an acidic form of this testicular protein.

The *t*-complex is therefore extremely interesting from the dual perspectives of population genetics and genome evolution. The basis of the developmental effects, however, remains elusive. The *t*-complex contains analogues of several genes within the comparable region of the wild-type chromosome. yet homozygosity for particular sets of these *t* gene produces products distinct developmental aberrancies. In some sense, the gene activities of the *t*-complex are mismatched with the developmental activities that arise from the rest of the genome of *M. musculus,* in effect a form of hybrid lethality. Such lethality is usually believed to reflect a failure of the transcriptional regulatory apparatus to cope with gene products or regulatory sites that are slightly different.

Many questions about the *t*-complex and *t* mutants remain. Solutions to the puzzle of how particular *t* lethals affect critical early transitions, in the mouse embryo may prove highly informative about these transitions, but the concept that the complex consists of a set of switch genes that operate in normal development is probably incorrect.

TERATOCARCINOMAS: MODEL SYSTEMS FOR DIFFERENTIATION

Some of the most interesting events in postimplantation development are those of tissue allocation and determination and the beginning of tissue differentiation. Unfortunately, these are also among the events most inaccessible to the experimenter. The methods of clonal analysis furnish some information about tissue foundation, but the conclusions are necessarily indirect and tentative and, by their very nature, cannot elucidate the underlying mechanisms in these early cell assignments. On the other hand, direct biochemical analyses of the first commitments and tissue differentiations in primitive streak embryos are rendered difficult or impossible because of the limited amounts of material available.

In recent years, an alternative approach to early development has been made possible by the realization that EC cells, a class of neoplastic stem cell, many provide an analogue of the early totipotent cells of the mouse embryo. EC cells were initially identified as the stem cells of certain tumorous growths called "*teratocarcinomas.*" As the name implies, these growths are both tumorous and "*monstrous.*" First discovered in humans, they appear to be highly disorganized embryos consisting of a nearly random mixture of tissues and structures such as teeth and hairs. The first class of teratocarcinomas discovered in mice was reported by *L.C. Stevens* and his colleagues in the 1950s. These were testicular tumors arising at a high frequency in the inbred mouse line 129. Analysis of these growth revealed that they certain two classes of cells, the various differentiated cell types and a core of cancerous stem cell, the EC cells. Culturing of the tumors showed that the differentiated cells are benign (non-cancerous), while the EC cells are neoplastic. In general, serial culture of teratocarcinomas in vivo leads to a loss of virulence expressed in the production of benign growths called "*teratomas.*" However, the EC cells can be maintained as cell lines by serial culture in the intraperitoneal space. Such in vivo ascitic cultures maintain their malignancy and differentiative capacities indefinitely.

The first teratocarcinomas to be described were testicular tumors. However, EC cells can be derived from any mouse strain by transplantation of early (3- to 7-day) embryos to nonuterine sites, such as the kidney or testis capsule of the male adult. It is also possible now to derive EC lines directly from early embryos using in vitro culture. The frequency with which embryos give rise to teratocarcinomas is a function of their own genotype—different inbred lines giving different results—and of their maternal source, suggesting that there is a maternal effect in the potentiation to form EC cell. Furthermore, spontaneous ovarian teratocarcinomas occur with measurable frequency in one strain; these apparently arise from early oocytes. Teratocarcinomas can therefore develop from either male or female germ line cells, although always from diploid germ line precursors, or form early embryos. When the source is an early embryo, the neoplastic stem cells presumably originate either from the normal totipotent cells of the ICM or forth the primary EmE. Regardless of source, all EC lines can be cultured in vitro. Depending upon its history and a variety of little understood factors, each line comes to be characterized by a very broad, intermediate, or narrow range of differentiative capabilities.

When cultured intraperitoneally, EC cells typically give rise to two kinds of embryonic bodies. The first, simple embryoid bodies, consist of a core of EC cells surrounded by a layer of endoderm; superficially, they resemble the ball of ICM cells covered by endoderm, typical of the early blastocyst stage. The second class, cystic embryoid bodies, arise after more prolonged culture of EC cells and probably develop from simple embryoid bodies. They resemble the embryonic region of late egg cylinder embryos, but are substantially more disorganized and exhibit a mixture of differentiated tissue regions. Certain in vitro EC lines also differentiate in suspension, under the appropriate conditions, and produce both kinds of embryoid bodies. Other EC lines typically show differentiation in vitro in monolayers on the inner surface of the culture vessel.

Although EC cells resemble early embryonic cell and can be derived from them, there has been some debate as to which specific embryonic cell type they most resemble. From their patterns of synthesized polypeptides, they appear closer to primary EmE in developmental state than to ICM.

The most productive approaches to the characterization of early embryonic differentiation through EC cells have been immunological. In one procedure, rabbits are injected with EC cell s to produce an anti-EC serum, and the purified antiserum is then applied to staged early embryos or to other EC lines. The presence or absence of the target antigen on the embryos or cells being screened is determined either by the direction of a biological effect or by using fluorescein-conjugated reagents that recognize and bind to the first antibody. [The latter approach was used by *Kemler et al.* (1976) to detect *t* specific antigens.] A different strategy is to prepare antibodies to purified components found only incertain differentiated cells and then to monitor the time course of appearance of these antigens, usually by means of the fluorescence assay, in differentiating cells lines.

An example of the immunological strategy for identifying events in early development using EC cells involved the production of an antiserum to one nullipotential line of cells, cell line F9 ; the F9 antiserum has served to identify a cell component required for compaction of the early embryo. This antiserum is complex, but its specific gamma globulin component detects a cell surface molecule present throughout early development. It is present on the surface of all cells of premorula embryos and in early and late blastocyst stages, becomes localized subsequently to the EmE, and finally disappears on the ninth day of development. In adults. The F9 antigen is found only on cells of the adult male germ line, including sperm cells. When early embryos are treated with the antigen-binding fragment of the gamma globulin fraction, compaction of the eight-cell embryo is prevented. Both the anti-F9 serum and several monoclonal antibodies that prevent compaction bind to a specific glycoprotein termed "uvomorulin." Uvomorulin undergoes a calcium ion-mediated conformational change that is essential for the process of compaction; the antibodies that block compaction evidently prevent this conformational change. Although the structural gene for uvomorulin has not been identified, two lethal *t* complementation groups block the appearance of F9 antigen. These *t* mutations presumably interfere with either the synthesis or surface localization of uvomorulin.

From the standpoint of genetics, the interest of EC cells lies in their potential use as vehicles for genetic alteration of mouse development. This possibility exists because the cells of certain EC lines can participate in normal embryonic development when injected into blastocysts, contributing descendant cells to all tissues and organs. This ability also shows that the neoplastic state of the EC cell is reversible when the cell is placed in a normal embryonic environment.

If normal EC cells can be reinserted into the developmental program, EC cells that are mutant but viable in culture may be similarly capable of reintegration. If so, the way is open for the construction of new lines of mice bearing mutations induced in vitro. The only conditions for success are that the introduced mutant cells be capable of participating in somatic development and giving rise to germ line tissue in the viable chimeras. This procedure, if successful, would be considerably more efficient than standard mutant screening of whole animals because vastly more cells can be examined in culture for mutant clones than whole animals screened for mutant pheno-types.

The first demonstration that EC cell can participate in whole animal development was reported by *Brinster* (1974). His experiment involved the injection of *agouti* cells into blastocysts of an albino strain. The source of EC cells that *Brinster* used was the ascitic fluid of a tumor line derived from strain 129, serially propagated over many years by intraperitoneal injection. Of 60 surviving animals, each derived from a blastocyst injected with two to four EC cells, one animal showed a coat with partial gray striping. The source of this agouti colour must have been the donor EC cells.

This particular animal did not sire agouti progeny when mated with an albino female—the dominant *agouti* allele would have been detected in any progeny from agouti germ line cells—and was therefore probably not a germ line chimera. However, a germ line chimera was subsequently obtained by *Mintz* and *Illmensee* (1975), who used the same ascites tumor as donor. They obtained three chimeric animals, detected by patches of donor coat colour, which were chimeric in their red and white blood cell-forming tissues and in their livers (as shown by the presence of donor cell GPI. One of these mice, a male nicknamed *"Terry Tom,"* sired progeny bearing tumor cell line markers and was therefore a germ line chimera.

The 129 cell line is an in vivo line of EC cells, using EC cells to transmit new mutations requires the utilization of EC cells from individual cell clones, and such clones can be obtained only from *in vitro* EC cell lines Somewhat surprisingly, the great majority of such cell lines, when tested, including those with large pluripotential somatic developmental capacities and lacking any detectable chromosomal abnormalities, have failed to give germ lines chimeras. Either these cell lines are genetically deficient in one or more ways or the PGCs derived from them fail to migrate to the gonadal ridges in sufficient number.

However, successful transmission of an in vitro cultured cell line METT-1 (for "mouse euploid totipotent teratocarcinoma-1") has been reported by *Mintz* and *Stewart* (1981). The strategy is diagrammed in Figure 5.16. The cell line, also derived from strain 129, shows a normal karyotype and is chromosomally female ; it was repeatedly passaged in vitro before being injected into blastocysts of a host non-agouti (black) strain. Measured either in terms of the percentage of visible chimeras (as scored by the frequency of agouti contribution to the costs of the surviving animals) or by the extent of the contribution of donor genotype cells to internal tissues (as shown by the GPI marker), this cell line is significantly more normal than other previously tested in vitro lines. Of nine coat colour female chimeras, one proved to have strain 129 oocytes, giving rise to three agouti F_1 progeny. These three animals in turn gave rise to agouti F_2 progeny.

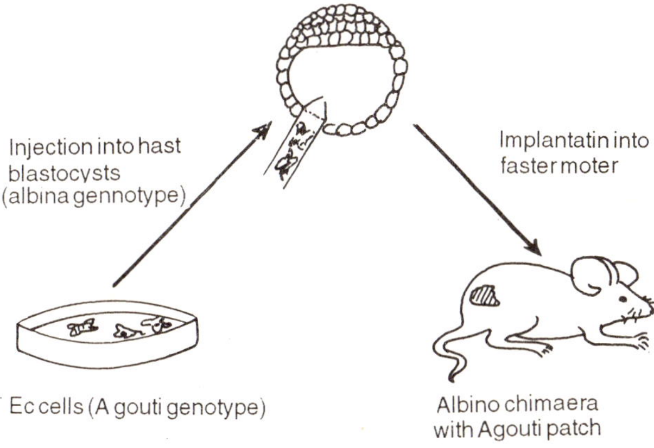

Fig. 4.20. Transmission of *agouti*-genotype of in vitro EC cells into the normal fetal development pathway.

A similar approach with embryo-derived pluripotential (EK) cells may be even more efficient. Bradley et al. (1984) report that over 50% of surviving blastocysts injected with EK cells are visibly chimeric. Of these, approximately one-third proved to be germ line chimeras. Induction of new mutations in EK cells for passage into the germ line promises to be superior to the use of EC cells in this respect.

It may soon be possible to transmit mutations, induced in EC or EK cells, into new strains of mice. The principal challenge may be in devising selection procedures for isolating mutant cell lines in vitro that show interesting developmental phenotypes when integrated into developing embryos. However, the development of direct modification of the germ line through transformation of the early embryo, including the ability to induce mutations with the inserted genes, provides a simpler method of producing genetically modified offspring than the EC cell route.

COAT COLOUR

The hereditary basis of coat colour is the oldest problem in mammalian genetics. Genetic investigation of coat colour predates the rediscovery of Mendelian genetics in 1900, but the modern analysis of coat colour genetics begins with the work of *Cuenot*, *Castle* and *Wright* in the first two decades of this century. The work of *Wright* on the genetics of guinea pig coat colour was particularly important; it was the first attempt to use genetic differences to probe the biochemistry of development. The reason for this early attention to the coat colour of rodents is its ready accessibility to genetic analysis. Although the mammalian coat still conceals many of its secrets, the broad outlines of its biology are now visible.

The mammalian coat serves two functions. The primary one is insulation. By constituting an efficient air-trapping layer next to the skin, the packed hairs of the coat provide substantial protection against thermal loss. (The features of birds furnish similar protection.) The coat typically consists of two classes of hair: the shorter and thinner underharis—about 80% of the total hair on a mouse—and the longer, thicker overhairs. Within each category, several different hair types exist. The underhairs collectively provide most of the direct insulating capacity of the coat, while the chief function of the overhair layer is to protect the underhair coat. In addition, some overhairs have a specialized sensory role, as do the vibrissae ("whiskers") of the mouse.

The second major function of the coat is to serve as a visual signaling device; in this role, colour becomes crucial.

The colour pattern serves as a recognition device for other members of the species, a function that is particularly important in mate selection. In some species, for example, the skunk, it serves additionally as a warning signal to other animals. In some mammalian species, colour serves the function of concealment through camouflage, as the snowshoe rabbit.

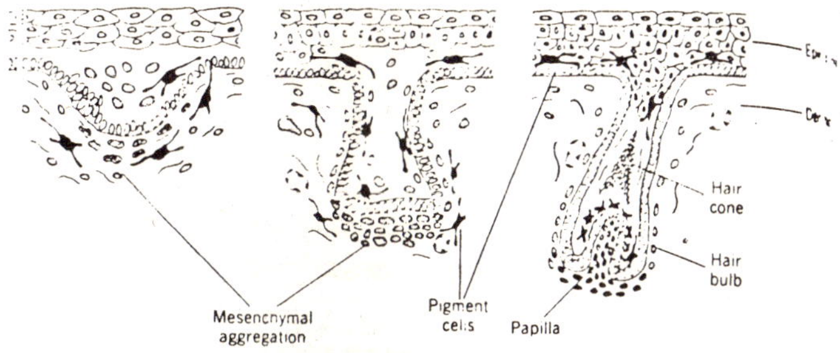

Fig. 4.21.   Sequences of hair follicle development in the mouse coat.

All hair shafts develop from multicellular structures termed *"follicles,"* which arise late in fetal development as inpocketings of the epidermis into the dermis. As the epidermal cells push down to form an inverted blunt-ended cylinder, mesenchymal cells of the dermis aggregate at the base of hair bulb. These mesenchymal cells from a thickened papilla at the end of the follicle that is partially enclosed by the epidermal cells of the hair bulb. It is from the hair bulb that the shaft originates. Each hair consists of an outer cylinder of thin, compressed cells, the cortex, and an internal set of cells, the medullary cells, separated by spaces within the shaft of the hair. Both cortex and medullary cells, but especially cortex cells, are rich in the structural protein keratin. Both cell types are also repositories of pigment. These melanic pigments are secreted directly into the cells of the hair shaft by melanocytes located in the region of the hair bulb. Pigment is transferred in the form of small pigmentary granules termed "melanosomes" to the basal cells of the shaft throughout the period of hair growth.

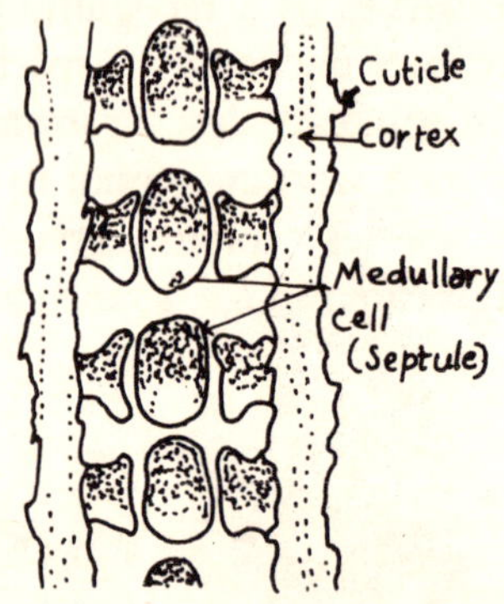

Fig. 4.22.   Internal structure of a hair shaft in the mouse coat.

The developmental biology of melanocytes is complex. As discussed earlier, the melanocytes of the skin are derived from the migratory melanoblasts of the neural crest. They arrive at the dermis—epidermis interface at about day 12 of development and form a network of interconnected cells throughout the epidermis. Each active melanocyte eventually transfers its melanosomes either into cells of a developing hair shaft or into neighbouring epidermal cells that are not part of a follicle. The transferred melanosomes are complex entities in their own right: each consists of fibrous protein granule attached to a melanic pigment. The pigments are of either of two kinds: eumelanin and pheomelanin. Eumelanin is the prototypical black melanic pigment, while pheomelanin is yellow. Both are synthesized from tyrosine in a series of sequential oxidations carried out by tyrosinase. Pheomelanin differs chiefly from eumelanin in having attached cysteinyl groups. Maturation of the protein moiety of the melanosome takes place gradually, and only mature melanosomes are transferred to epidermal cells. The final colour of a hair of hair subregion is function of which pigment is deposited, the number and spatial arrangement of a melanosomes within the cells of the hair shaft, and the shape and size of the melanosomes.

Given the developmental and biochemical complexity of

hair construction, it is not surprising that many genes directly or indirectly exert effects on the colour, colour pattern, or structure of the coat. To date, more than 50 loci have been identified on the basis of their mutant effects on coat colour; a large proportion of these loci have pleiotropic effects and for these, the relationship between colour and other mutant phenes is unknown.

Only a minority of the loci identified on the basis of their coat colour mutant phenotype are directly involved in the synthesis of the melanic pigments. Of the few genes in this category, the albino or c locus is the most important. The mutants of the c locus form a phenotypic series whose effects range from only a mild diminution of the pigments to a complete absence of melanin. This last is the original c mutant, characterized by a completely white coat and pink eyes. In standard genetic backgrounds, the wild-type C allele is dominant to all other members of the series. However, in the presence of certain nonallelic genetic differences that affect coat colour, C/c animals can be observed to make less pigment than C/C homozygotes.

The *albino* locus is almost certainly the structural gene for tyrosinase, the key enzyme in melanin synthesis. Not only is there a correspondence between the severity of the phenotypic defect and the amount of tyrosinase in the allele series, but a particular c mutant, the Himalayan variety (c^h), which is distinguished by a phenotypic temperature sensitivity in pigmentation—such that the extremities of the animal, which are always cooler, are darker—possesses a correspondingly temperature-sensitive tyrosinase activity (Siamese cats, no near relation of the Himalayan mouse, have a similar problem). Unlike the mutants of several other genes whose hair follicles lack melanocytes, the c mutant has normal numbers of melanocytes within the follicles, but they are devoid of melanosomes.

A second locus, *brown (b)*, is also involved in the biosynthesis of melanosomes. Mutants of this gene cause the replacement of the black granules with brown melanosomes. The effect of mutations in *b* appears to be on the proteinaceous granules that bind the melanic pigments. When examined microscopically, the melanosomes of *b* mutants are seen to be smaller than those of the wild-type; the visible alteration in colour is a direct consequences of this change in melanosome size.

However, the great majority of coat colour mutants produce their effects on colour indirectly, either through interference with melanocyte development or morphology or through subtle influences on the interactions of melanocytes with the other cell types in the hair follicle. Examples of two genes that affect colour through an effect on melanocyte morphology are the loci *dilute (d)* and *leaden (In)*. Mutants of both genes exhibit an apparent diminution of pigment intensity, the reduction depending upon the allelic composition of the other pigment genes. Thus, homozygosity for *d* causes mice that would otherwise be black to be Maltese blue and those that would be dark brown to be light brown. The dilution effects for mutants of both genes stem from a similar change in melanocyte morphology, although the precise biochemical basis of the change probably differs between *d* and *In* mutants. Normal melanocytes have a highly branched, or dendritic, morphology that facilitates the transfer of melanosomes. In the mutants, the melanocytes are much more compact. In consequence, melanosome transfer takes place in an irregular, bottlenecked pattern of release. The individual hairs of both mutants show large aggregations of pigment that are unevenly spaced within the hair shafts. Although the macroscopic impression is one of a reduction in pigment, there is as much pigment present as in the wild-type.

A second group of mutants that are indirectly affected in pigmentation are the various white-spotting mutants. Some are dominants, producing the spotting phenotype in heterozygotes; others are recessive. Two of the white-spotting mutants show severe anemias and partial sterility. Although while spotting looks like regional albinism, the cause is fundamentally different; hair follicles in the white areas lack melanocytes. In some mutants, the fundamental defect may be in the neural crest, involving a failure of formation or of migration of melanoblasts. In others, the melanoblasts migrate but die en route. The latter defect has been studied in chimeras and found to occur in the melanoblasts of the mutants, suggesting that the mutation causes autonomous cell death.

Perhaps the most interesting locus affecting coat colour is the *agouti* gene because of the range and patterns of colour associated with its various alleles. Seventeen alleles of the locus have been identified, ranging from two causing dominantly yellow expression (one, the classic lethal, the other a homozygous viable) to various black or dark brown recessive alleles. Thus, in an otherwise wild-type background, A^y and A^{vy} produce a yellow coat; the *A*, agouti, standard allele, produces a gray coat: and a^e, extreme nonagouti, produces a completely black coat in homozygotes. The *agouti* locus evidently governs, in some fashion, the balance between phenomelanin and eumelanin synthesis. A diagram of several of the *agouti* allele hair colour phenotypes is shown in Figure 5.19. A general feature of these phenotypes, except for the extreme black-pigmented form, is that hairs on the ventral side are always lighter than those on the dorsal side.

The puzzle of the *agouti* locus action is revealed by considering the colour pattern of individual hairs in the standard *A* strain. These hair shafts have black tips, yellow shafts, and black bases. The apparent grayness of the coat in the *A* strain results from this internal distribution of the two pigments. The transitions between black and yellow in the hair are not abrupt but take place over the length of three to four medullary cells. Indeed, some medullary cells in the transition zone contain both pheomelanin and eumelanin granules. This finding suggests that the gene somehow affects the synthetic behaviour of the melanocyte rather than its intrinsic synthetic capacity for the two melanic pigments.

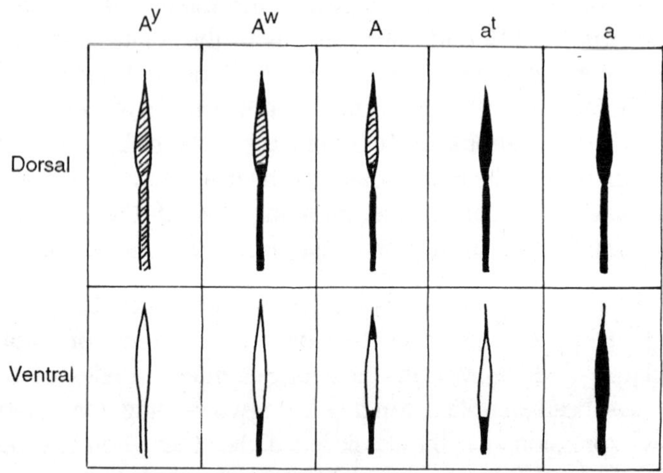

Fig. 4.23. Hair color phenotypes corresponding to representative alleles of the *agouti* locus.

In fact, the *agouti* locus acts in the epidermal or dermal cells of the follicle rather than in the melanocytes, as shown by the creation of skin chimeras in grafting experiments. When embronic or neonatal skin is grafted between strains, the developing follicles are populated by host melanoblasts. In all cases, the phenotype of the developing hairs, with respect to *agouti*, is characteristic of the genotype of the grafted skin and not of the host melanoblasts.

Thus, if one grafts skin of nonagouti, *ala* (black) to A^y (yellow) hosts, the developing skin graft produces a patch of black hair, although the hair follicles of the graft are populated with A^y genotype melanoblasts. The results do not reflect the action of an indigenous population of melanoblasts carried along in the skin grafts because the same result is obtained when the skin graft is taken from white patches of white-spotting mutants (which lack melanoblasts: the colour phenotype of the developing hair on the graft still follows that of the *agouti* allele of the graft.

Is the instruction affecting pigment synthesis given by the epidermal or dermal cells of the follicle? One can answer this question by making reciprocal dermis-epidermis hybrid grafts using embryonic skin. In such experiments, skin is taken from 13- to 14-day embryos of both *agouti* genotypes and separated into its two components by light trypsinization. The reciprocal hybrid grafts are then made by recombining the layers between the two genotypes, allowing a day for the grafts to heal, and then transplanting them individually to the testis of newborn males (a particularly favourable location for growth of the new grafts). If the hair phenotype follows the dermal genotype,one concludes that the gene acts in the dermis. Alternatively, if the phenotype follows the genotype of the epidermis, the latter must be the site of gene action.

The observations have been surprisingly complex. For some *agouti* allele combinations, the dermis seems to be crucial, but for two others, which produce a *"mottling"* phenotype (A^{vy} and a^m), the hair phenotype follows the genotype of the epidermis. *Poole* (1980) has suggested that the immediate instructions to the melanocytes are given by the epidermal cells of the follicle but that the epidermal cells receive their orders from the dermis. In those genotypic combinations in which the epidermal cells govern the hair phenotype, the epidermis may have "set", in other words, lost its competence to respond to a new dermal signal.

The complete set of results shows that the final pigmentation pattern is the product of a subtle interaction between two very different cell types, dermal cells and melanocytes. As we have seen, these cell types have distinctly different developmental origins. In fact, chimera patterns can reveal such differences in origin and sites of gene action. The melanocytes are ectodermal and derive from the melanoblasts of the neural crest: the major part of the body dermis is mesodermal and derives from the somites. Because the melanocytes and somites have different tissue origins, genes expressed in one of these hair follicle components but not the other show very different spatial patterns of expression from one another in chimeras. All the melanocytes of skin, as we have seen, may ultimately trace their origins to 34 melanoblast clones in the body; what is certain is that the coat patterns of melanoblast chimeras can be reduced to a basic archetypal pattern of 17 stripes on each side. In contrast, agouti-nonagouti (gray-black) chimeras show a different and much more finely striped pattern. Examination of large numbers of these animals suggest that there may be as many as 85 basic stripe widths per side, including about 18 in the head region, altogether a much larger number than seen in the comparable melanoblast-generated patterns.

Two interpretations of the **agouti** chimera pattern have been offered. The first is that each unit stripe derives from a hair follicle progenitor clone. Apart from the 18 or so in the head, there are about 65 per side for the body and tail, a number that matches the somite number for these regions. In effect, each somite may contain one hair follicle progenitor clone. (In this view, the source of the stripes on the head raises the interesting possibility that there are rudimentary "invisible" head somites, a subject that has been debated by mammalian developmental biologists). The other interpretation of the fine grained pattern is based on the observation that there is much greater regularity of striping in these hair follicle chimeras than in melanocyte chimeras. *McLaren* (1976a) suggested that this regularity signifies a systemic patterning of some kind that is superimposed on the chimeric skin rather than an alternation of genetically determined clones. On the basis of random clonal placement, such regularity is unexpected. That there are *some* systemic influences that regulate *agouti* locus expression is apparent from the consistent dorsoventral differences in pigmentation within single genotypes.

Solution of the various puzzles associated with the *agouti* locus ultimately requires determination of what the gene product does. Earlier results were interpreted to mean that the gene regulates the pheomelanin-eumelanin decision through control of the synthesis of sulfhydryl (SH)-containing compounds, with higher levels of SH compounds favouring eumelanin synthesis. However, this hypothesis has received little subsequent support. The action of the *agouti* gene product remains a puzzle.

Apart from the systemic effects on the expression of *agouti*, the idea that there are general systemic influences on the development of coat colour pattern is inescapable from the fact that many mammals show a species-specific coat pattern of spots or stripes, or sometimes both. Some common examples are zebras, giraffes, and many members of the cat and squirrel families. In contrast to the colour patterns of genetic chimeras, constructed in the laboratory, or those of X inactivation mosaics, each of these species-specific coat patterns is produced in animals whose cells are essentially identical in genotype. The question of the origins of such patterns is therefore one of developmental physiology rather than developmental genetics (although the comparative genetics of the different species' patterns is an interesting matter). In formal terms, these patterning processes can be thought of as morphogen systems acting throughout the presumptive field of pigmentation in the embryo or fetus. The morphogens may either suppress or enhance pigment formation.

SEX DETERMINATION

Sex Chromosomes and Gonadal Differentiation

The ultimate source of the differences between male and female mammals is the sex chromosomes. As in the fruit fly, females are the homogametic sex, being XX, and males are the heterogametic sex, being XY. Unlike *Drosophila*, the crucial difference is the possession or absence of the Y rather than the number of Xs. Thus, female XO mice are phenotypically female and fertile (in humans, the XO condition produces sterile females) and XXY animals are phenotypically male, the reverse of the fruit fly situation. Evidently, in mammals, the Y carries one or more sex-determining factors, and maleness is the dominant sexual phenotype (in the biological sense); in the absence of the Y, femaleness results.

There is a further difference in the sex determination mechanisms. In *Drosophila*, sexual phenotype is determined autonomously, cell by cell, whereas in the mouse, gonadal sexual determination is primary, with all other aspects of sexual phenotype following from it. Secretion of the hormone testosterone by the male gonad triggers male genital development; in the absence of testosterone, the gonad secretes estrogens and the secondary sex characteristics that develop are female. This mechanism is general in mammals.

The first differences in gonadal differentiation become apparent at about 12.5 days p.c., 2 days after the arrival of the first PGCs. The male gonad begins to organize itself into testis cords, consisting of solid strings of spermatogonial cells encased in mesodermal somatic cells. The early female gonad retains a compartmented appearance, with groups of oogonial cells surrounded by a matrix of mesodermal cells. The urogenital system undergoes an accompanying differentiation. Initially, both sexes possess both kinds of urogenital structures, the female Mullerian ducts and the male Wolffian ducts. In each sex, one duct system develops while the other degenerates.

Although the primacy of gonadal differentiation in setting sexual phenotype is clear, it is not known with certainty whether the initial events occur in the gonadal soma or the germ line. In the chick, it has been possible to show that the soma of the gonad plays the crucial role: when PGCs are prevented from colonizing the genital ridge, the gonad nevertheless develops according to the sex chromosomal composition of the (somatic) ridge. In mammals, this kind of unequivocal demonstration has not yet been possible. What can be done is to construct chimeras, and analyze for correlations between phenotypic sex and the respective chromosomal composition of the somatic and germ line tissues of the gonad. Because this procedure almost never produces an animal that has reverse compositions for the two components, the interpretations are always uncertain.

However, the chimera results show two things. The first is that most the XY/XX chimeras (which comprise about half of the experimental animals) are male, a consequence of the dominance of maleness. Above a certain threshold of male (XY) cells, perhaps as little as 30%, a chimeric gonad tends to develop as a testis, while around that threshold, an ovotestis (a mixture of testicular and ovarian material forms. Development of a testis then brings in train the development of secondary male sex characteristics.

The other finding of interest is that in XY/XX chimeras, regardless of the phenotypic sex of the gonad, there is little or no conversion of germ line cells of the "wrong" composition into functional gametes. For XX germ line cells developing within a testis, the prohibition seems to be absolute and may reflect an absolute block of spermatogonial development in cells with two X chromosomes. For XY cells, transfer of germ line cells out of the somatic gonad before these cells encounter the normal meiotic block typical of the developing male gonad can sometimes permit oocyte development. Possession of the Y chromosome, therefore, is not an intrinsic block to oocyte development.

The probable sequence of events therefore is an initial commitment in the somatic portion of the gonad created by the sex chromosome composition of that gonad, followed by a distinctive course of gamete maturation, and finally by the entrained development of secondary sex characteristics by the secreted male or female sex hormones.

Mutants and the Role of the Y Chromosome
Sexual development in animal systems with sex chromosomes involves two stages, an initial phase of sex determinative decisions and an ensuing period of sexual differentiation. Both sets of processes are subject to genetic

derangement. Because of the temporal separation between the initial event (gonadal determination) and the subsequent differentiation events, it is possible to male informed guesses as to which process has been genetically altered.

In mammals, the differentiation period is dependent on the continued presence of the secreted sex hormones: if androgen production is stopped by removal of the testes in male fetuses, the further course of somatic sexual development becomes shifted toward the production of female characteristics. Similarly, mutations that interfere with either gonadal sex hormone production or hormone binding would affect the later stages of sexual development.

Although mutants deficient in hormone production have not been reported, a sex differentiation mutant, characterized by insensitivity to testosterone, has been described. It is the X chromosomal testicular feminization (*Tfm*) mutant, and its phenotype is that of partially feminized males. (The mutation has no effect in females). XY mice that carry *Tfm* develop testes, but these are small and underdeveloped; externally, the animals resemble females, although not perfectly. The animals have normal testosterone levels, and administration of neither testosterone nor dihydrotestosterone rescues the male phenotype. Evidently the defect is not in testosterone production but in the response to testosterone. Similar genetic syndromes have also been reported in rats and humans. In all cases, the specific defect appears to be a deficiency of the androgen receptor protein, which is normally distributed in all or most cell types.

The most dramatic effects of *Tfm* are on secondary sexual characteristics, but there is also a block to spermatogenesis. In *Tfm*/Y individuals, spermatogonial are prevented from giving rise to spermatocytes, the immediate precursor cells of the sperm. To determine whether the effect is a direct one in the germ line or a secondary effect of an inappropriate gonadal soma, *Lyon et al.* (1975) constructed chimeras of *Tfm*/Y and +/ Y embryos and then mated the mature chimeric males with *Tfm*/ + females. Some of the progeny obtained were found to be *Tfm/Tfm* embryos, showing transmission of some *Tfm*-carrying sperm from the male chimeras. The result indicates that the presence of some *Tfm*⁺ in the gonadal soma can rescue spermatogenesis in some of the *Tfm* spermatogonia, and that the effect of the mutation on spermatogenesis is in the soma rather than the germ line.

Sex determination mutants should differ from sex differentiation mutants in having transformed gonads as well as transformed secondary sex characteristics. Although hunts for sex determination mutants in mammals are infeasible, because large numbers of animals cannot be screened, a number of putative autosomal and X chromosomal sex determination mutants have been identified in mice, goats, wood lemmings, and humans. In mice, several genetic conditions that affect sex determination have been identified. However, the precise genetic basis of most of these conditions remains ill-defined, including the number of loci involved. For several, the effect involves an aberrant interaction between the Y and the autosomes, where the Y has been introduced from other subspecies or lines, resulting in a partial feminization of XY individuals. The genetic basis of these syndromes is obscure, but their existence shows that simple possession of a Y is insufficient to guarantee normal male development—that Y-autosome interaction in some manner is essential for this development.

Of the several murine genetic conditions that alter sex determination, that of *Sex reversal (Sxr)* is the best understood. It was first described by *Cattanach et al.* (1971) as a dominant mutation that transforms Xx animals into sterile males. Because of this sterility the initial mapping had to be carried out in crosses between normal females and XY *Sxr* carrier males; the results indicated that the condition was inherited independently of both the X and Y chromosomes. In consequence, the authors concluded that the trait was autosomal. However, all subsequent attempts to localize it to a particular autosome met with failure.

The solution of the *Sxr* puzzle was made possible by the use of a molecular marker specific to the Y chromosome. This marker is a particular satellite DNA that is unique to the Y in mammals but was first isolated not from mice or humans but from a snake, the banded krait. More surprisingly, it was isolated from the heterochromatic *female*-specific W sex chromosome of snakes. In the banded krait, as in most snakes, sex determination is carried out by the ZW-ZZ sex chromosome system. In this system, male are the homogametic sex, with two euchromatic Z chromosomes, and females are the heterogametic sex, with a ZW constitution. By looking for a satellite DNA species unique to the heterochromatic W chromosome, *Singh* and *Jones* (1982) isolated such a satellite, which they designed Bkm (for "banded krait, minor satellite").

The Bkm satellite is found throughout eukaryotes, from yeast to humans although in varying amounts and different locations. The fact that it is found on the female-specific chromosome of snakes and on the male-specific chromosome of mammals implies that it does not determine femaleness or maleness per se but that it is a marker of sex chromosomes. When radioactively labeled Bkm satellite is hybridized in situ to metaphase spreads of chromosomes from XX *Sxr* animals, it is found to be located at the distal tip of one of the X chromosomes. The result indicates that the *Sxr* condition derives from the translocation of Y chromosomal material to the distal tip of the X, presumably by recombination during meiosis. The transfer of Y chromosomal material to the X by a high-frequency recombination event explains the absence of genetic linkage between the Y and the *Sxr* condition found in the earlier studies.

Eicher (1983) has proposed a hypothesis for the event. The idea is that the *Sxr* condition is characterized by a Y chromosome carrying two sex-determining, Bkm-containing regions, and that transfer of the more distal region from one chromatid to the X invariably occurs during meiotic pairing in the XY *Sxr* animals. Note that the process, as diagrammed, produces equal numbers of X^{Sxr} and normal Y chromosomes. Indeed, XY *Sxr* carriers give rise to normal males in addition to transformed XX females and carrier XY *Sxr* sons (who bear the non-recombinant, duplication-bearing chromatid).

The central question about mammalian sex determination concerns the manner in which the Y chromosome confers maleness. The general belief is that the Y carries one or more genes that actively promote testicular development. Much speculation has focused on the possibility that one of the known male-specific antigens might play that role and be encoded by genes on the Y. One of these is detectable as a transplantation antigen, causing the rejection of male skin in females of the same inbred strain. This antigen has been dubbed the "H-Y antigen (for the Y histocompatibility antigen) and is both highly conserved and male specific throughout the vertebrates. The possibility that the H-Y antigen is encoded by a gene on the Y and is the male-determining factor was originally proposed by *Ohno* (1976). A second antigen that is male specific was subsequently discovered in sera and designated SDM. For a long time, it was believed that SDM and H-Y were the same antigen. It is now apparent that they are not. More importantly, the current evidence shows that neither H-Y nor SDM is invariably associated with phenotypic maleness, nor is their absence invariably correlated with femaleness. These antigens are therefore probably consequences rather than causes of male determination. The nature of the relationship of the putative male-determining factors to the Bkm satellite also remains to be elucidated. The Bkm sequence may contain protein coding functions but these are presumably not directly male-determining because Bkm is not invariably associated with maleness in vertebrates. If there are male-determining, protein-coding function on the Y, they are presumably closely linked to the Bkm satellite.

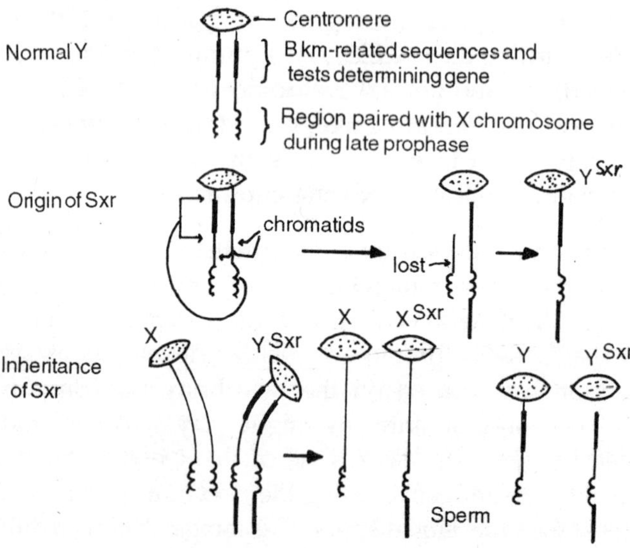

Fig. 4.24.   Model of the origin of *Sxr*-carrying X chromosomes. See text for discussion. (Reproduced with permission from Eicher. 1983; copyright Animal Reproduction Laboratory, Colorado State University.)

A different view of the role of the Y in setting maleness has been proposed by *Chandra* (1984), who suggests that it is purely passive: to absorb repressor molecules, thereby permitting the function of an X chromosomal gene essential for testicular determination, designated *Tdm*. In this view, the Bkm satellite might be the male determinant itself by functioning as the repressor binding sequence. Chandra's specific proposition is that an autosomal locus encodes a repressor that binds with high affinity to the repeated Y chromosome sequence and with lower affinity to *Tdm*. In animals possessing one or more Y chromosomes, the repressor is bound by these chromosomes and *Tdm* expression is turned on; in XX animals, there is no sink for repressor and *Tdm* is repressed. Although this model is at best an approximation, it fits most of the known facts and provides a provocative alternative to the conventional view of the Y chromosome's role in sex determination.

Superficially, sex determination in mammals appears to operate by very different rules than in fruit flies or nematodes. The controvertible difference is that in fruit flies, the decision is made autonomously in all cells capable of a sexually dimorphic response, while in mammals the key sex determination events occur in one organ, the gonad, with all of the remaining sexual dimorphism following from this first step. There may, however, be certain underlying unities. In both organisms, the decision is a binary one in the cells that make it—either a "yes" or "no" to development along the pathway of the dominant sex. In addition, the key sex chromosome (the X in fruit flies, the Y in mammals) interacts with one or more autosomal factors in making the sex determination decision. These autosomal genes have been identified in *Drosophila*, while those of mammals are still obscure. However, the general logic of such an interaction, if not the precise molecular details, may prove to be surprinsingly similar. The future of this problem is certain to be interesting.

5

Human Development

Scarcity of Early Human Embryos
Because of the inaccessbility of the fertilized human ovum, its cleavages have never been observed. Human embryos can be obtained only through accidental or surgical abortion from the uterus, and such embryos are past the cleavage period. Even later stages are comparatively rare, since they are extremely small and are usually lose or injured in scraping the uterus and searching for them. However, of late two authors, *Hertig* and *Rock* have published a great deal about early human embryos. One of these concerns an embryo nine days old of which a reconstruction has been made, and one is even younger. These two men have added greatly to our knowledge of human embryology.

Development of Blastocyst
Since the early cleavages of mammalian ova have been found to be uniform in their cleavage pattern, it seems reasonable to assume that the human egg is no exception in this respect. At any rate, one may safely say that its cleavages are like those of the macaque monkey. The arrival of the blastocyst in the fundus of the uterus has been variously estimated to be from four to seven days after fertilization, but it has been established that the tubal journey of the macaque monkey takes only three days and that would confirm the lesser estimate of four days in man. From the study of the estrous cycle it is safe to assume that it arrives within 4 to 6 days after ovulation. Basing our conclusion on the development of the blastocyst in other

mammals, the human embryo at that time is probably slightly past the morula stage. The inner cell mass, or embryonic knob, has been formed and the trophectoderm consists of a single cellular layer functioning as an envelope of the blastodermic vesicle. After adhering to the uterine wall for a few hours, it begins to sink into the utterine mucosa or endometrium by localized cytolysis. Apparently the erosion of the tissue is due to the cytolytic action of the trophectoderm cells. This destructive action of the blastodermic vesicle has been observed in several mammals, as for example the guinea pig. Further more, it appears that the blastocyst elongates in its descent into the endometrium, thus temporarily losing its spheroidal shape. Having disappeared beneath the surface of the uterine mucosa, it readjusts itself, resuming more or less its former contour.

The presence of the blastodermic vesicle within the endometrium has a destructive effect on the uterine mucosa immediately surrounding it. Indeed, this deleterious action of the trophectoderm is even more pronounced than the one taking place during its entrance into the endometrium. The cells of the trophectoderm multiply at a rapid rate and push into the adjoining uterine tissue. The latter is broken down under the cytolytic action of the trophectoderm cells and forms cellular fragments, broken-down capillaries, fibrin, degenerating blood cells, and other cellular debris. This material is called *embryo-troph* and is absorbed by the blastodermic vesicle to supply the necessary food for growth. This embryotrophic nutrition lasts until the embryo has developed its own vascular system and perfected its placental attachment. One of these early human embryos was described by *Hertig* and *Rock* in 1945 and is represented in which is based on a plastic reconstruction of the uterus and its glands.

The original trophectoderm has now considerably increased in size. Its inner-most layer remains distinctly cellular and is from now on referred to as the *cytotrophoderm (cytotrophoblast).* However, the cells which have been proliferated from it and have invaded the uterine tissue from syncytia and loosely organised spongy tissue called the *plasmotrophoderm (plasmotrophoblast)*. The latter acts as does a parasite, disintegrating and living on the tissue of the host. Spreading and increasing its destructive effect, the plasmotrophoderm comes in contact with uterine blood vessels : it disintegrates their walls and maternal blood comes in direct contact with the plasmotrophoderm.

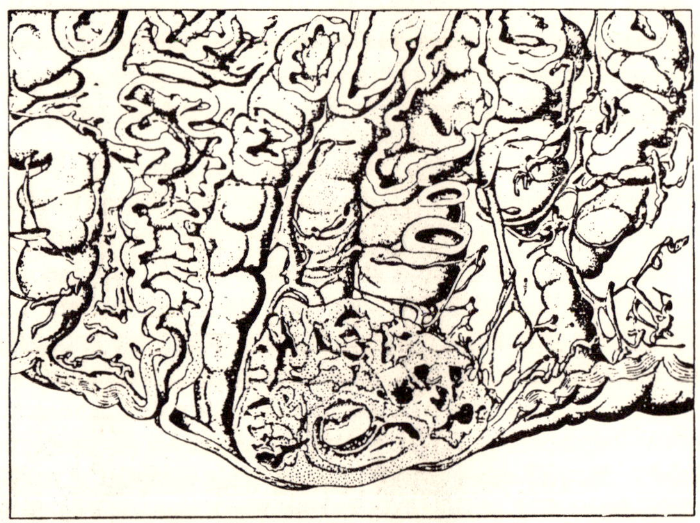

Fig. 5.1.   Reconstruction of the 9-day ovum, sectioned through the middle, viewed at right angles to the cut surface.

At that stage of its development, the cytotrophoderm attains a second period of rapid cell proliferation which is followed by a corresponding diminution of the plasmotrophoderm. This new cellular multiplication produces minute projections or *chorionic villi*. As the plasmotrophoderm shrinks, the chorionic villi come in direct contact with the blood and form in this manner the foundation of the discoidal, deciduate placenta.

Early Human Embryos

Hertig and Rock (1945) described a human embryo, the age of which they estimate to be seven days. If one assume that the segmenting egg arrives in the fundus of the uterus after four days of moving down the oviduct, then this 7-day embryo is not much older than the blastocyst.

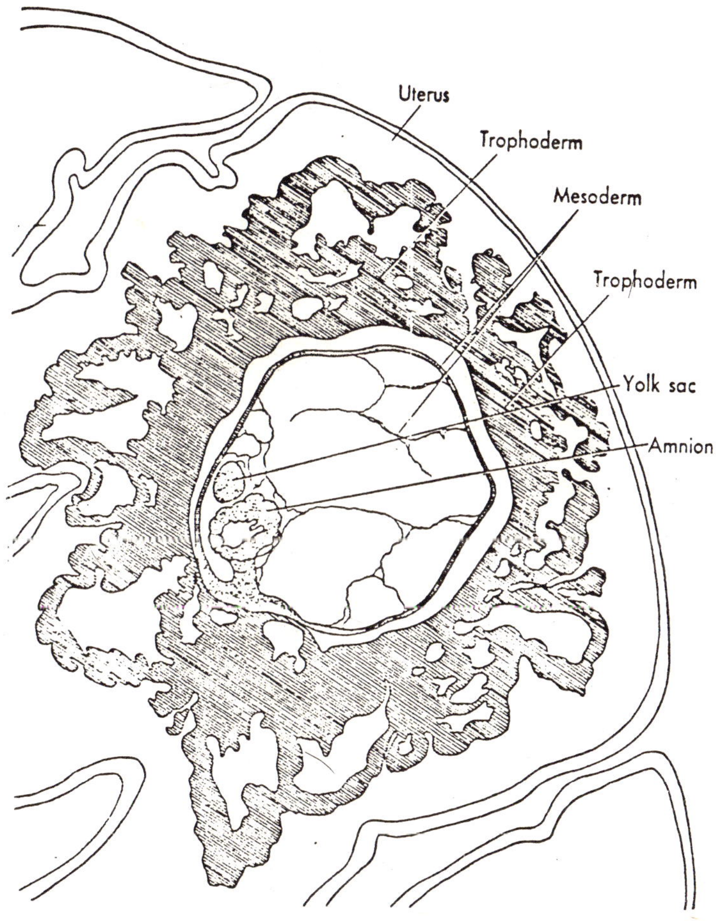

Fig. 5.2.   A human embryo.

A blastodermic vesicle of a slightly more advanced stage has been described by Streeter in the Miller ovum. It has been estimated to be from ten to twelve days old. The cytotrophoderm and plasmotrophoderm are indicated by the heavily-shaded area. The mesodermal layer is shrunk away from it, and fine mesodermal strands are seen all over the vesicle. The amnion and the yolk sac have been differentiated, but the allantois is not apparent as yet.

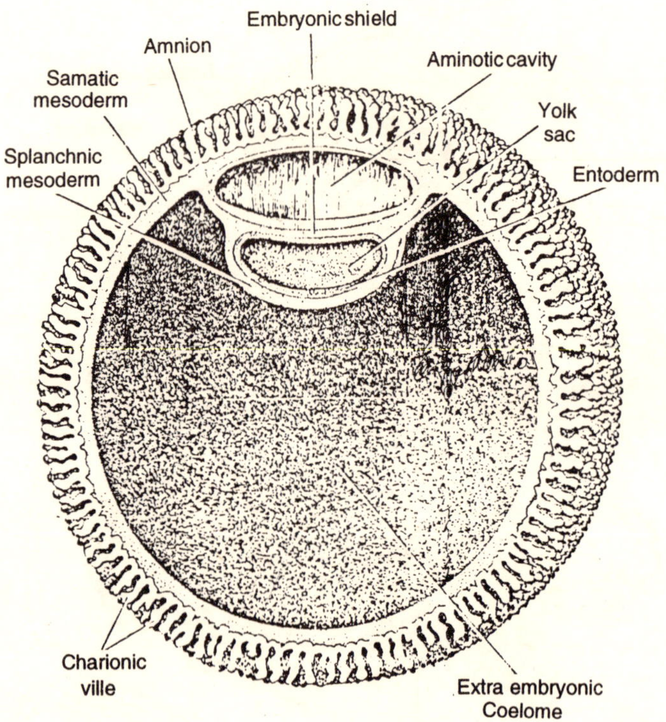

Fig. 5.3.   Stereogram of a human blastcyst contianing an embryo of approximately fourteen to fifteen days.

At this stage of development the future embryo can be allocated to the area represented by the embryonic shield. It appears that the blastocyst has grown extensively by the increase of the cytotrophoblast and its chorionic villi. The condition of the young embryo indicated in the diagram is realized in the Peters embryo as well as in von Mollendorff's embryo. The Peters embryo was probably the best preserved as well as the most interesting of all the human embryos up to the description of such embryos by *Hertig* and *Rock*, and even these two writers give it adequate credit in their pages. The embryo is estimated to be fifteen days old, measuring 0.19 mm. in length. It appears that in this embryo the plasmotrophoblast has become reduced and that the cytotrophoblast has begun its second period of cellular multiplication. Chorionic villi extend peripherally into blood lacunae which are in direct contact with blood vessels. These lacunae become the large blood sinuses of the placenta. The chorionic villi with their central cores of mesoderm from nipple-like projections from the cytotrophoderm. Here and there one may be seen to form secondary beaches which will finally culminate in the complex chorionic villi of the mature placenta. Though the extent of the plasmotrophoderm has been reduced, the destructive action against the uterine mucosa is still continued and is now carried on by the cytotrophoderm by means of the rapidly extending and branching chorionic villi. Nutrition is still embryotrophic, but the beginning of the hemotrophic form of nutrition is in preparation.

The triangular space on the left side has been tentatively diagnosed by Peters as the extraembryonic coelome. He is not certain whether this identification is correct. It may merely represent a vacuolized area surrounded by a more distinct mesodermal stand, of which many are visible in the cavity of the large extraembryonic coelome.

Von Mollendorff's embryo OF is probably a day older than that of Peters. Its chorionic villi are larger and more branching, and their rapid growth has pushed them much deeper into the blood lacunae.

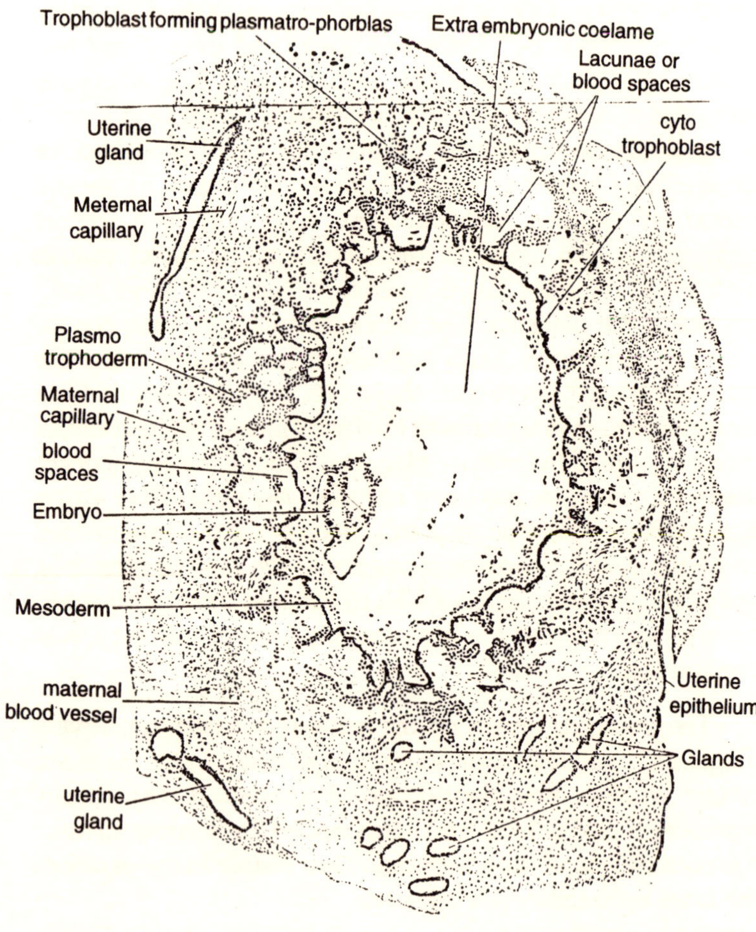

Fig. 5.4.   Human embryo estimated to be fifteen days old and 0.1 mm. long.

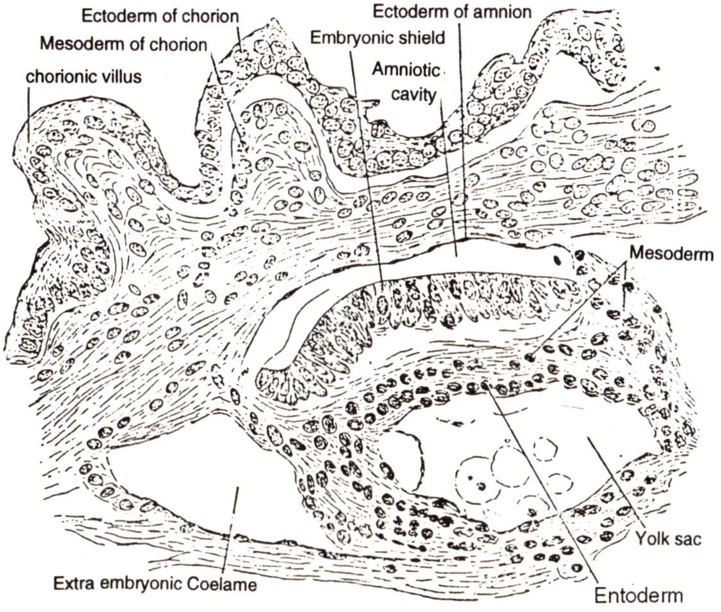

Fig. 5.5.   Human embryo.

In the next stage of development a radical change occurs in the position of the embryonic mass in that it detaches itself partially from the inner surface of the cytotrophoderm, of chorion, as it may now be termed. At the same time, it increases rapidly in size and forms a tabular outgrowth from the upper region of the yolk sc into the body stalk. This tubular extension is the diminutive allantois, which grows toward the chorion but does not spread over it as it does in the ungulates or carnivores. The Mateer embryo described by *Streeter* represents the initial stages of this change. This Mateer embryo is estimated to be seventeen days old and has a dimension of 0.92 mm. It is therefore five times larger and only two days older than the Peters embryo. It is just detaching itself from the chorion, some of its connecting fibres being still intact between the amnion and the chorion. Viewed from the dorsal aspect with the amniotic membrane removed, the Mateer embryo shows a primitive streak on the embryonic shield contianing an anteriorly located primitive pit.

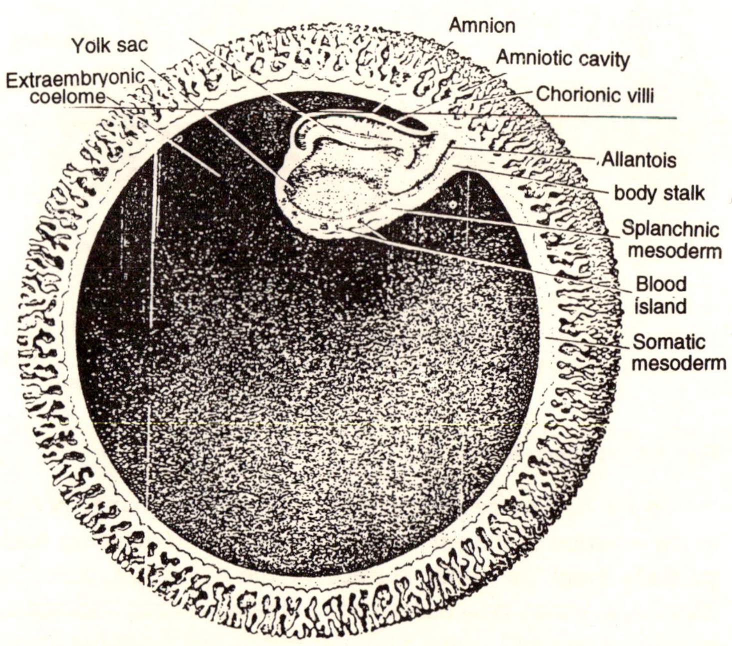

Fig. 5.6.   Stereogram of a human blastocyst containing an embryo of approximately twenty to twenty-two days.

The embryo of *Graf von Spee,* which he terms embryo Gle., is estimated to be twenty-one days old and is 1.54 mm. long. It has a well-developed yolk sac containing blood islands for the organization of the vitelline circulation. The fact that it is nearing the end of its embryotrophic nutrition is indicated by the profusely branching chorionic villi and their close association with the blood sinuses of the endometrium, in which they are suspended and surrounded by maternal blood. The chorionic villi are becoming vascular and will soon join the umbilican circulation which is established within a few days. The primitive pit observed in the Mateer embryo and a few other human embryos has disappeared in the Graf von Spee embryo, as was established in later investigation.

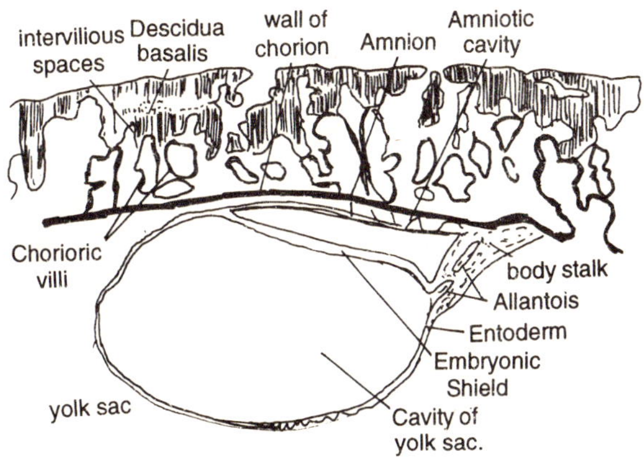

Fig. 5.7.   Human embryo estimated to be seventeen days old and 0.92 mm. long.

An instructive and illuminating contribution by *Hertig* and *Rock,* featuring human embryos of different ages, is contained in the accompanying illustration. The legend is self-explanatory and includes several of the embryos discussed in this chapter. They are estimated to be from 11 to 15-16 days old.

Between the sixth and seventh week of gestation the embryo grows so rapidly that the blastodermic vesicle forms a bulge on the surface of the endometrium. It pushes the uterine lining (decidua) ahead of itself in its extension into the uterine cavity. During this period the amnion also increases in dimension so that the extra-embryonic coelome becomes almost entirely replaced by it. This expansion of the amnion is the cause for the change in the position of the embryo. It rotates almost 180° from its original position, and now faces the body stalk with its venal side. When the expansion of the amnion is completed its envelops the body stalk and becomes opposed to the inner side of the chorion. The body stalk is from then on known as the umbilical cord.

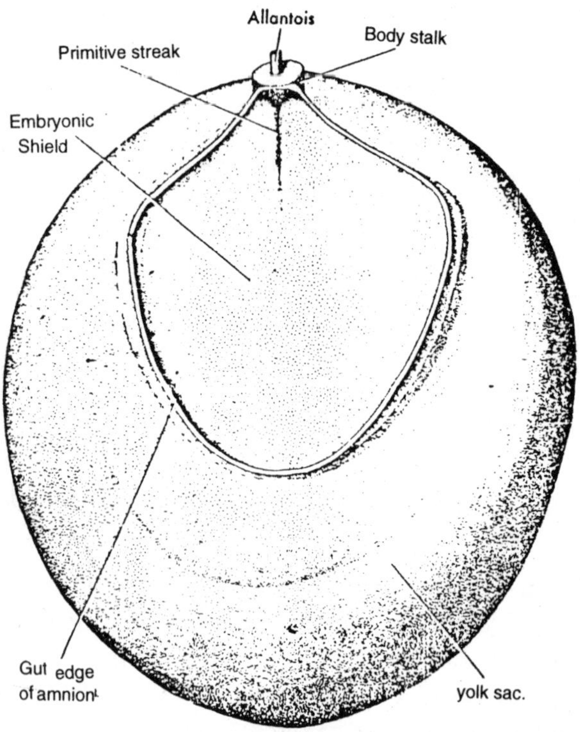

Fig. 5.8.   The same embryo as shown in dorsal aspect. The amnion has been cut away to expose the embryonic shield.

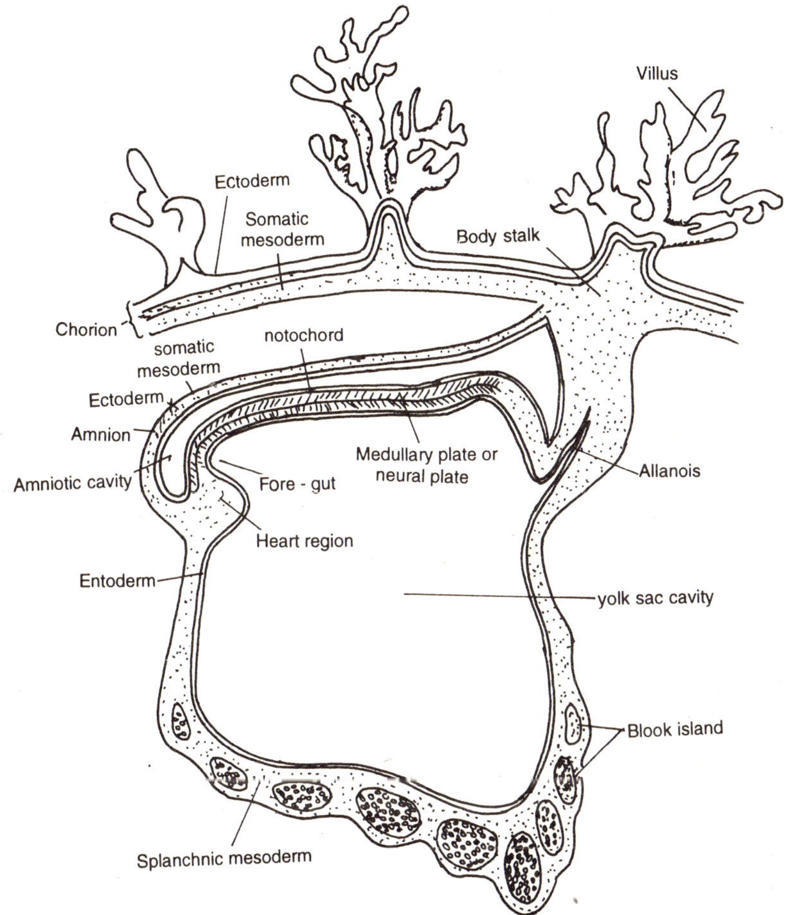

Fig. 5.9.   Human embryo described by Graf von Spee, estimated by him to be twenty-two days old and 1.54 mm. long, and referred to as "Ovum Gle".

The membranes have been cut away to expose the embryo. The chorionic villi have progressively diminishes in size and number on the side facing the uterine cavity. In later stages the chorionic villi are relegated to the uterine side of the chorion to form the disc-shaped or discoidal placenta. As the embryo grows away from the placenta, the two remain in communication by the umbilical cord. The yolk sac and the allantois have remained vestigial organs throughout the entire development; they have become secondary in their importance in the development of man.

Prenatal Influences

It was formerly assumed that the mammalian embryo develops entirely independently of its mother, that its embryology was the function of the embryo alone, and that the parent merely supplied the "raw materials" in the form of food, protection and warmth for its development. This belief is no longer held, for with the development of modern chemistry and the better understanding of semipermeable membrance, much progress has been made in prenatal care of the human embryo. These investigations show that a good many substances can be passed from mother to fetus or from fetus to mother in the ordinary metabolic processes of the embryo. Thus, it has been demonstrated that certain substances, such as salts, vitamins, hormones and other, can be diffused through the placenta into the fetal blood. This is to be expected since, in the natural course of metabolism, gases and food substances and excretory products in solution have to pass through the placenta. These facts have been established by feeding the mammalian mother certain substances and examining the blood of the fetus afterwards. Some materials in solution do not pass because of the selective action of the placental membrane, but a great number do enter the fetal circulation.

However, there are also many instances of large molecules, as for example, proteins and even blood cells passing through the placenta when it has been injured by inflamation or weakened mechanically. Thus a uterus that is infected can be very troublesome for the embryo and may cause its abortion or death. Certain infectious and communicable diseases, as for example syphilis, may easily pass from mother to offspring and do considerable damage there through an infected placenta. Within the last few years, a considerable number of such cases have been recorded, one type of which, the cause for certain still-births, is best established. This is the well-known fetal *erythroblastosis*, a rather common cause of death of the infant before birth which is due to the destruction of its blood and the blood forming centres. To understand the cause of this disease, with all its ramifications, involves the knowledge of immunity reactions and also the laws of Mendelian inheritance.

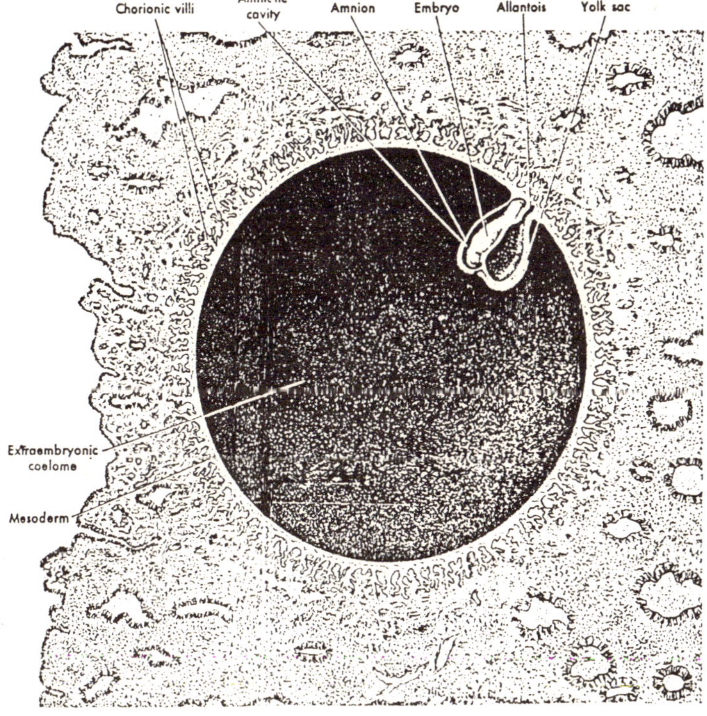

Fig. 5.10.   Stereogram of a blastocyst containing a human embryo, twenty-five days old and 2.4 mm. long, imbedded in the endometrium.

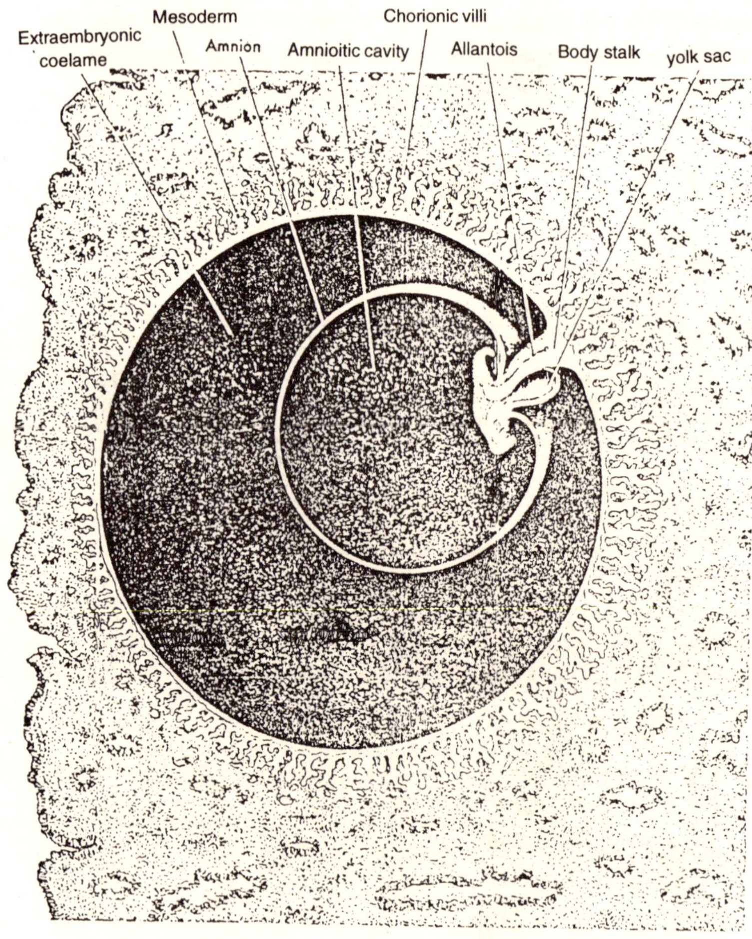

Fig. 5.11. Stereogram of a blastocyst containing a human embryo, twenty-nine days old and 4.3 mm. long. The blastocyst is bulging slightly over the uterine mucosa into the uterine cavity. The embryo has rotated and is facing the body stalk. The extraembryonic coelome has been partially replaced by the amniotic cavity.

The greater majority of human carry a factor in their blood which agglutinates the corpuscles of certain other human blood. This is due to the presence of the Rh factor, Rh standing for Rhesus monkey, in which this factor was first discovered. Subsequently, the same factor was demonstrated by *Landsteiner* and *Wiener* in human blood. With regard to the latter, there are two types : the Rh+, having the Rh factor, and the RH-, having none of it. Being a dominant Mendelian factor, the Rh factor shows individuals who are heterozygous with regard to the Rh factor. An Rh—mother may conceive form an Rh+ father a child which is Rh+ (Rh+ being dominant). When the placenta is defective for one reason or another as shown above, or is so for some other cause not known to us as yet, a limited but continuous amount of fetal blood will enter the mother's circulation and cause the development of anti-Rh substances in the mother, which then pass back to the embryo by means of the fetal circulation. In the fetus these anti-Rh substances will harm its blood. If there are not too many such substances generated and passed over by the mother, the child may live; but in the second and further pregnancies, there is an increasingly high danger that the child will be still-born, because with every pregnancy more of the anti-Rh substances are developed so that eventually there will be enough to kill the infant.

Multiple Births

Within recent years the interest of the public has been centred on twinning and multiple births. They are of common interest because they are considered to be comparatively rare in man, and because the similarity of certain twins is so striking that it arrests the attention of the crowd.

It is a familiar fact that many mammals give birth to more than one young at a time, and that others, usually the larger ones, being forth only one at birth. One may make the arbitrary rule that this phenomenon is due to the size of the animals. For example, such mammals as the horse, elephant, and the

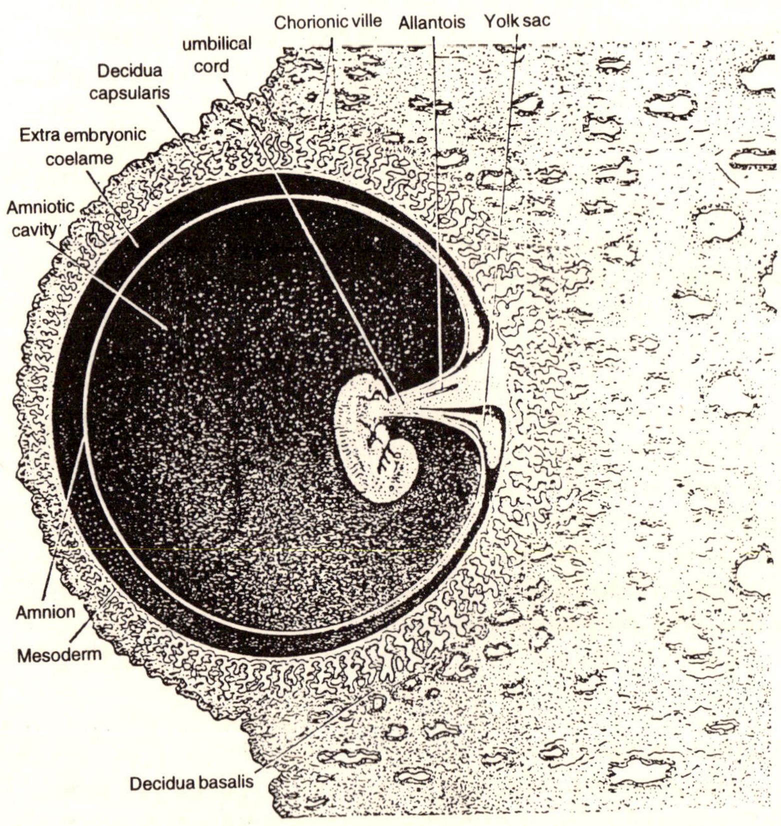

Fig. 5.12. Stereogram of a blastocyst containing a human embryo, thirty-three days old and 5.00 mm. long. The blastocyst has formed an elevation on the surface of the uterine mucosa. The extraembryonic coelome has been almost entirely replaced by the amniotic cavity. The body stalk has been converted into the unbilical cord by being covered with the amnion.

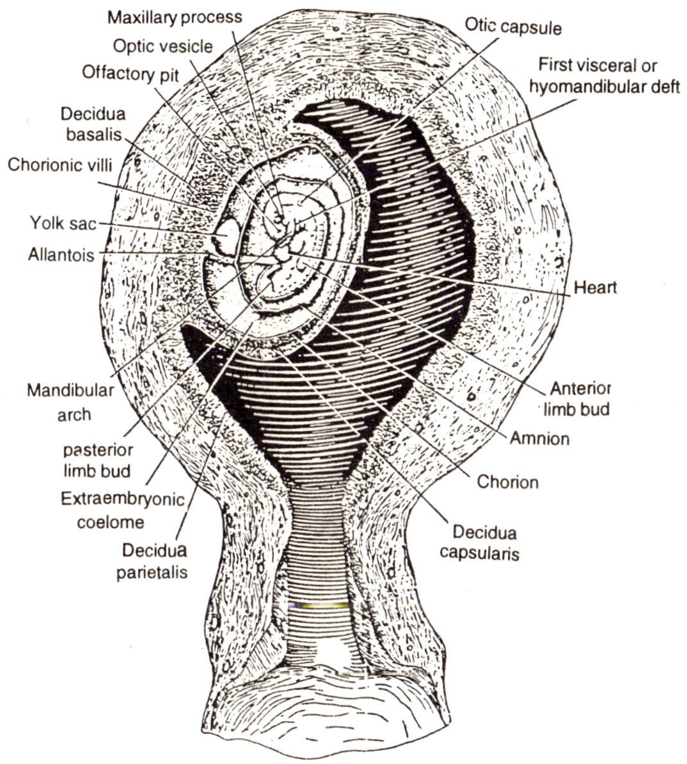

Fig. 5.13. Human embryo in the eterus, thirty-eight days old and 7.5 mm. long. (Adapted from Coste and other sources.) The chorion and amnion have been cut away to expose the embryo to view. The yolk sac extends into the extraembryonic coelome. The vestigial allantois may be seen next to the yolk stalk. The umbilical arteries and veins are not shown.

giraffe give birth to only one young, while dogs, cats, rabbits, mice, and others produce litters of many individuals. That this division is not universally true in exemplified by the lion, which is larger and heavier than man and may have a litter of several cubs, or by the pig, which may bright forth a dozen young at one time. Large litters are chiefly confined to mammals having a bicornuate or bipartite uterus, but even this condition has its exceptions.

Twins, triplets or other multiples may be formed in three different ways in man : they may arise from two independent ova, or they may develop from a single one, or they may be formed by a combination of the two. In the first instance they are referred to as *fraternal* twins because they are not any more closely related to each other than they are to other brothers sisters in the family. Having their origin from two different ova, the resulting blastodermic vesicles are entirely independent and develop their separate fetal membranes. In later development the placentae may approach each other and may even grow together, but their separate identify can always be demonstrated by the independence of the umbilical blood circulation. Fraternal, or more accurately *biovular*, twins have a much chance to be of the same sex as to be of different sexes. Their origin may be due to the simultaneous ripening of two Grafian follicle and the liberation of two ova for fertilization. Another cause of their origin can be ascribed to the presence of two oocytes within one Graafian follicle. In either case biovular twins will be formed, and the possibility of producing triplets and more individuals is determined by the increase of these causes or their combination.

Those of the second type, called *identical or monovular* twins, are always of the same sex, and they resemble each other so closely that it is sometimes difficult to tell them apart. Monovular twins are derived from the same zygote. Their duality is established early in the formation of the blastodermic vesicle. From evidence gathered in other mammals, such as

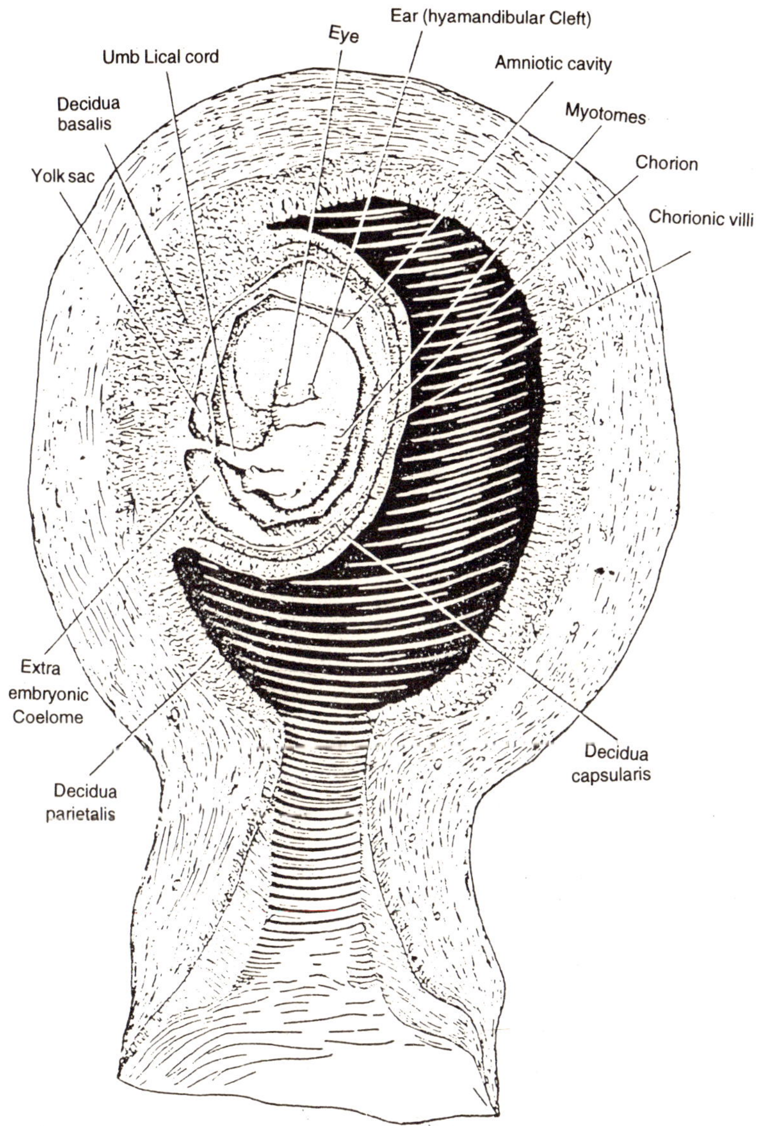

Fig. 5.14. Human embryo in the uterus, seven weeks old and 17 mm. long.

the pig and e specially the armadillo, it seems that the inner cell mass or possibly the embryonic shield divides into two parts, each developing a separate amnion, but retaining the same yolk sac, chorion and placenta. In the formation of triplets, quadruplets or quintuplets of this second type, the inner cell

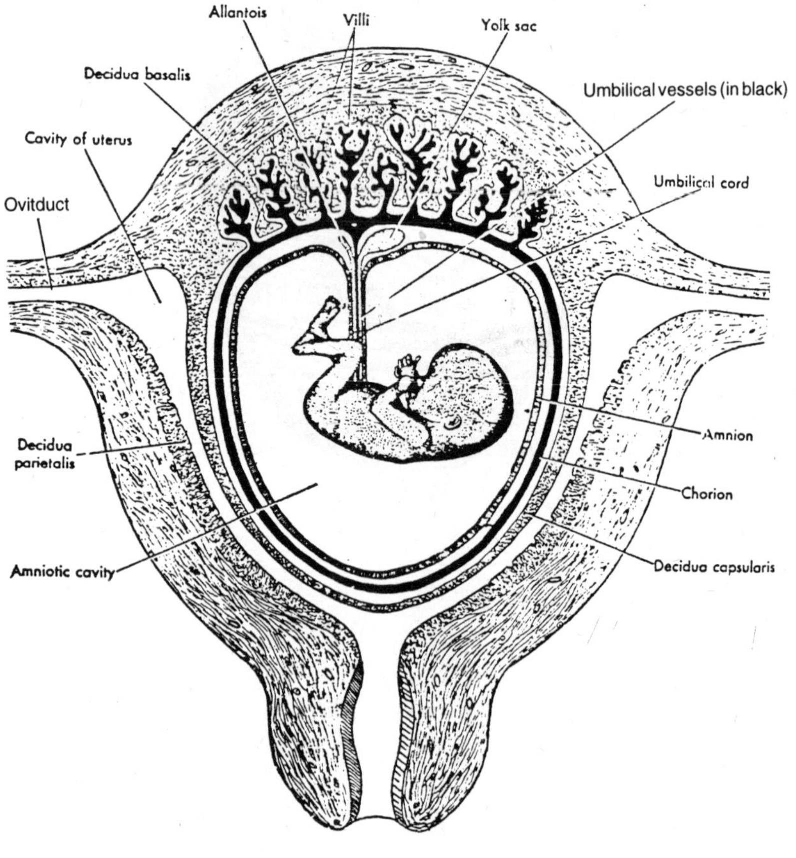

Fig. 5.15.   Sectional diagram of human uterus with fetal membranes and their relationship to the uterus and the embryo.

mass separates into the corresponding number of units. Since monovular twins have identical chromosomes and genes, they are identical in their physical and mental traits and characteristics. There is no closer human relationship than between identical twins. They are more closely related to each other than to their father and mother or any other member of the family.

A third and more common method by which triplets and other multiples are produced is by the combination of the two processes discussed above. Such multiples show this dual formation usually in the sex ratio, placenta, and in their general appearance.

Twinning is not quite as rare as it is generally assumed. One birth out of every eighty-five will have twins. Triplets occur in one out of 85 2(=7.225), quardruplets in one but out of 85 4 = 32,2000.625)> Mortality of infants of multiple births increases in direct ration of multiplicity. The higher the multiple, the less chance there is for the individual to survive. The survival of quintuplets is a biological phenomenon.

6

Ectodermal Derivatives

The Integumentary System

The contributions of ectoderm to the development of the teeth, tongue, palate, salivary glands, hypophysis and anal canal are described in earlier chapters. Here will be presented the histogenesis of the skin and the development of its specialized derivatives, all of which make up what is known as *integument*.

The Skin

The skin is an organ of double origin. Its superficial component is a stratified epithelium, called the *epidermis*, that specializes from the general ectoderm not involved in making the nervous system. The epidermis lies upon a fibrous *corium* of mesodermal origin. Beneath these two layers of the skin is the loose, fatty *subcutaneous layer*.

The Epidermis

The embryonic ectoderm is originally a single sheet of cuboidal cells but in the fifth week it begins to add a second layer. The outer cells make up a distinct, transient layer named the *periderm*. Its cells flatten and later spread to several times the diameter of the deeper cells. The basal cells, cuboidal in shape, are the reproducting elements that presently give rise to new layers above them. During the third and fourth months the epidermis is typically three-layered, an intermediate stratum being gradually interposed between the basal and periderm cells.

After the fourth month the epidermis becomes highly stratified and specialized. The lower layers consist of living cells, whereas the upper layers constitute 'dead skin'. The deepest stratum (basal cells), and its immediate descendants in the layers next above (pickle cells) constitute the definitive *stratum germinativum*. It contains the actively dividing cells of the epidermis. Daughter calls of this layer are crowded upward by still newer and ones and eventually reach the free surface. As a cell attains higher and still higher levels it undergoes progressive changes, culminating in cornification. Thus, directly above the germinative cells is the thin *stratum granulosum*, containing keratohyalin granules. Next higher lies the thin and clear *stratum lucidum* whose content is a fluid eleidin, supposed to represent softened and fused keratohyalin granules. Still nearer the surface, the epidermal cells flatten steadily and comprise the many-layered *stratum corneum*. The peripheral cytoplasm of cells in this layer becomes cornified in a way not well-understood, an the epidermis thereby loses its primitive transparency. More centrally in the cytoplasm of these cells a wax-like substance collects that is considered to be transformed eleidin (para-eleidin).

It is important to understand that only in the thickened epidermis of the palm and sole are all the layers, just mentioned, distinguishable; over the general body surface the granular and lucid strata are not clearly represented. In a few regions, like the red margin of the lip and the anus, cornification is slight. Pigment granules appear soon after birth in the cells of the stratum germinativum; these granules are obtained from *melanoblasts* that migrate from the primitive neural-crest tissue and specialize in pigment formation. Negro infants are quite light in colour at birth, but begin to darken within a few days; at six weeks their integument approaches the final degree of pigmentation.

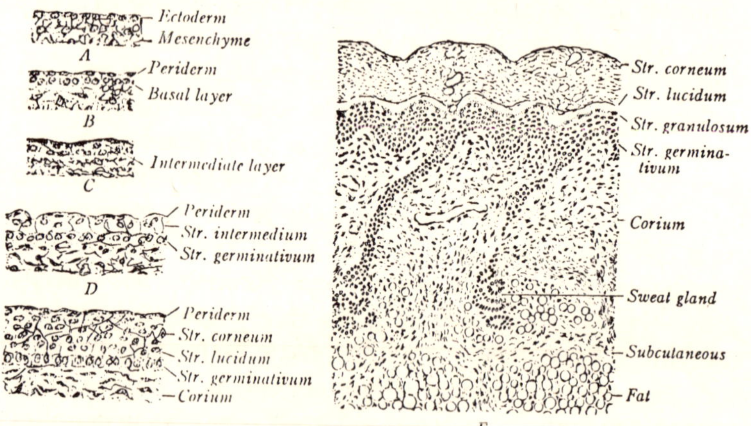

Fig. 6.1.   Development of human skin, shown in vertical sections. A, at 4 mm.; B, at 12 mm.; C, at two months; D, at three months; E, at five months; F, at birth. Epidermis of the palm and sole (F) is more highly differentiated than that of the body surface in general.

When the hairs emerge, at about the sixth fetal months, they do not penetrate the toughened periderm of the epidermis but loosen or break it. Hence in mammals this layer, which is characteristic of land vertebrates, is known also by another name, the *epitrichium* (i.e., upon the hair). Desquamated epitrichial and epidermal cells mingle with cast-off lanugo hairs and sebaceous secretions to form the pasty *vernix caseosa* which smears the fetal skin. This material is alleged to protect the epidermis against a macerating influence which otherwise would be exerted by the amniotic fluid. It also prevents chafing-injuries from the amnion as the growing fetus becomes progressively confined in its fluid-filled sac.

The plane of union between epidermis and corium is smooth until early in the fourth month when epidermal thickenings grow down into the corium of the palm and sole. About two months later corresponding elevations first appear on the skin surface. These epidermal ridges complete their permanent, individual patterns in the second half of fetal life.

The Derma or Corium

The fibrous layer of the integument is customarily traced to cells proliferated from the lateral walls of the paired somites. In consequence, this region of a somite has received the name *dermatome*, cutis plate. Evidence in support of this claim is not plain in mammals and it has been urge that the so-called dermatome really belongs to the myotome. In any event, the dermatome would only supply connective tissue in the vicinity of the somites. Much of the corium must differentiate from non-specific mesenchyme subjacent to the epidermis, most of which comes from the lateral sheets of somatic mesoderm.3

Collagenoous fibers appear in the third month and *elastic fibers* in the sixth month; their manner of differentiation has been described previously. Only gradually does a distinction between the compact corium proper and the looser, subcutaneous tissue become recognizable. Columnar papillae project upward from the corium into the germinative stratum; the dermal papillae are of two kinds, depending on whether they contain blood vessels or nerve endings. Some of the corial cells acquire pigment granules. In the lower sacral region deep-lying pigment tends to give local areas a bluish to brownish colour. They occur more commonly in infants and young children of the dark-skinned races and are known 'Mongolian spots'. Fat develops in the *subcutaneous layer*, but does not become abundant until the later months of fetal life.

The skin is innervated by segmental spinal nerves that supply, successive zones of the integument known as *dermatomic area*. The implication, however, that all of the corium derives from portions of paired somites and that each girdling dermatome carries with it a sensory nerve, just as myotomes maintain their original motor innervation, is not at all secure. It will be noted in that the radial (outer) surface of the arm and tibial (inner) surface of the leg receive nerves from higher levels that do the respective inner and outer surfaces. This is in agreement with the primitive cranial and caudal surfaces of the limbs before rotation occurred in oppositive directions.

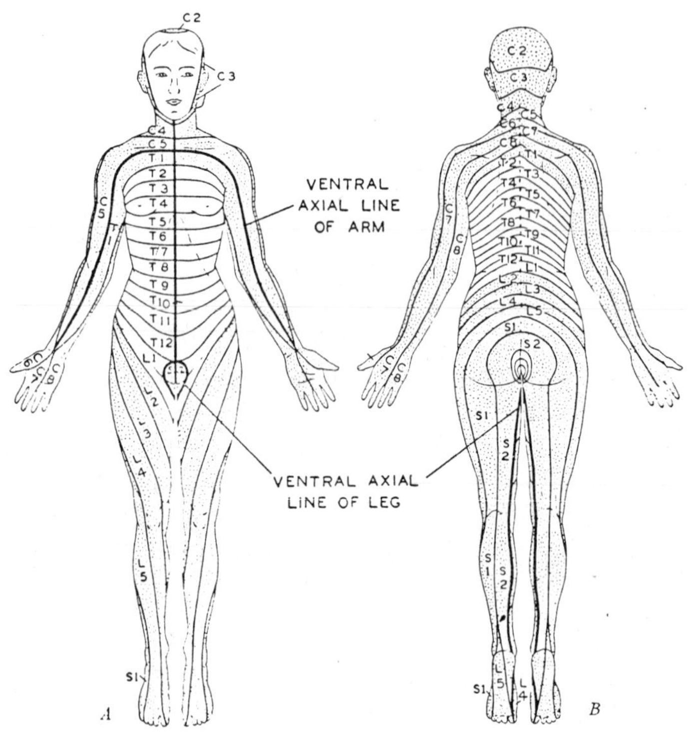

Fig. 6.2.   Distribution of the segmental apinal nerves to 'dermatomic' cutaneous areas.

Causal Relations. Presumptive epidermis is for a time a highly plastic tissue. When pieces from an amphibian gastrula are transplanted into the flank of a neurula, it will develop into various things: brain or cord; sense organs; ganglia; branchial cartilages; notochord; myotomes; pronephros. The ability of neural-crest cells to induce the epidermis to differentiate into cartilage has been mentioned previously. After determination occurs, skin can self-differntiate in culture media.

Anomalies. Skin may fail to complete normal differentiation, thereby retaining its fetal characters. Rarely it is astonishingly elastic. The deposition of pigment in the epidermis and elsewhere sometimes fails (albinism) or is over-abundant (melanism). Such atypical pigmentation may affect local areas only. Naevus is a name given either to a pigmented spot ('mole') or to a red purple discolouration caused by a cavernous, vascular plexus in the corium ('birthmark'). Ichthyosis designates a rough, scaly skin due to abnormal cornification of the superficial layer in extreme cases the epidermis shows thick plates, separated by cracks ('alligator skin'; B).

The Nails

Nails are modifications of the epidermis that correspond to the claws and hoofs of lower mammals. The first indication of a nail is foreshadowed at ten weeks by a thickened area of

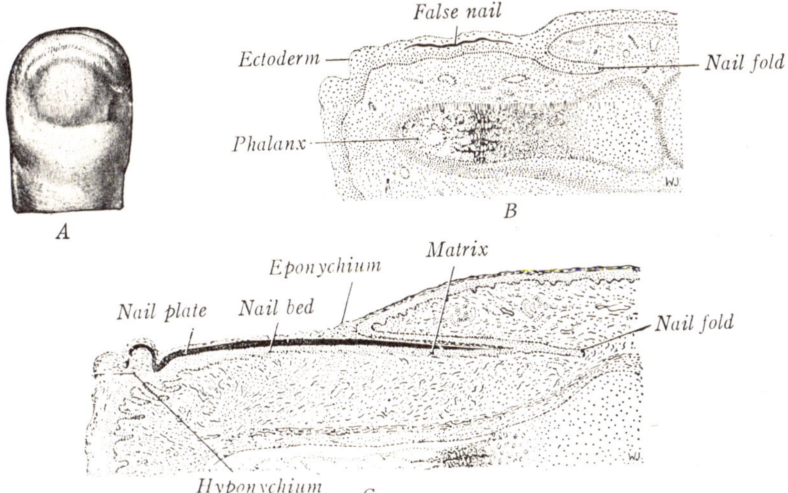

Fig. 6.3. Development of the human nail. A, Dorsum of finger, at ten weeks. B, Longitudinal section, at fourteen C, Longitudinal section, at birth.

epidermis (*nail field*) on the dorsum of each digit (Fig. 399 A). The adjoining territory, on each side and at the base, tends to overgrow this field, thereby giving rise to shallow *laternal nail folds* which continue into a much deeper *preoximal nail fold* that extends nearly to the proximal end of the terminal phalanx (B).

Although the primitive nail field undergoes some local cornification ('false nail'; the material of the true nail is developed within the under layer of the proximal nail fold. This layer is accordingly named the *matrix*. During the fifth month specialized keratin fibrils differentiate in the matrix layer, without having passed through a keratohyalin or eleidin stage as in the ordinary method of cornification. The keratinized cells flatten and consolidate into the compact tissue of which the *nail plate* is composed. In this manner the nail substance differentiates in the proximal nail fold as far distad as the outer edge of *lunula*, which is the whitish crescent at the base of the exposed nail. Beyond the lunula, the nail plate merely shifts progressively over the *nail bed* and reaches the tip of the finger one month before birth. As might be expected, the nails of the toes are begun and completed slightly later than thee finger nails. The corium beneath the nail is thrown into parallel longitudinal folds which are said to produce the characteristic ridging and grooves.

The stratum corneum and periderm of the epidermis for a time cover completely the free nail and are jointly termed the *eponychium* (i.e., upon the nail. In late fetuses this layer is lost, except for horny portions that continue to adhere to the nail plate along the curved rim of the nail fold. Underneath the free end of the nail the epidermal cells also accumulate to constitute a piled-up epidermal mass known as the *hyponychium*, or substance beneath the nail this region is much more important in a claw, and still more so in a hoof where it forms the 'sole'. The opacity of the lunula has been interpreted variously.

Anomalies. Misshapen nails occur and absence of nails

(*anonychia*) may accompany other failures or defects among epidermal derivatives.

The Hair

Hairs are specialized epidermal threads that rest upon a sunken papilla of the corium. They are produced only by mammals and are a distinctive characteristic of that vertebrate group. The comparative hairlessness of man today is a feature acquired within relatively recent times. Since it is similar to the condition found in late fetuses or anthropoid apes, this reduced hairiness is regarded as an example of arrested development. The primary insulating that heat-conserving function of hairs is compensated for in man by a heavy deposit of fat beneath the skin. Hairs of a fetus tend to be grouped in threes or fives, with the central one larger, and also to be arranged in lines. These relations are interpreted as the survival of a primitive mammalian condition in which the hairs stood in definite relations to scales which covered the skin, after the manner still seen in certain living forms.

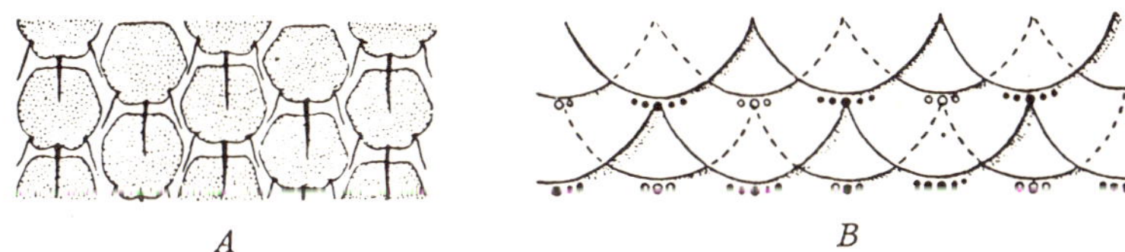

A   *B*

Fig. 6.4. The relation of hair groups to scales. *A*, Arrangement on the oppossum's tail, *B*, Arrangement in the human fetus, with hypothetical dermal scales drawn in.

Hairs begin to develop early in the third month on the eyebrows, upper lip and chin; those of the general integument make a start one month later. The first evidence of a future hair is the crowding and elongation of a cluster of germinative cells in the epidermis. Their bases sink root-like into the corium, and active proliferation soon produces a cylindrical, epithelial peg. At this stage the hair follicle consists of an outer wall of columnar cells, continuous with the basal layer of the epidermis, and an internal mass of polyhedral cells. About the whole is a mesenchymal investment (the later *connective-tissue sheath*), and at the clubbed base the mesenchyme condenses into a mound-like *papilla*.

As development proceeds and the hair peg pushes deeper into the conum, its base enlarges into the *bulb* which becomes molded like an inverted cup over the papilla. The actual hair substance is a proliferation from the basal epidermal cells lying next the papilla. These cells give rise to an axial core, destined to become the *inner epithelial sheath* and *shaft*; the latter grows progressively upward toward the surface. Quite distinct are the peripheral cells on the sides of the original downgrowth, which comprise the *outer epithelial sheath*.

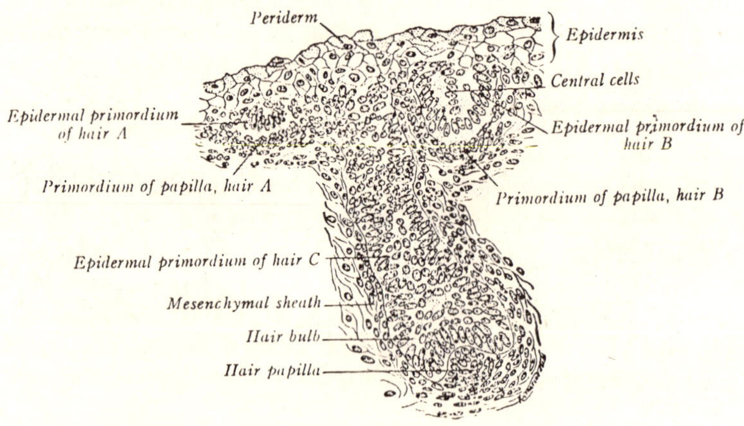

Fig. 6.5.   Human hair follicles, at three months, shown in longitudinal section.

The young 'hair cone' grows by the steady addition of new cells in the bulk. In this manner it is pushed up through the central cells of the solid, primordial follicle and is molded into shape by the organizing inner sheath. The shaft finally reaches the epidermis, follows along a *hair canal* in it, and erupts at the surface. Above the level of the bulb, the cells of the hair shaft cornify and differentiate into an outer *cuticle*, middle *cortex*, and central (inconstant) *medulla*. Two swellings of the outer epidermal sheath appear on the lower side of the obliquely directed follicle. The upper of these becomes the *sebaceous gland*, which will remain permanently associated with the hair; the deeper swelling is the *epithelial bed*, a region of rapid mitosis that also contributes to the growth of periodically regeneration hair follicle. Mesenchymal tissue near the epithelial bed transforms into the smooth fibers of the *arrector pili* muscle, which attaches to the side of follicle. Pigment granules appear early in the basal cells of the beginning hair; they are acquired from wandering melanoblasts that originate in the neural crest. Such cells are carried upward along with other hair cells and cause the characteristic colouration.

The first generation of fetal hairs is a downy coat termed *lanugo*. It constitutes a dense covering to the body, especially on the back and limbs, prominent by the fifth month. Lanugo hairs are short-lived, all being cast off either before birth or soon afterward. The replacing hairs develop, at least in part, from new follicles; they are likewise fine and constitute the so-called *vellus* (i.e., fleece) of the prepuberal years. Thereafter hair is shed and formed a new periodically throughout life. At the termination of any growth cycle the hair is carried upward by its shortening, regressive follicle. After a time the follicle reorganizes and begins to elaborate a new hair in the same manner as already described.

Some hairs remain permanently of the vellus type. In the female such occur on the face, neck and trunk; in the male they remain on the face (except beard), the flexor surface of the upper arms and various regions of the trunk. The replacing hairs of the brows, eyelashes and scalp of children are progressively larger and coarser than the first set. Under the influence of hormones, and especially those of the gonads, coarser and darker hairs appear at puberty on the pubis and axillae of both sexes and on the face and trunk of the male.

The hair coat shows definite, directional patterns (streams; whorls). These are established by similar angular slants of the hair follicles at their first development in any local region.

Casual Relations. Mouse ectoderm of the tenth day, grown in culture medium, develops into epidermis and normal hairs. Probably, as with feathers, it is the presence of the derman component of a papilla that induces the differentiation of the epithelial component. Ectoderm and hair follicles have no inherent power to form pigment. Only those cultures containing presumptive neural crest can differentiate melanoblasts — and hence pigment.

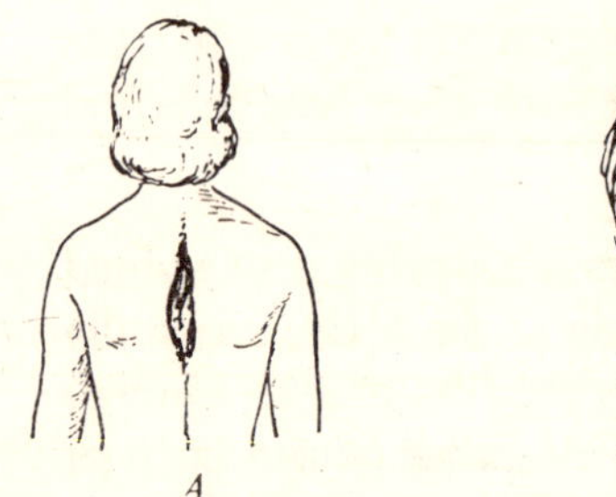

Fig. 6.6.   Hypertrichosis, or excessive hairness. A, Local tuft; B, local area, normally 'downy'; C, general hairness.

Anomalies. Hypertrichosis refers to excessive hairness which be localized or general, as in exhibited 'hairy monsters' (C). It is undecided whether this is due to an augmented development of the later hair follicles or to a persistent overgrowth of the lanugo set. In the rare *hypotrichosis* the congenital deficiency of hair may be complete (*atrichia*); the latter is usually associated with defective teeth and nails.

Sebaceous Glands

Most of the sebaceous glands accompany hairs. However, some independent ones, such as those on the genitalia, anus, nostrils and upper eyelids, develop from the general epidermis. Many of these do not organize until after birth, and some (e.g. upper lip) not until puberty.

Gland primordia appear first in the fifth month as swellings on the outer epithelial sheaths of the hair follicles. The swelling becomes a lobulated, flask-shaped sac whose lumen arises by the fatty degeneration of the central cells. The resultant oily secretion is an important constituent of vernix caseosa it is usually credited with helping of preserve the fetal skin from maceration. Cells in the neck of a sac are the reproducting elements. Throughout life they supply new cells, which are forced centrad and disintegrate in the process of oil elaboration. Such a gland is *holocrine* (i.e., the secretion consists of altered, disintegrated gland cells themselves).

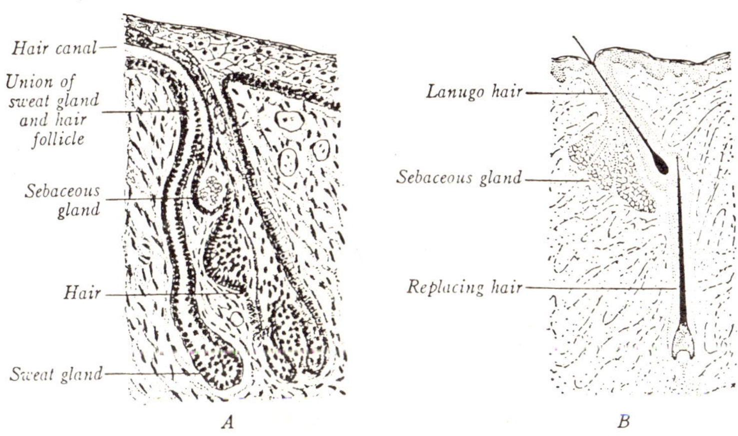

Fig. 6.7. Human cutaneous glands. A, Early stage in the development of a sebaceous gland and an apocrine sweat gland, both in association with a hair follicle. B, Later sebaceous gland, connected to the follicle of a lanugo hair which is being replaced by a coarser, vellus hair.

Anomalies. Skin glands (sebaceous or sweat glands) may be overdeveloped in local regions. Contrariwise they may be underdeveloped and even lacking, either regionally or generally.

Sweat Glands

Sudoriferous glands first begin to develop at four months

from the deep epidermal ridges of the finger tips, palms of the hands, and soles of the feet. They are formed as solid, cylindrical ingrowths, but differ from hair primordia in being more compact and in lacking mesenchymal papillae at their bases. During the sixth month the simple cords coil, and at seven months internal clefts that arose earlier by hollowing now unite into a continuous lumen. An inner layer of cells about the lumen constitutes the gland cells. By contrast, the more peripheral cells transform into flattened elements that are usually considered to be smooth muscle fibers; this interpretation is of special interest since such muscular elements would then be ectodermal. The duct portion of the gland at first ends blindly at the epidermis, but later, as cells are replaced during the course of growth in the stratified epithelium, a canal is left which continues the duct lumen to the surface. Glands appear to be capable of functioning at seven months, but probably are not needed until after birth.

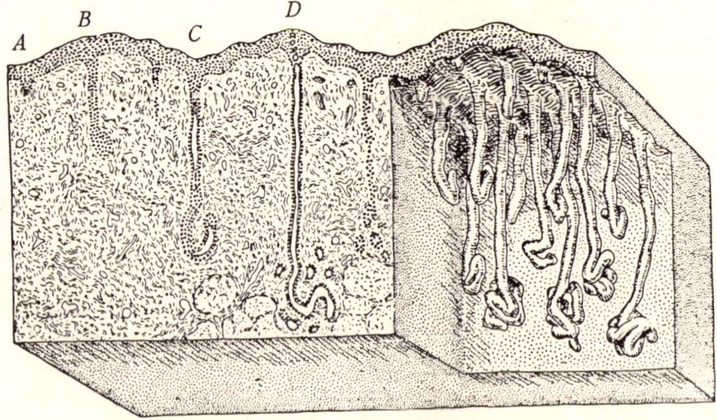

Fig. 6.8.   The development of human sweat glands, shown in a model of the skin. Left half, gland stages (A-D; four to seven months) in longitudinal section. Right half, epidermis and glands at similar stages isolated from the corium.

In certain regions of the body supplied with coarse hairs (pubis; axilla; areola; eyelids) there are large, specialized sweat glands. Many of these develop on the sides of the hair follicles and move upward until they acquire separate openings on the epidermis. An association with hair follicles is characteristic for sweat glands in general in most mammals, but this is not true of the ordinary kind in man. Human glands of the axillary type are *apocrine* (i.e., the top of their secretory cells break away along with the secretion). They have a thicker secretion and exhibit cyclic activity.

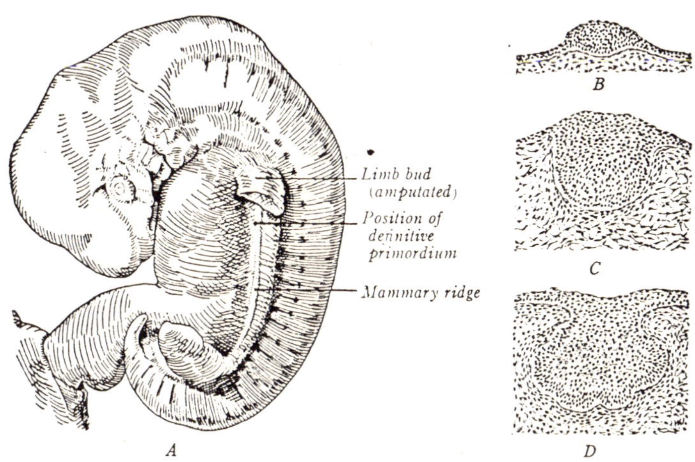

Fig. 6.9.   Early development of the human mammary gland. A, Unusually prominent mammary ridge, at 13 mm.. *B-D*, Vertical sections of gland primordia, at six weeks, nine weeks and four months, respectively.

Mammary Glands

Mammary glands are peculiar to mammals. It is remarkable that they appear so early in development, not only since they are of use to adults alone but also because they are a late acquisition among vertebrate organs. Early in the sixth week of human development an ectodermal thickening extends on each side as a longitudinal band between the bases of the limb buds. At about 9 mm. it makes a distinct linear elevation that has been called the *mammary ridge*, or *milk line*. In man this usually is inconspicuous except in the pectoral region, and in any event all but the cranial third normally vanishes quickly. By contrast, lower mammals with serially repeated glands, like the hog, have a prominent milk line extending from axilla to groin.

Each human mammary gland begins as a localized thickening on the corresponding epidermal milk line in the region of the future breast. At first lens-shaped the primordium gradually becomes globular and then bulbous and lobed. During the fifth month from 15 to 20 solid cords begin to bud inward, pushing aside the corial connective tissue as they advance. These primary *milk ducts* continue to grow and branch throughout fetal life. The duct system also slowly acquires *lumina* by hollowing. The originally elevate free surface of the primordium flattens and cornifies; hollowing produces a pit into which the ducts open. About the time of birth, or even considerably later, this sunken area elevates into the *nipple*. The *areola* is first recognizable as a circular area, free of fair primordia but acquiring branched *areolar glands* (of Montgomery) in the fifth month.

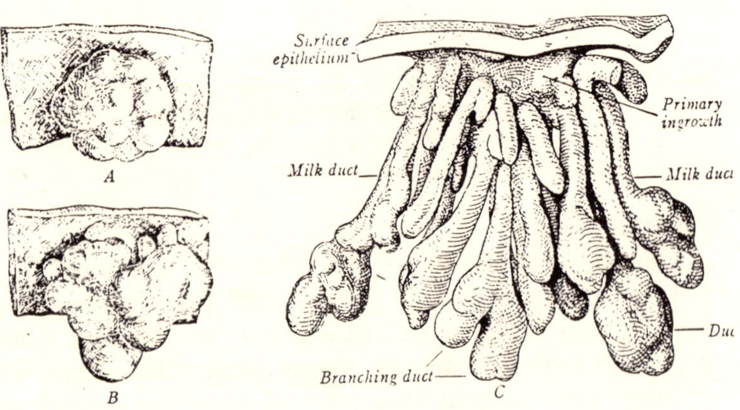

Fig. 6.10.   Intermediate development of the human mammary gland, shown by models. *A*, At fourteen weeks *B*, at five months *C*, at six months.

In the female the areolar region becomes elevated before puberty, whereupon this stage is followed by a rapid enlargement (through fat deposition about the growing ducts)

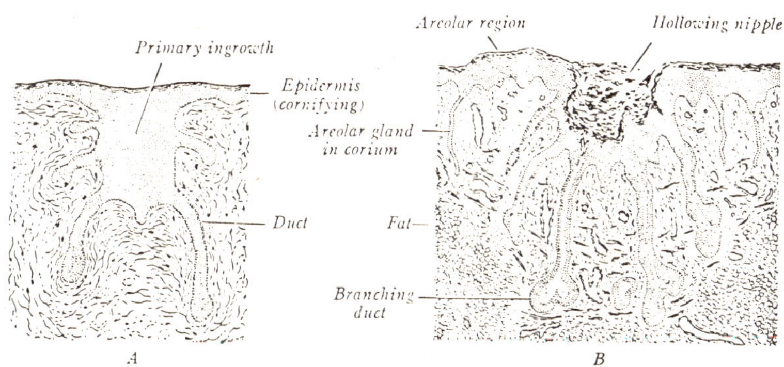

Fig. 6.11.   Later prenatal development of the mammary gland, in vertical section. *A*, At five months *B*, at eight months.

until the breast is a hemisphere bearing the areola and nipple at its apex. During the puberal growth-period the ducts continue to branch and it is then that *acini* first become definitely recognizable as swollen end-pieces. The mammae are further augmented during pregnancy, the system of epithelial ducts and acini advancing greatly both in bulk and structural differentiation. Two or three days after parturition the glands become functionally active. This culmination of development is a response to hormonal stimulation. In this process, the ovarian secretions excite the preliminary changes, whereas the anterior lobe of the hypophysis is the final activator responsible for actual lactation. The mammary glands of the newborn of both sexes also yield to little secretion ('witch milk') within a few days after birth. Their activity at this time depends upon changing hormonal relations similar to those that bring about lactation in the mother. The early mother's milk is a thin fluid called *colostrum*.

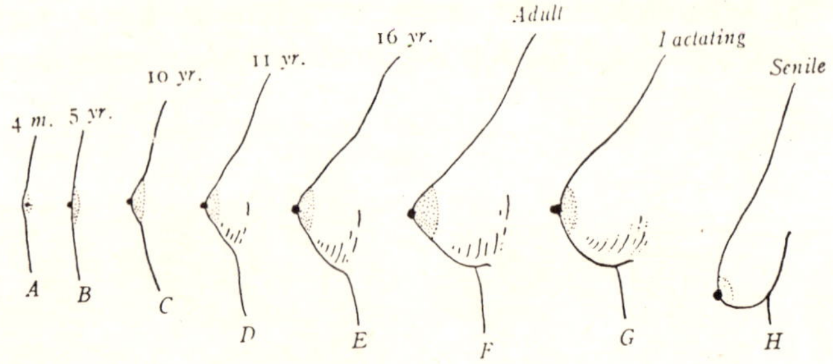

Fig. 6.12.   Profiles of the female breast throught posnatal life.

The gland of both sexes show equal development throughout fetal life and until late childhood. The male gland reaches its full development at twenty years; at this time is resembles the female gland at an early stage of puberty. After the menopause the mammary glands of the female undergo a regression that parallels a similar decline of the internal genitalia. All were built up and maintained by ovarian hormones; when the ovary declines, they correspondingly suffer regression.

The mammary glands are regarded by most authorities as modified sweat glands of the apocrine type. The homology is made because their development is similar and because in the lowest mammals their structure is the same. Moreover, rudimentary mammary glands (the areolar glands), which also resemble sweat glands, occur about the nipple. In many mammals several pairs of mammary glands are developed along the milk line (hog; dog); in some a single pair occupies the pectoral region (primates; elephant) or even axilla (fruit bat; flying lemur); in others they are confined to the inguinal region (sheep; cow; horse) or even occur near the genitalia (some whales). The human gland on each side develops from one of several local thickenings along the ridge; this multiple appearance of potential mammary sites possibly represents an expression of atavism.

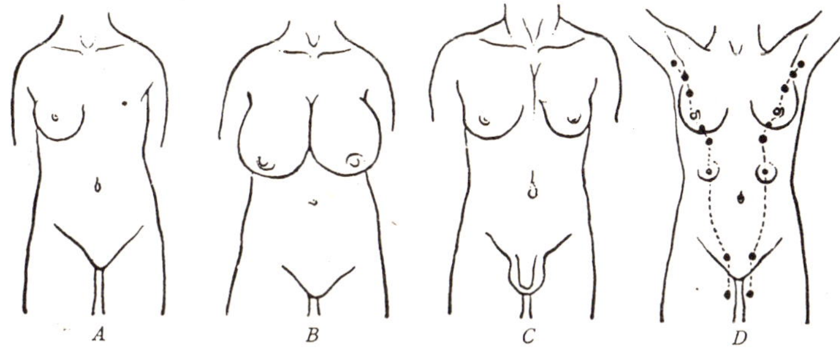

Fig. 6.13. Anomalies of the breasts. *A*, Amastia (except for a nipple rudiment); *B*, macromastia; *C*, gynecomastia; *D*, accessory pair of glands on abdomen. The commoner sites of accessory glands (or nipples) are indicated in *D* by dots along the course of the former milk lines.

Anomalies. Absence of one or both mammary glands (*amastia*), retention of the prepuberal condition (*micromastia*), and the attainment of abnormal size (*macromastia*) are all well-known. In some instances the male develops a breast most or less of the female type (*gynecomastia*. This condition is dependent on a disturbed androgen-estrogen balance and is frequently present in hermaphrodites. Two examples of actual milk secretion by an adult male have been recorded. Supernumerary mammary glands (*hypermastia*) are quite rare, but accessory nipples (*hyperthelia*) are fairly common in both sexes. It is aid that at least 1 per cent of large populations may show trace of them. They occur chiefly between the axilla and groin, but have been recorded on the arm, back and thigh. Such structures represent independent differentiations, usually along the primitive milk line, such as occur normally in lower animals.

FORMATION OF NERVOUS SYSTEM

The Neural Tube and Neural Crest

Chorda-mesoderm, represented by the notochorda plate

and paraxial mesoderm, exerts neutralizing and mesodermalizing effect on the overlying ectoderm and induces the formation of *neural plate* and *neural crest*. The epithelium of the neural plate is made up of mitotically active cells, so-called *ventricular cells*, which are characterized by abundant free polysomes and a poorly developed cytomembrane system. By the *neural tube* stage, the epithelium has become pseudo-stratified and has three nuclear zones. Next to the lumen and separated from it only by an *internal limiting membrane* is a *mitotic zone* containing the nuclei of rounded-up dividing cells. After division, the daughter cells extend their cytoplasm peripherally and their nuclei move away from the internal limiting membrane and enter a *resting zone*. As a preliminary to the next division, the nuclei move further from the internal limiting membrane and enter an outer zone where they synthesize DNA. Meanwhile, the cytoplasmic processes of the ventricular cells have extended beyond this *synthetic zone* and formed a nuclei-sparse *marginal layer* bounded by an *external limiting membrane*. The nuclei with replicated DNA, however, re-enter and pass through the resting zone, approaching the internal limiting membrane before they divide.

These undulating movements of nuclei within the *ventricular layer* are commonly referred to as *elevator movements or interkinetic migration*. Some of the progeny of dividing cells leave the ventricular layer without synthesizing DNA and become *neuroblasts* as they begin to develop the rough polysome-bearing endoplasmic reticulum which is characteristic of neurons. These neuroblasts accumulate in an *intermediate layer*, between the venetricular and marginal layers. Later, other daughter cells accumulate in a *subventricular layer* between the ventricular and intermediate layers. Some of the subventricular cells divide to form neuroblasts, others to form *glioblasts* which differentiate into the large-celled connective tissue of the central nervous system — the *macrolia* —*astrocytes* and oligodendrocytes. In contrast, the small-celled connective tissue—the *microglia*—is derived from mesenchyme, probably of neural crest origin.

The cells which remain in the ventricular layer gradually reduce their mitotic activity and differentiate into ependymal cells.

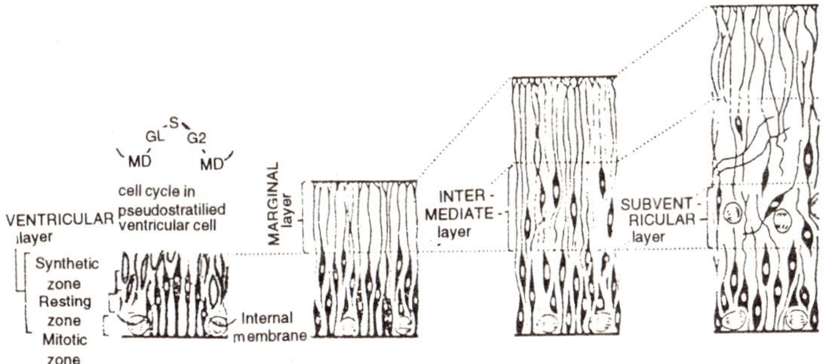

Fig. 6.14. Neural Tube Histogenesis

The derivatives of the intermediate and subventricular layers form most of the *grey matter* of the CNS. The *white matters* is formed from myelin-forming glioblasts associated with certain neurites (processes of neuroblasts) of the marginal and subventricular layers. The wall of the neural tube remains thin at the *roof* and *floor plates*. Laterally it thickens as the intermediate layer forms two distinct masses — a *dorsal (alar) lamina* and a *ventral (basal) lamina*. As they enlarge, the masses project into the central canal and a longitudinal groove   the *sulcus limitans* — appears between them. Within the laminae, neuroblasts become multipolar as they develop a number of neurites. Most become dendrites but one elongates to form an axon. The tip of the axon divides and makes cell-membrane contacts —*synapses* with other cells. Axons from alar neuroblasts remain within the nervous system and synapse with other neurons in both laminae. Axons from some basal neuroblasts are induced to grow beyond the confines of the neural tube, as the ventral roots of spinal nerves, to synapse with the subneural apparatus of muscle fibres or with developing autonomic neurons.

Neural Crest

In the process of neural tube formation the *neural ridges*

meet and form a midline wedge which roofs in and closes the tube. The remaining neural ridge tissue on each side forms a *neural crest* which is initially a continuous longitudinal column in the angle between neural tube and surface ectoderm. Migration of neural crest cells then occurs. Some cells destined to form *neural tissue* stop dorsolateral to the neural tube and form a series of primordia for the *spinal* and *cranial sensory ganglia*; others stop ventrolateral to the neural tube and form *sympathetic ganglia* and the *superarenal medulla*; yet others pass into head mesenchyme or into the mesentery with viscera, and in those sites form *parasympathetic ganglia* or network of *ganglionic neurons*. Whether a particular cell becomes adrenergic or cholinergic depends upon the microenvironment in which it finds itself.

In the *spinal ganglia*, neural crest cells form bipolar neuroblasts whose peripheral and central processes complete the dorsal roots of spinal nerves. The peripheral processes reach and run with the outgrowing ventral root fibres to form mixed spinal nerves. Thence, the peripheral processes extend out and from receptor endings in the various body tissues. The central processes enter the neural tube and travel in the marginal layer to synapse with neurons at different levels in both laminae. The central and peripheral processes of the primitive bipolar neuroblasts fuse near their cell body to form the unipolar neurons of spinal ganglia. Neurons of some of the cranial sensory ganglia are also derived from neural crest cells but the neurons of others are derived from placodes.

Neural crest tissue, which accompanies the migrating autonomic neuroblasts, differentiates into masses of sympathochromaffin cells which flank the abdominal aorta. Of these, the largest forms the suprarenal medulla while others persist for a while as *paraganglia* but many degenerate soon after birth. The *satallite cells* of ganglionic neurons (capsular cells) and of peripheral axonal processes (Schwann cells) are derived from the neural crest. When a Schwann cell is invaginated by a single axon of a certain diameter and with one of several types of peripheral termination, its plasma membrane spirals around the axon and forms an early segment of myelin sheath. (Myelin sheaths in the central nervous system form in a similar manner from the plasma membranes of processes of macroglial cells. Myelination begins near the cell body of a neuron and spreads peripherally. Growth in axonal diameter and myelin sheath thickness is generally completed within a system at about the same time as it becomes fully functional but growth in length continues until the innervated part ceases to grow.)

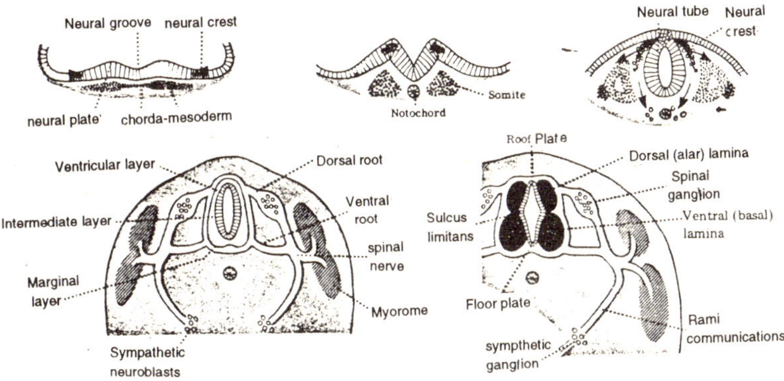

Fig. 6.15.   The Spinal Cord and Nerves

Neural crest tissue is also held to form the intracapsular cells of encapsulated sensory nerve endings, skin elements, including melanocytes and pigment cells, and the dental papillae and odontoblasts. It contributes to head mesenchyme in general but particularly to the facial prominences and pharyngeal arches. There it forms connective tissue elements as diverse as the skeleton, the stroma of endocrine and exocrine glands and the musculoadventitial walls of derivatives of the aortic arches. The calcitonin producing cells of the ultimobranchial body and type I and type II cells of the carotid body are also derived from the neural crest. The optic vesicles, nasal placodes and athke's pouch may represent it anterior extremity and give rise to neural crest cells during development. Finally, the neural crest gives rise, in part at least, to the leptomeninges.

The Meanings

Surrounding the developing neural tube is a loose tissue— the *primitive meninx*. In the spinal region it is derived mainly from the sclerotomes, but partly from neural crest. In the cranial region it is derived from mixed head mesenchyme. The innermost cells of the primitive meninx, probably of neural crest origin from the vascular pia mater. The outermost cells are initially indistinguishable from those of the blastemal vertebrae and membranous vault. Later, vascular channels appear and separate the spinal dura mater from perichondrium and, at the sites of the dura venous sinuses, the meningeal layer from the endosteal layer of the cerebral dura mater. The intermediate tissue cavitates and forms arachnoid trabeculae and the ligamentum denticulatum. In the *spinal region* the arachnoid mater is formed by delamination from the inner surface of thee dura mater. In the *cranial region* it is more closely associated with the pia mater. At three sizes the roof plate and overlaying pia mater of the hindbrain break down, thus establishing communications between the cavities of the brain and subarachnoid space; cerebrospinal fluid can then circulate. At first the spinal cord is co-extensive with the vertebral canal but in later antenatal and postnatal development, due to differential growth rates and caudal degeneration, it retreats up the canal. Cavitation of the intermediate tissue leaves an extensive subarachnoid space in the wake of the retreating cord.

The Spinal Cord

The early CNS—*the neuraxis*—is co-extensive with the notochord and, from within out, consists of ventricular, subventricular, intermediate and marginal layers. At this stage and throughout the length of the neuraxis, the cells bodies — *somata* — of neuroblasts are deeply placed in the subventricular and intermediate layers while their processes—*neurites*—are superficially placed in the marginal layer. The somata of neuroblasts are arranged in alar and basal laminae separated by a sulcus limitans. While it is modified in the subsequent development of the brain, this generalized arrangement persists in the spinal cord so that *spinal grey matter* (neuronal somata, dendrites, axon terminals, synapses and glia) is confined to the subependymal position and *spinal white matter* (axon bundles, myelin sheaths and associated glia) is subpial in position.

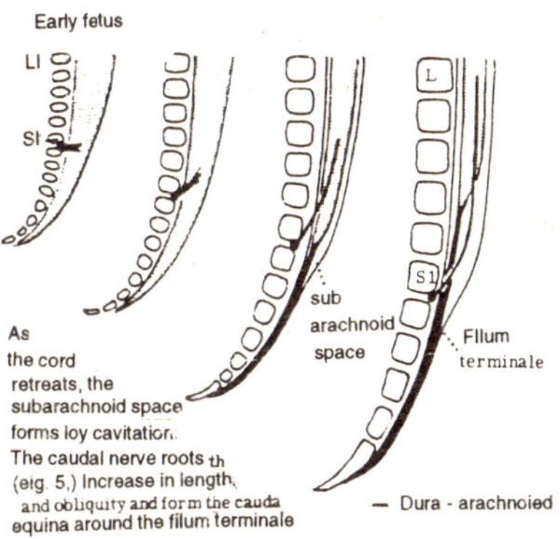

Fig. 6.16. The Caudal Subarachnoid Space

On each side of the spinal cord four longitudinal, initially cylinderical, columns of neuroblasts are formed — two derived from the alar lamina and two from the basal lamina. With continued differentiation, growth and maturation, their cylinderical form is lost but their main derivatives can be recognized in the mature cord. Each of these columns is given a name referring to the functional attributes of neurons derived from its neuroblasts. It should be noted that these neurons are usually in the minority (the majority forming interneurons of great variety), that the attribute may not be exclusive to the column and that in some situations alar neuroblasts migrate into basal laminae. Nevertheless, with these reservations, the classification is useful.

Spinal Alar Lamina

Each spinal alar lamina divides into a somatic afferent

column adjoining the roof plate and a visceral fferent column adjoining the sulcus limitans. These columns are called 'afferent' because some of their neurons receive the synaptic terminals of some of the fibres (central processes) of primary sensory neurons which enter the CNS. Other primary afferent fibres project beyond the alar and reach the basal lamina while yet others ascend in the dorsal part of the marginal zone, providing collateral branches to the spinal grey matter as they ascend to their final destination—the gracile and cuneate nuclei in the hindbrain.

Somatic Afferent Column

The dorsal alar neuroblasts which form a somatic afferent column of interneurons gradually mature into most of the dorsal column (or 'horn') of the adult spinal grey matter. These neuroblasts receive the synaptic terminals of sensory neurons which convey information from somatopleuric structures (exteroceptive information from skin, proprioceptive information from striated muscle, tendon, joint capsule and fascia, and nociceptive information from all these tissues).

Visceral Afferent Column

The ventral neuroblasts of the alar lamina form a smaller visceral afferent column of interneurons. These neuroblasts receive the terminals of sensory neurons which convey interoceptive, including nociceptive, information from splanchnopleuric (visceral) structures and from the blood vessels and glands of the somatopleure. This column contributes to the intermediate grey matter of the mature cord.

The axons of some alar neuroblasts are involved in intra- or intersegmental local reflex loops, others invade the marginal layer and form intersegmental fibres and polysynaptic tracts to higher centres, while other long ascending or descending fibres from tracts in the developing nervous system. Many of these fibres acquire myelin sheaths so that the marginal layer is transformed into while matter. Some cross the midline in the floor plate and form the ventral white commissure of the cord. Some grow into the basal lamina.

Spinal Basal Lamina

Each spinal basal lamina, like an alar lamina, differentiate into two columns. These are the somatic efferent column adjoining the floor plate and the visceral efferent column adjoining the sulcus limitans. These columns are called 'efferent' because they are the source of fibres which leave the CNS.

Somatic Efferent Column

The ventral cells of the basal lamina form a somatic efferent column characterized by motor neurons. The axons of their neuroblasts grow out into the ventral roots and innervate striated muscle of myotomic and somatopleuric origin. Some of these neuroblasts form large multipolar neurons with large diameter axons which supply extrafusal, somatic muscle fibres. Others form small multipolar neurons with axons of small diameter which supply the intrafusal muscle fibres or muscle spindles. The majority, however form interneurons including the inhibitory neurons of Renshaw. Collectively, the somatic efferent cells develop into the ventral grey column (or 'horn') of the adult cord. The large neurons become grouped into subcolumns, particularly in the limb enlargements : these are related to different groups of musculature.

Vesceral Efferent Column

The dorsal cells of the basal lamina form a visceral efferent column of interneurons. The axons of these neuroblasts grow out as preganglionic fibres which leave a ventral ramus in a white (myelinated) ramus communicans to reach a developing sympathetic ganglion, where they synapse with sympathetic neuroblasts either re-enter a spinal nerve in a grey (unmyelinated) ramus communicans, to be distributed to smooth muscle and secretory cells of somatopleuric origin, or pursue a direct (perivascular) course to smooth muscle and glands of splanchnopleuric origin. Axons pursue both of these courses to innervate adipose tissue. With the development of the limb enlargements of the spinal cord, the visceral efferent column becomes confined to the thoracolumbar sympathetic outflow (first thoracic to second lumbar spinal segments) and the sacral part (second to fourth segments) of the craniosacral parasympathetic outflow. Together the afferent and efferent visceral columns form much of the intermediate grey matter of the adult cord, the surface of which projects as a lateral 'horn' in the thoracic region of the cord.

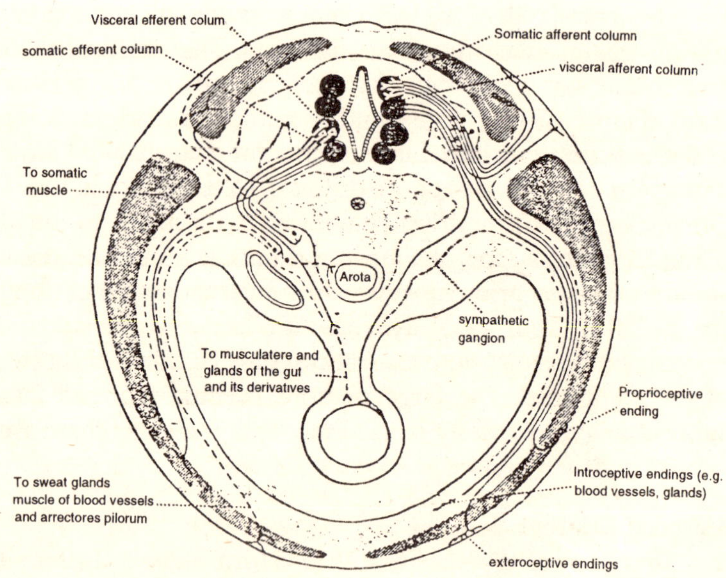

Fig. 6.17.   Efferent Fibres (primary sensory neurons ending on alar lamina cells.

The Cranial Neural Tube and Neural Crest

Whereas the generalized arrangement of the neuraxis persists in its caudal part which forms the spinal cord, it is modified in its cranial part which forms the brain. There, while some subventricular cells remain deeply placed (subependymal), others, mainly from the alar lamina, migrate through the intermediate layer into the marginal layer and thus become superficially placed. Those not migrating beyond the middle of the marginal layer differentiate into nuclei which lie between or are traversed by interweaving tracts of white matter. The vast majority, however, invade the roof plate of the brain, become virtually subpial in position and form a primitive *cortical plate*, a new zone of cell division and differentiation from which is derived superficial grey matter. Thus in the forebrain it forms the cerebral cortex, in the mid-brain the colliculi (corpora quadrigemina), and in the hindbrain the cerebeller cortex.

The primary sensory neurons of the cranium are derived from neural crest or placodal tissue. The cranial neural crest forms trigeminal, acousticofacial, glossopharyngeal, vagal and transient occipital primordial ganglia. It also forms the primordial cranial autonomic ganglia (parasympathetic neuroblast and supporting cells). It is probably also represented by Rathke's pouch, by the nasal placodes, by the optic vesicles and by a group of cells incorporated in the hindbrain which extend rostrally to form the mesencephalic nucleus of the trigeminal nerve. All the supporting cells of the definitive sensory ganglia are derived from neural crest, as are all the neurons of the superior ganglia of the glossopharyngeal and vagus nerves and some of the neurons of the trigeminal and geniculate gangila (general somatic afferent). Other neurons are derived from the epibranchial placodes (V, VII, IX and X) and the otic placode of each side. Thus neuroblasts in neural crest and placodal tissues; and in the olfactory placode which forms olfactory epithelium, differentiate into the primary sensory neurons of the cranium. Their central processes provide the input at the cranial level and discharge on interneurons. The olfactory bulbs and optic nerves are forebrain derivatives and thus part of the CNS.

The Brain Stem

The midbrain and those parts of the hindbrain called the pons and medulla oblongata together constitute the *brain stem,* a term which thus excludes the cerebellar part of the hindbrain.

The alar and basal laminae and sulcus limitans are continued throughout the brain stem. As in the spinal cord, the laminae differentiate into somatic and visceral columns. However, in addition to general columns, which are continuous and homologous with those of the spinal cord, special columns differentiate in association with the pharyngeal arch nerves and the vestibulocochlear nerve. These special columns are limited to the medulla oblongata and pons of the hindbrain.

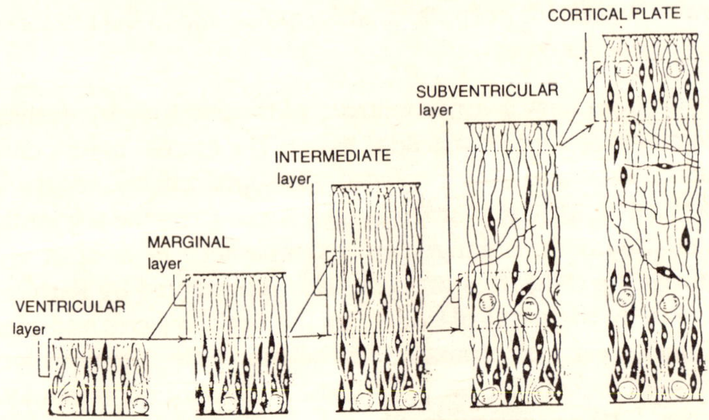

Fig. 6.18. Neurohistogenesis.

In the hindbrain, the roof plate is expanded and the laminae lie alongside each other in the floor of the fourth ventricle. Furthermore, cells migrate from the lateral part of each alar lamina, ventrally and rostrally as a bulbopontine extension. The derivatives of this extension are found ventrolateral to the basal lamina derivatives and comprise the olivary, reticular and arcuate nuclei in the medulla oblongata, and the scattered pontine nuclei in the ventral part of the pons. Finally, alar lamina cells migrate to form the cortex of the cerebellum and the nuclei in its roof. The remaining parts of the alar laminae and the basal laminae of the hindbrain form the nuclei in the floor of the fourth ventricle, including the dorsal part of the pons.

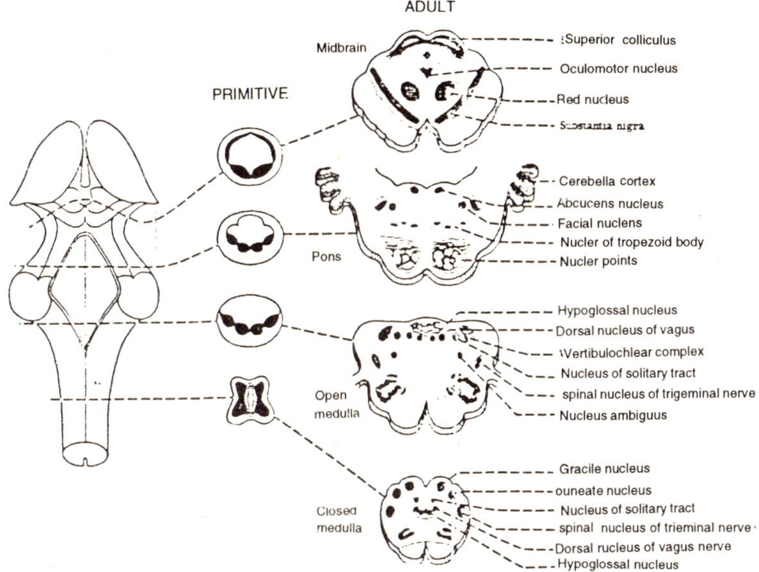

Fig. 6.19. Midbrain and Hindbrain Derivatives of Alar and Basal Laminae

In the midbrain the alar laminae form the nuclei of the dorsal tectum and invade the roof plate, while the basal laminae form some nuclei in the ventral tegmentum and invade the floor plate. In the gegmentum, however, the red nuclei, the substantia nigra and reticular nuclei are formed from alar lamina cells which have migrated ventrally.

On inspection, the sulcus limitans appears to continue into the forebrain as the hypothalamic sulcus. On this basis some consider that the hypothalamic nuclei are derived from basal lamina. However, with some neuroendocrine exception, hypothalamic neurons behave like alar lamina cells and their axons remain within the nervous system. Others thus consider it preferable to regard almost all forebrain grey matter as derived from alar lamina and the sulcus lamitans as ending with the midbrain.

THE BRAIN STEM

Cranial Alar Lamina

As in the spinal cord, the cells of the alar laminae of the brain stem form interneurons. Each alar lamina differentiate into four columns. General somatic and visceral afferent columns continue upward from the olumns of the spinal cord and subserve similar functions. Between these columns a *special visceral afferent* column differentiates in the centre of the alar lamina. Its neuroblasts receive the terminals of sensory neurons which convey information from the taste buds (pharyngeal endoderm). Between the general somatic afferent column and the roof plate a *special somatic afferent* column differentiates in the most lateral part of the alar lamina. Its neuroblasts receive the terminals of sensory neurons which convey information from the organs of equilibration and hearing.

Cranial Basal Lamina

Three columns differentiate in each cranial basal lamina in the brain stem. Somatic and general visceral eferent columns continue upwards from the columns of the spinal cord and subserve similar functions. Between these columns a *special visceral efferent (branchiomotor)* column differentiates in the centre of the basal lamina. The axons of its neuroblasts grow out of the brain and innervate the striated special visceral (branchial) muscles of the pharyngeal arches.

Columns in the Brain Stem

At first, the various cell columns are continuous throughout the brain stem. Later, they break up into serially arranged masses of cells (nuclei) associated with individual cranial nerves.

Somatic Efferent Column

The upward continuation of the somatic efferent columns of the spinal cord form the cranial somatic efferent columns on either side of the midline floor plate. The nuclei derived from these columns contain motor neurons whose axons grow out in nerves, homologous with ventral roots, to supply striated muscle derived from myotomes or their homologues. Axons from the hypoglossal nucleus grow out to occipital myotomes which form the musculature of the tongue. Axons from the abducent nucleus grow out to the ventral part of the maxillomandibular condensation of mesenchyme, which at this stage lies lateral to the pons. The ventral parts forms the lateral rectus muscle. At first the trochlear nucleus is also in the pons but later it migrates into the midbrains. Its axons grow out to the dorsal part of the maxillomandibular condensation of mesenchyme which forms the superior oblique muscle. The rostral part of the column forms most of the oculomotor complex in the midbrain. Its axons grow out to the premandibular condensation of mesenchyme which forms the other extrinsic eye muscles.

Special Visceral Efferent Column

The cells of the basal lamina between the somatic efferent and general visceral efferent columns form a special visceral efferent (*branchiomotor*) column. The nuclei derived from it contain motor neurons whose axons supply striated special visceral (pharyngeal arch) muscle via nerves V, VII, IX, X and XI. The motor nuclei of the last three nerves form a complex. The rostral part of the complex migrates laterally and forms the nucleus ambiguous. The caudal part retains its primitive position and forms the spinal nucleus of the accessory nerve. Axons from this complex grow out in the glossopharyngeal, vagus (cranial root of the accessory) and accessory (spinal root), nerves. They supply the third and subsequent pharyngeal arch mesenchyme which forms the musculature of the pharynx, larynx and cranial half of oesophagus, and the postpharyngeal mesenchyme which forms the sternomastoid and trapezius muscles. Axons from the facial nucleus grow out to supply the second pharyngeal arch mesenchyme which forms the muscles of facial expression. Subsequently, the facial nucleus migrates around the abducens nucleus into its definitive position so that the root of facial nerve is looped around the abducens nucleus. Axons from the motor nucleus of the trigeminal nerve grow out in the motor root and run with the mandibular division of the nerve to supply the first pharyngeal arch mesenchyme which forms the muscles of mastication. Palatine muscles' supply in mixed.

General Visceral Efferent Column

The upward continuation of the visceral efferent column of the spinal cord forms the cranial general visceral column which adjoins the sulcus limitans. The nuclei derived from the general visceral efferent column contain interneurons whose axons grow out as preganglionic fibres to synapse in parasympathetic ganglia, or mural visceral networks, with parasympathetic neurons of neural crest origin. The postganglionic fibres supply smooth muscle or cardiac muscle of grandular tissue. Axons from the large-celled part of the dorsal nucleus, of the vagus grow out in the vagas nerve and cranial root of the accessary nerve to synapse with parasympathetic neurons in or near the walls of the foregut, the midgut and their derivatives and the heart. Axons from the inferior salivatory nucleus grow out in the glossopharyngcal nerve to synapse with parasympathetic neurons in the otic ganglion which supply the parotid gland. Axons from the superior salivatory nucleus grow out in the focial nerve to synapse with parosympathetic neurons in the submandibular and the pterygopalatine gunglia which supply the submandibulr and sublingual and other oral glands and the nasal. Edinger-Westphal nucleus in the midbrain migrates to the head of the oculomotor complex. Its axons grow out in the oculomotor nerve to synapse with parasympathetic neurons in the ciliary ganglion which supply the smooth intrinsic muscles of the eye (iris and ciliary body).

General Visceral Afferent Column

The upward continuation of the visceral afferent column of the spinal cord forms the cranial general visceral afferent column which adjoins the sulcus limitans and is represented by the small-celled part of the dorsal nucleus of the vagus. Its neuroblasts receive the synaptic terminals of primary sensory neurons in the inferior ganglia of the glossopharyngeal and the vagus nerves which convey general interoceptive information from the foregut, mudgut and their derivatives, and the heart. The nucleus of the solitary tract is probably also involved.

Special Visceral Afferent Column

The cells of the alar lamina between the general visceral and general somatic afferent columns form a special visceral afferent column which is represented by the nucleus of the solitary tract. Its neuroblasts receive the terminals of sensory neurons in the geniculate ganglion of the facial nerve and the inferior ganglia of the glossopharyngeal and vagus nerves. They receive information from taste buds derived from the first, third and subsequent arch endoderm of the tongue, palate, oropharynx and epiglottis.

General Somatic Afferent Column

The upward continuation of the somatic afferent column of the spinal cord forms the cranial general somatic afferent column which is represented in part by the main sensory nucleus and the spinal nucleus of the trigeminal nerve. Their neuroblasts receive the terminals of sensory neurons in the trigeminal ganglion, the mesencephalic nucleus of the trigeminal nerve, the superior ganglia of the glossopharyngeal and vagus nerves. They receive proprioceptive information from the muscles of mastication(first arch), of the tongue (occipital myotomes), of the orbit (premandibular and maxillomandibular mesenchyme), and, possibly, of facial expression (second arch) via the mesencephalic nucleus. They receive exteroceptive information from the skin and ectodermal mucous membrane of the head, via the ophthalmic division (eye and frontonasal prominences) the maxillary division (maxillary prominence) and the mandibular division (mandibular prominence) of the trigeminal nerve. Additional exteroceptive information from the external ear comes via the diminutive auricular branch of the vagus and the communicating branch of the glossopharyngeal nerves. The large gracile and cuneate nuclei, which receive (amongst many others) the synaptic terminals of dorsal white column fibres conveying a massive, highly discriminative, exteroceptive and proprioceptive input from spinal levels must also be classified as general somatic afferent.

Special Somatic Afferent Column

The most laterally placed cells of the alar lamina form the special somatic afferent column which is represented by the vestibulocochlear complex. Its neuroblasts receive the terminals of bipolar sensory neurons in the vestibular ganglion and the spiral ganglion of the cochlea.

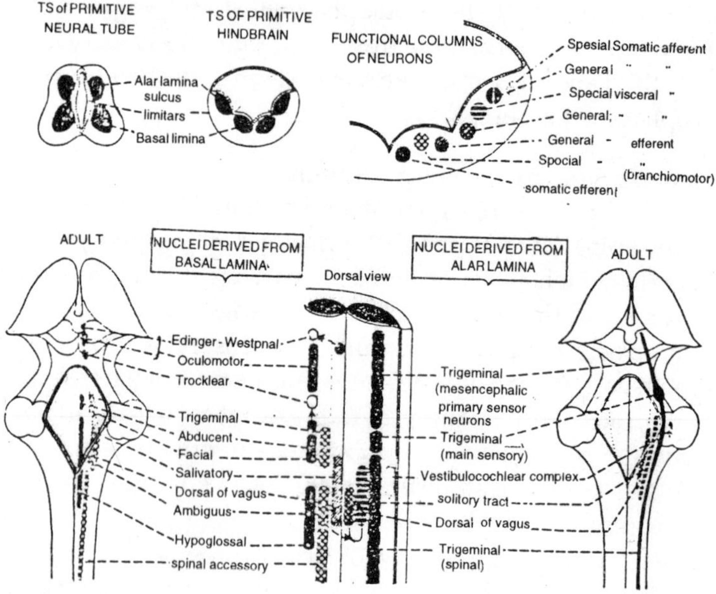

Fig. 6.20.   The Brain Stem.

The Brain

By the 7-somite stage, the neural groove has closed between the fourth and sixth somites. Cranial to the fourth somite, the expanded neural plate forms the fiddle-shaped primordial brain. The wide anterior segment is the future *prosencephalon* (forebrain). A depression near its lateral edge is the optic sulcus. The intermediate narrow segment is the future *mesencevhalon* (midbrain). The posterior wide segment, continuous with the neural tube, is the future hindbrain or *rhombencephalon* (named from the rhomboidal shape which its cavity — the fourth ventricle—assumes). By the 20-somite stage, closure of the neural groove has spread and the anterior neuropore has closed, completing the primary brain vesicles. The prosencephalon, mesencephalon and rhombencephalon are now tubular structures, each containing a dilatation of the central canal. On each side the cavity of the prosencephalon opens into an optic vesicle which has grown out from the optic sulcus. At this stage the neural tube conforms with the general 'comma' shape of the embryo and has a marked primary embryonic curve. This becomes compound when a ventral flexure (cervical) appears between the rhombencephalon and the spinal cord. Another ventral flexure appears in the mesencephalon, and a dorsal flexure (pontine) follows in the rhombencephalon. Later another dorsal flexure (telencephalic) appears in the prosencephalon.

Rhombencephalon

The hindbrain of a late somite embryo shows a series of transient surface elevations (the neuromeres) which contain its differentiating efferent nuclei (IV, V, VI, VII, IX and X). At this stage the facial nucleus is rostral to the abducens nucleus but later it migrates dorsally, caudally and then ventrally into its definitive position. The trochlear nucleus migrates rostrally from the region of the isthmus—the narrow segment adjoining the mesencephalon. Meanwhile, as the pontine flexure develops, the roof plate becomes thinned, and the cavity dilates to form the diamond-shaped fourth ventricle. Its caudal and cranial angles continue into the central canal and the aqueduct of the midbrain. Its laternal angles wind around the side of the brain stem and form the lateral recesses. The ependymal roof is attached peripherally to a ridge of alar lamina—the rhombic lip. The caudal half of the rhombencephalon, laying between the level of the alteral recesses and the first cervical nerve roots becomes the medulla oblongata. Its ependymal roof is reinforced by vascular pia mater and infolded rostrally to form the choroid plexus. The median and lateral apertures are formed by local resorption of these layers and permit cerebrospinal fluid from the ventricle to prevade the subarachnoid space. The cranial half of the rhombencephalon forms the pons. The cells of the cranial part of the rhombic lip and of the adjoining alar lamina proliferate to form the cerebellar rudiment.

The midbrain vesicle grows much more slowly than adjacent parts, becoming relatively greatly narrowed to form the aqueduct which extends between the isthmus and the forebrain vesicle. The roof and floor plates thicken as they are invaded by lamina cells and from the midline parts of the tectum and tegmentum. Most of the nuclei differentiate within the midbrain, but the trochlear nucleus migrates from the isthmus, and the mesencephalic trigeminal nucleus extends from the pons. The tectum becomes subdivided by surface depressions into superior and inferior pairs of colliculi, which are reflex, correlation and relay centres for visual and auditory information, but mudulated by inputs from the spinal cord, reticular formation and cerebellar and cerebral cortices. The ventral part of the marginal layer is invaded by massive tracts of fibres which descend from the forebrain on each side and from the cerebral peduncles.

Prosencephalon

Rostral to the optic cups, the forebrain develops massive bilateral evaginations which are the primitive cerebral hemispheres. Each contains a lateral ventricle which communicates with the prosencephalic cavity (third ventricle) by an interventricular foraman. The rostral wall of the third ventricle is a thin sheet—the lamina terminalis—in which the optic chiasma and major cerebral commissures develop. The primitive hemispheres, the lamina terminalis and the brain wall between them constitute the telencephalon (end-brain).

The rest of the third ventricle and its wall constitute the diencephalon (between-brain).

Diencephalon

The third ventricle becomes laterally compressed as epithalamic, thalamic, hypothalamic and subthalamic nuclear masses appear in its lateral walls. These masses form reflex, correlation and relay centres; the epithalamus for olfactory information and the maintenance of some biorhythms mediated by the pineal gland; the thalamus for all sensory information except olfactory; and the hypothalamus for visceral and olfactory information. The hypothalamic mass contain into the floot of the ventride, where median diverticulum forms the rudiment of the neurohypophysis. The roof remains thin except caudally where the pineal gland, habenular nuclei and related commissures form.

Telencephalon

The corpus striatum (a primitive motor control centre) develops as a nuclear mass in the floor of each lateral ventricle. Elsewhere, the walls constitute the pallium or cortex. The hemispheres expand dorsally caudally and to a lesser extent, rostrally. They become dorsiflexed on the diencephalon and gradually grow over it and the mesencephalon. The rostral part of each hemisphere forms the frontal lobe and has an olfactory bulb on its under-surface. The caudal part curves downwards and forwards behind the developing eye and forms the temporal lobe. Later, a further backward extension forms the occipital lobe. The lateral ventricle extends into these lobes as its anterior, inferior and posterior horns. At first, the surfaces of the hemispheres are smooth, but, as growth proceeds, the pallium is thrown into a number of complex folds of gyri with intervening sulci. The groove between the frontal and temporal lobes forms the stem of the lateral sulcus. As the hemisphere grows, the cortex covering the corpus striatum remains relatively static, and is overgrown by a series of opercula from the surrounding cortex. The burried cortex

The Brain

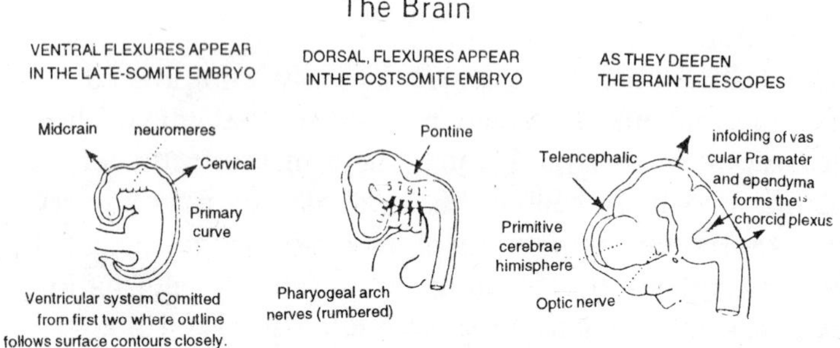

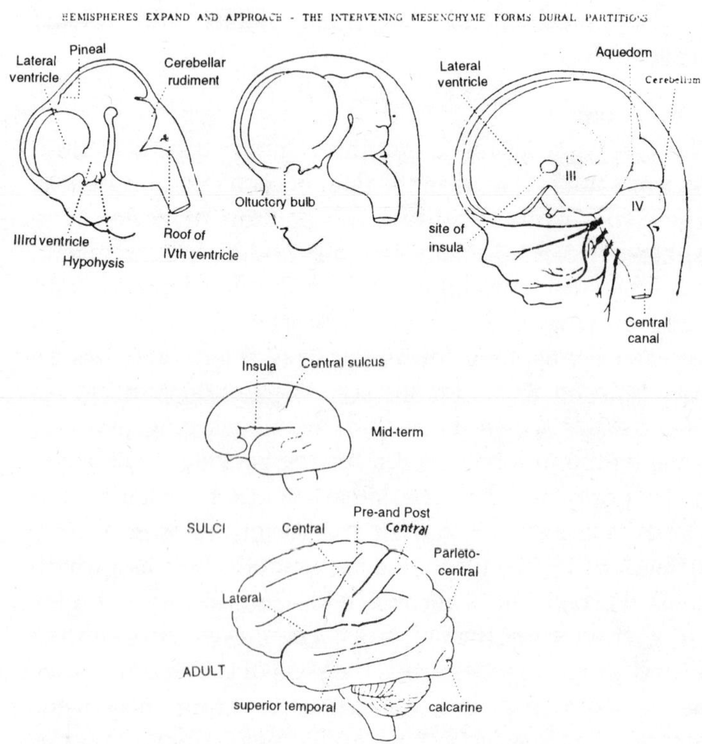

Fig. 6.21.   The Brain.

forms the insula, and the opercula meet at the posterior ramus of the lateral sulcus. Other sulci may be limiting (infolding around the 'margins' of a so-called functional area) or axial (infolding within a 'functional area') or merely reflect growth within a confined space and have not functional significance.

It should be noted that while the concepts of 'centres' and 'functional areas' are useful at an introductory level, they do not stand up to sophisticated analysis.

The Forebrain

Like the roof of the primitive fourth ventricle, the roof of most of the third ventricle and a curved strip in the medial wall of the lateral ventricle do not develop nervous tissue, but become membranous. In these regions vascular pia mater overlies the ependymal lining of the ventricle and forms the *primitive tela choroidea.* Invaginations into the ventricles form *choroid plexuses.* The site of invagination into a lateral ventricle is termed the *choroid fissure.* At this stage a horizontal section through the diencephalon and telencephalon, caudal to the interventricular foramen, shows the diencephalon with the developing thalamus and hypothalamus in its lateral wall, and the telencephalon separated from it by the primitive meninx. In the floor of the telencephalon is the developing corpus striatum. Apart from these subependymal nuclei, neuroblasts from the subventricular layer invade the marginal layer and form a cortical plate which differentiates into three main areas of superficial cortical grey matter. Of these, two are initially associated with olfaction. The central processes of primary sensory neurons in the olfactory muscous membrane become grouped into bundles—the olfactory nerves. These grow into the telencephalon at the future alfactory bulbs and make synaptic contact with mitral and other varieties of neuroblast. The axons of the mitral cells grow back in the olfactory tract which divides into lateral and medial roots. The medial root runs to a strip of cortex immediately dorsal to the choroidal fissure—the *archipallium* (hippocampal cortex). The lateral root runs to an area of cortex lateral to the anterior part of the striatum — the *palaeopallium* (piriform cortex). Between these primitive olfactory cortices is the *neopallium* which deals with motor control, with sensory information relayed from the diencephalic nuclei and with higher thought processes. It expands enormously during subsequent development. The olfactory input to archi- and palaeopallial structures should not be over-stressed. Both develop massive interconnections with neopallial cortical areas, the hypothalamus, thalamus, reticular formation and other centres in the brain stem. Thus, they become involved in polydimensional perception, total environmental orientation, the establishment of memory traces and the planning of executive commands for complex behavioural responses.

The fibres of the interneurons which make up much of the white matter of the forebrain are of three kinds; *commissural fibres* cross the middle and bring the two sides into communication, *association fibres* connect different areas of the same side and bring them into communication, while *projection fibres* connect the forebrain with the brain stem and spinal cord.

Commissural and Association Fibres

Cortical and other areas become connected with the opposite side by *commissural* axons. Many of these cross the midline in the *lamina terminalis*. Its lowest part is invaded by decussating axons from the ganglion cells of the retina and forms the *optic chiasma*. Above this, axons connect the piriform cortices and olfactory bulbs and from the primitive *anterior commissure*. Above, again, axons connect the hippocampal cortices and form the primitive *commissure of the fornix*. Axons connecting the neopallial cortices run mainly in the upper part of the lamina, where they form the *corpus callosum*, but they also reinforce the posterior part of the anterior commissure. Other commissures develop in the caudal part of the roof of the third ventricle in relation to the stalk of the pineal gland. Above it, axons connecting the two epithalami form the *habenular*

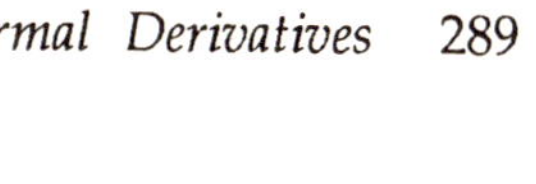

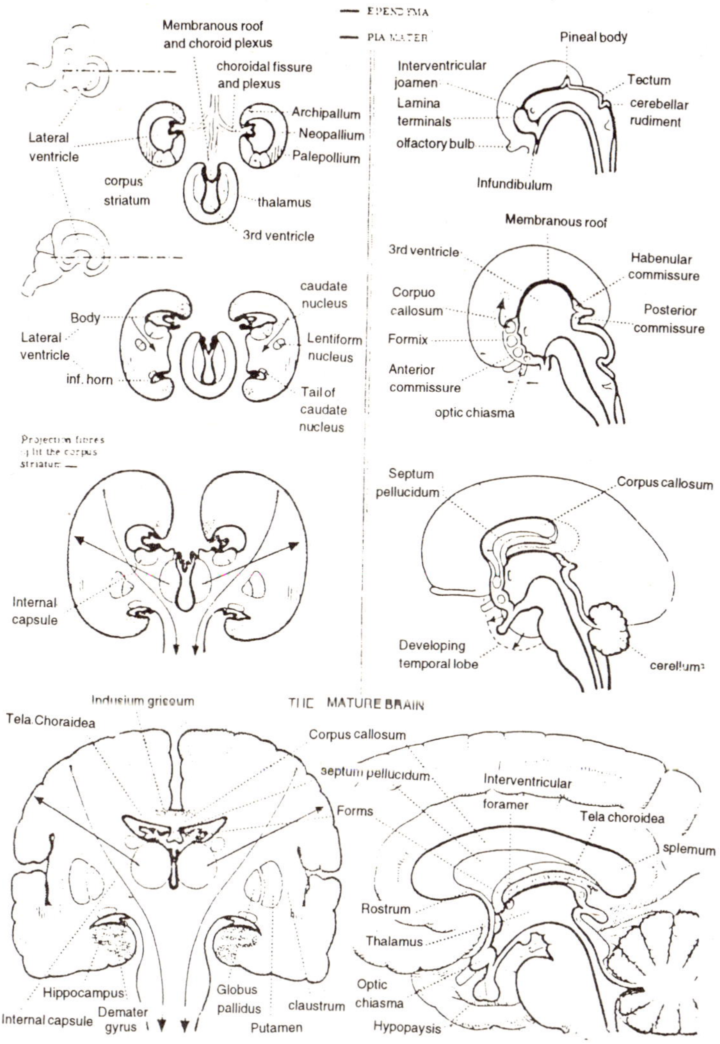

Fig. 6.22.   The forebraina and its commissures.

commissure, collateral branches probably invading the pineal gland as the habenulopineal tract. Below it, axons connecting certain diencephalic and mesencephalic nuclei form the *posterior commissure*. Different cortical areas in the same hemisphere become connected by long and short arcuate *association* axons.

The Limbic System

The formation of the temporal lobe brings the caudal ends of the two 'primitive' cortices (the archipallium and the palaeopallium) into secondary continuity as the piriform cortex and the hippocampus (cornuammonis and dentate gyrus). These structures are derived from neuroblasts that *border* the convex external margins of the interventricular foramen and its curved caudal extension,the choroid fissure; hence, the collective name *limbic lobe* (L. *limbus* + border) or *limbic system* when its numerous connections and functional associations are included. Traced radially from the outer lip of the choroid fissure, the derived tissues show increasing complexity of structures, and this remains evident even after the complicated infolding of layers has occurred. Near the fissure, and most elementary in structure, lies the *denate gyrus*, followed by the more highly ordered *cornu ammonis*, then progressing through the still further differentiated regions of the *subiculum*, to finally merge with fully developed *neopallium*.

At first, the *primitive fornix* consists only of commissural fibres connecting the hippocampal cortices. Soon, other axons from hippocampal neuroblasts traverse the lateral part of the fornix to reach diencephalic and other nuclei of the same side. When the corpus callosum grows backwards, the commissural part of the fronix becomes confined to an area above the ependymal roof of the third ventricle. With the formation of the temporal lobe, the backward growth of the corpus callosum and the reduction of the dorsal part of the hippocampal formation, the lateral fibres pursue a long, curved course. On each side they leave the medial edge of the hippocampal formation, in the floor of the inferior horn, and form a

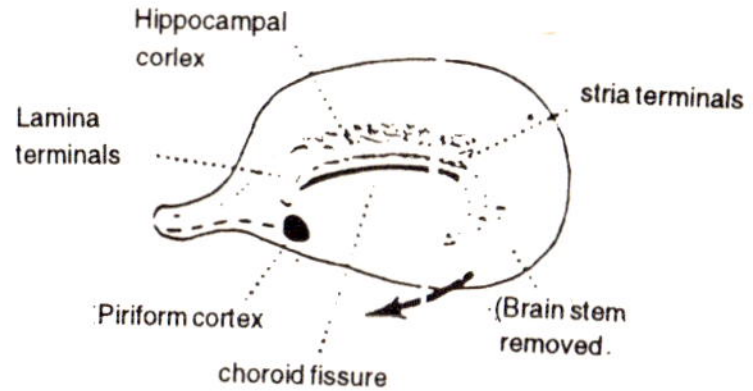

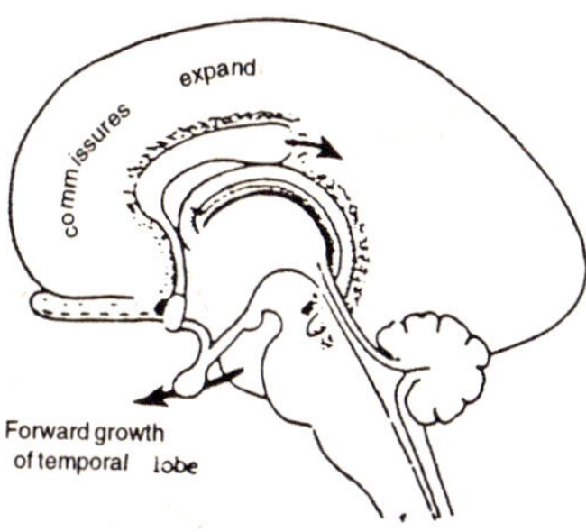

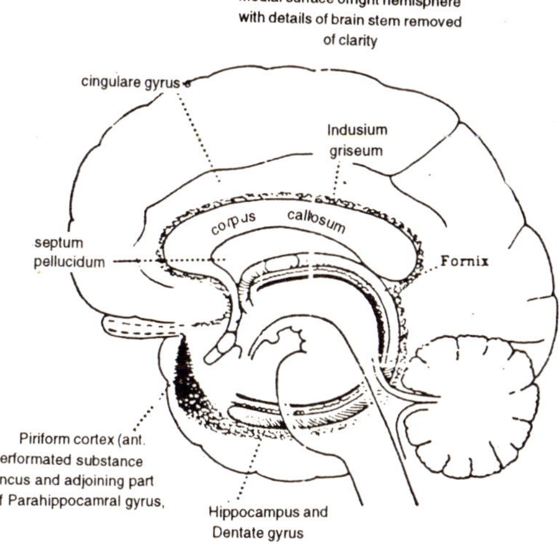

Fig. 6.23.   Limbic Lobe.

hippocampal fimbria. From the fimbriae they traverse the right and left crura of the fronix. The crura are joined across the midline by the commissural fibres and fuse at the body of the fornix. The fornix divides again above the interventricular foramina, and homolateral fibres descend in the right and left columns to reach and terminate in septal, hypothalamic and reticular nuclei. Between the expanding corpus collosum and for mix-lamina terminalis tissue and paraterminal archipallial septal cortex become attenuated and form two thin paramedian laminae—the *septum pellucidum*.

After fusion between the telencephalon and diencephalon, the expanding thalami enroach on the third and lateral ventricles. In the third ventricle, fusion produces a variable interthalamic adhesion. In each lateral ventricle, the thalamus forms the medial part of the floor of the body and, in it, is separated from the corpus striatum by a bundle of axons which project from the amygdaloid body and pass rostrally to the piriform cortex and the hypothalamus as the *stria terminalis*. This, too, becomes C-shaped as the temporal lobe grows. In the angle between the lateral wall and roof of the third ventricle, axons pass between the piriform cortex and the epithalamus and habenular commissure in the *stria medullaris thalami*. Neither this tract nor the *medial forebrain bundle*, which interconnects piriform cortex, hypothalamus and reticular nuclei, is affected by the growth of the temporal lobe.

The parts primitively concerned with olfaction were for long grouped together as the so-called rhinencephalon (smell-brain). In man, however, only the piriform cortex (palaeopallium) seems to be concerned with conscious olfactory experience. With the archipallial component and amygdaloid body, projections to the hypothalamus, epithalamus, hypothalamic projections to the cingulate gyrus and the cingulate gyrus itself, the subiculum and parahippocampal gyrus, and their numerous neocortical interconnections, it forms the basis of the limbic system.

The cerebellum

The neuroblasts of the cranial part of the rhombic lip and the adjoining alar lamina proliferate to form bilateral cerebellar rudiments. At first, these project into the fourth ventricle, but soon they meet and form a dumb-bell-shaped swelling which projects externally. A *posterolateral sulcus* appears and demarcates the *flocculonodular lobe*. A *primary fissure* separates the *anterior* and *middle lobes*. Numerous less significant fissures appear and give rise to the folia of the cerebellum. The cerebellar cortex may be divided into three (overlapping) zones which receive afferent fibres from different sources. The neuroblasts of the *archicerebellum* receive equilibratory information, directly and after relay in the vestibular nuclei, from the bipolar neurons of the vestibular ganglion. The *palaeocerebellum splits the archicerebellum* into a small rostral part—the lingula—and a larger caudal part—the flocculonodular lobe. The neuroblasts of the palaeocerebellum receive tactile and proprioceptive information, after relay and perhaps directly from the unipolar neurons of the spinal and trigeminal nerves. The *neocerebellum splits the palaeocerebellum* into a rostral part, mostly the anterior lobe, and a caudal part the pyramid, uvula and tonsils. The neocerebellum forms mostly the middle lobe and receives the terminals of cerebral, tectal and olivary neuroblasts. The rostrocaudal division of the cerebellar cortex into main regions, depending upon their principal (but not exclusive) sources of afferent input, is only one method of classification. Alternative criteria result in different divisions.

Two distinct waves of migration occur from subventricular alar presumptive neuroblasts of the cranial part of the rhombic lips into the membranous roof of the fourth ventricle. One remains for a while subependymal and forms an *internal germinal layer* whilst the other forms a subpial cortical plate—the *external germinal layer*. Orderly sequences of further migration, differentiation and growth occur at both sites.

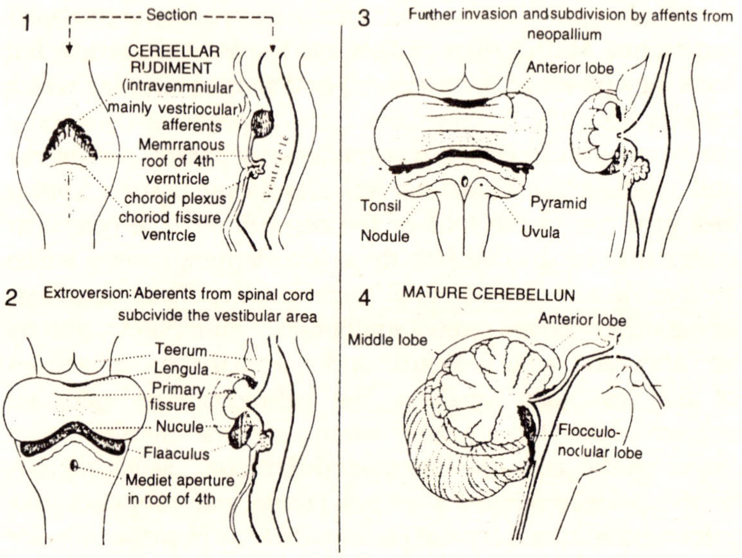

Fig. 6.24. The Cerebellum

The cells of the *internal germinal layr* form three main groups of definitive neuroblasts. *Nuclear neuroblasts* remain subependymal and form the deep cerebellar ('roof') nuclei (archicerebellar-fastigial; palaeocerebellar emboliform and globose; neocerebellar-dentate), their axons growing via the cerebellar peduncles to their destinations in motor control integrative centres throughout the brain stem and diencephalon. In contrast, *Purkinje neuroblasts* migrate superficially from the internal germinal layer to the junction of molecular and granular layers where they distribute in precisely patterned arrays, their flattened dendritic trees penetrating and molecular layer whilst their axons retain synaptic contact with neuroblasts of the deep roof nuclei. Somewhat later another superficial migration of *Golgi neuroblasts* occurs and they almost reach the Purkinje cells before developing their relatively vast dendritic and axonal arborizations.

The *external germinal layer* gives rise first to generations of *stellate* and *basket cell neuroblasts*, both differentiating in a subpial position in the future molecular layer. Soon this is followed by an intense proliferation of cortical plate cells to give a vast population of *granule cell neuroblasts*; these migrate deeply (centripetally), meeting and then passing through the Purkinje cell layer. Some remain dispersed between the somata of the Golgi neurons, but the majority form a densely packed granular layer deep to them. The granule neuroblasts migrate with their rudimentary dendrites in advance; ultimately these make synaptic contacts with incoming 'mossy' cerebellar afferent and Golgi cell axons terminals in the spherical complex cerebeller synaptic glomeruli that occupy the crevices between adjacent granule cell clusters. The tip of the elongating granule cell axon remains in the molecular layer and bifurcates at right angles, each branch growing a few millimetres in the long axis of its cerebellar folium. Collectively these branches form the parallel fibre bundles of the cortex.

Anomalies
While occurring less frequently than spine bifida occulta *spina bifida cystica* remains very common. Where the gap due to non union of paired vertebral arches is *primary* and is wide, normally developed meninges and spinal cord may extend through it so that a sac covered with skin is seen on the surface. In a *meningocele,* only the meninges bulge through the gap; in some cases of *meningomyelccele* a normally developed cord is in the sac and may be adherent to it. Where the non-union of vertebralarches is *secondary* to defective neural tube formation, the overlying skin or skin and meninges may be absent so that the thin membrane of the sac is exposed or absent. In cases of *meningohydromyelocele* in this category, the thin membrane and meninges are initially intact but may be easily torn and the extruded cord is abnormal; in a

meningohydromyelocele its central canal is distended. In a *myelocele* the neural tube has not closed and an open plate of nervous tissue is widely exposed to the surface. Lesions are most commonly associated with the posterior end of the tube (i.e. the lumbosacral region) but any part or the whole of the spine can be affected.

In the same way, contents of the cranium may herniate through large primary defects of the skull, the occipital region being the commonest site. In a *meningocele*, only the arachnoid is herniated, the dura being related to the periosteum and therefore defective and the pia following the contours of the brain. In the *meningoencephalocele*, brain and pia are also herniated, whilst in a *meningohydroencephalocele* a diverticulum from a ventricle is included.

Neurological signs are absent in uncomplicated (closed) cases of meningocele, meningomyelocele and encephalocele, and surgical repair is frequently successful. In other cases, particularly in myelocele where the exposed nervous tissue has undergone secondary degeneration and vascularization prenatally, neurological signs such as paralysis are present. Most such cases die before the age of 3 years, usually from infection of the exposed nervous tissue. Closure of the defect may be life-saving but there is no way of replacing damaged nervous tissue.

Iniencephaly is an uncommon variant of spina bifida, due to defective development of the cervical and occipital sclerotomes and the parachordal cartilages. As a result, there are defects of the squamous part of the occipital bone and of the cervical vertebrae, with prolapse of the brain through enlarged foramen magnum.

In some cases of spina bifida cystica the spinal core remains tethered to the skin and vertebrae at the site of the lesion and thus does not, as normally, retreat up the vertebral canal. As a result, the hindbrain as the foramen magnum. Cerebrospinal

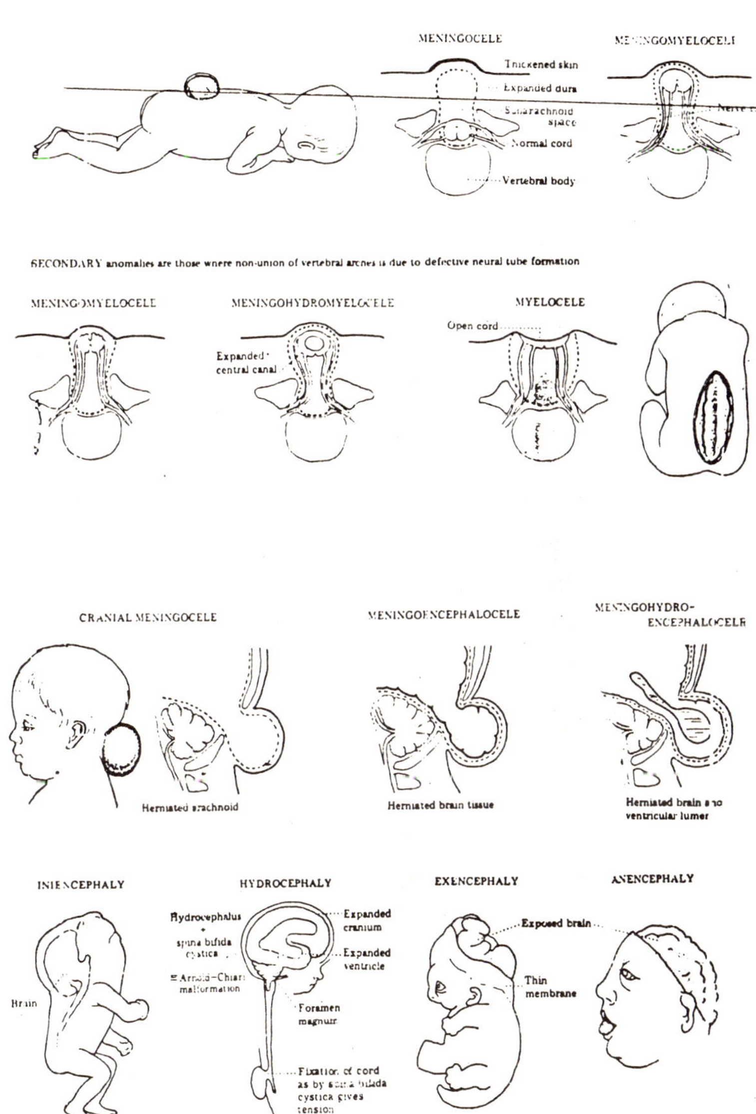

Fig. 6.25. Spina Bifida Cystica

fluid is thus unable to escape via the fourth ventricle and distends the brain, producing *hydrocephaly*. (In cases of hydrocephaly not associated with spina bifida the common site of obstruction is the aqueduct of the midbrain, so only the lateral and third ventricles are affected.)

Both anencephaly and exencephaly are due to defective closure of the anterior end of the neural tube. In *exencephaly* which is found only in the embryos and early fetuses, the scalp and vault of the skull fail to develop over an open but substantial mass of folded brain tissue. *Anencephaly*, which develops from exencephaly by secondary degeneration and vascularization of the exposed nerve tissue presents in later fetuses and newborn babies which may be stillborn or short-lived; an exposed flattened mass of vascular, cystic nervous tissue sits on the base of the skull and is frequently continuous with cervical myelocele.

In both anencephaly and open spina bifida, cerebrospinal fluid and vascular transudate are discharged into the amniotic fluid and cause hydramnios; in anencephaly the neural mechanism for swallowing is defective, which also contributes to the degree of hydramnios. The leakage of fetal blood or blood components into the amniotic fluid raises the level of α-fetoprotein above normal levels and may be detcted following amniocentesis. Howeveer, the α-fetoprotein level is raised in other conditions associated with leakage and is not raised in closed cases of spina bifida. The test is thus not particularly discriminatory and diagnosis by ultrasound is more reliable.

Spina bifida cystica and anencephaly together are amongst the commonest serious congenital conditions, in incidence being second only to congenital heart disease. Their inheritance is multifactorial. They are commoner in girls and in caucasians but regional variations in incidence exceed those of sex and race. The incidence of second affected babies in families with one affected is high, as is the incidence in the offspring of consanguineous marriages. The conditions have rarely been found in identical twin pregnancies and then only one twin has been affected. It is likely that pregnancies in which both twins have been affected have been abortive.

The Eye

The optic vesicle induces the formation of a lens vesicle from the overlying head ectoderm and forms a double-walled optic cup. This and the adjoining optic stalk are incomplete inferiorly, where they wrap around a strand of vascular mesenchyme at the choroidal fissure. The hyaloid artery forms in this mesenchyme and supplies the vascular capsule of the developing lens. Branches arise from the proximal part of the artery and supply the inner wall of the optic cup. The distal part and the posterior lens capsule subsequently degenerate. As the lips of the choroidal fissure fuse, the hyaloid artery becomes enclosed within the solid optic nerve as the central artery of the retina. The choroidal fissure may not close completely, resulting in a congenital cleft iris or *coloboma*. The hyaloid artery may persist and form an opacity in the vitreous body.

The margins of the optic cup extend and overlap the lens. Posterior to it, the inner wall of the cup forms the retina proper (rod and cone cells, bipolar cells, ganglionic cells, horizontal and amacrine cells and supporting elements) while the outer wall forms the pigmented layer of the retine. Anteriorly, the walls of the cup continue as the two-layered epithelium on the posterior surface of the ciliary body and iris. In the iris, optic cup cells (neural ectoderm) form the sphincter and dialator pupillae muscles. The mesenchyme surrouding the cup condenses to form an inner vascular coat—the choroid and the rest of the iris and ciliary body, including the ciliary muscle— and an outer fibrous coat—the sclera and the substantia propria of the cornea. Anterior and posterior chambers are at first separated by the anterior part of the mesenchymatous lens capsule—the pupillary membrane—which normally breaks down as term approaches but may persist. The margins of the

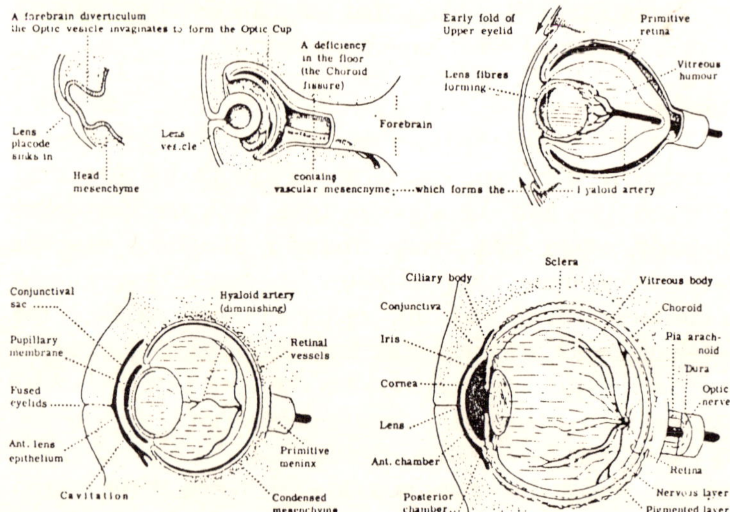

Fig. 6.26. A forebrain diverticulum the Optic vesicle invaginates to form the Optic Cup.

eyelid folds fuse and remain fused, forming a closed conjunctival sac, until the fetus is viable. The lacrimal gland arises as a series of ectodermal cords which grow from the outer part of the upper conjunctival recess (fornix) into the underlying mesenchyme. These ramify and later conalize, forming the secretory units and multiple ducts of the gland. The axons of the neuroblasts of the ganglionic layer of the retina converge on the optic papilla and grow through the neuroglial framework of the primitive optic nerve towards the lamina terminalis. There the fibres from the nasal half of each retina decussate and form the optic chiasma. The fibres of the optic tract are thus derived from the temporal half of the homolateral retina and the nasal half of the contralateral retina. These continue to grow round the side of the upper end of the midbrain (cerebral peduncles) and terminate in the developing thalamus (lateral geniculate body), the tectum (superior colliculus), and probably other diencephalic centres. The developing lenses are vulnerable to the rubella virus (German measles) and may become opaque—*congenital cataract*—resulting in blindness.

The Ear

The pinna arises by the fusion of tubercles around the orifice of the first pharyngeal groove. The lining of the external acoustic meatus and the outer epithelium of the tympanic membrane are derived from first groove ectoderm. The dorsal recesses of the first and second endodermal pouches combine to form a common tubotympanic recess which grows towards the first groove ectoderm. As the tubotympanic recess expands around the auditory ossicles and the chorda tympani, it invests them with endoderm, continuous with that lining the walls of the recess. The endorderm also clothes the inner aspect of the tympanic membrane, which thus represents the closing membrane between the first groove and the first pouch. In this way, the recess forms the tympanic cavity proper and the auditory tube. Later extensions form the epitympanic recess and the mastoid antrum. At term, the mastoid air cells begin to develop. Meanwhile, on each side, an ectodermal placode skinks below the surface and forms an optic vesicle—the primordium of the membranous labyrinth. An early outgrowth from its medial aspect is the endolymphatic sac. The out vesicle is related to and contributes to the acoustico-facial complex of neural crest material. After the geniculate ganglion of the facial nerve has separated from the complex, the optic vesicle and the remaining ganglionic tissue become subdidived into vestibular and cochlear parts. Three flattened divericula arise from the dorsal (vestibular) part. Only the peripheral rims of these diverticula remain patent. These form the semicircular ducts opening off a utriculosaccular chamber. At first the anterior and posterior semicircular ducts are in the same plane, but later the posterior duct swings laterally into its definitive position. The ventral (cochlear) part elongates, becomes coiled and forms the cochlear duct. The cells forming that part of the duct related to the basilar membrane differentiate into the rows of sensory hair cells and wide variety of supporting cells

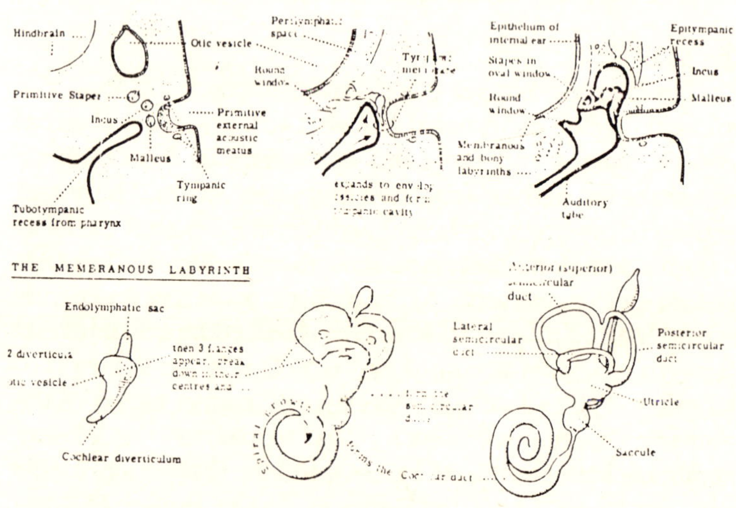

Fig. 6.27. The Membranous Labyrinth.

that characterize the organ of Corti. Subsequently, the connections between the various parts become narrowed. Particular patches of cells in the walls of the utricle and saccule differentiate and form polarized rows of sensory hair cells with their related otolithic membrane and supporting cells; they constitute the masculae of these endolymphatic cavities. The mesenchyme surrounding the middle and internal ears chondrifies and forms the cartilagnous otic capsule. The perilymphatic space develops between the membranous labyrinth and the cartilage, and approaches the tympanic cavity at the round and oval windows. The neuroblasts of the vestibular and spiral (cochlear) ganglia are bipolar. Their peripheral processes grow into the membranous labyrinth, and end in relation to cells with hair-like processes in the dilated end (ampulla) of each semicircular duct, the utricle, the saccule and in the spiral organ of the cochlear duct. Their central processes grow back to synapse with, and convey auditory and equilibratory information to the vestibulocochlear complex in the lateral part of the floor of the fourth ventricle. The rubella virus may affect the developing inner ear, later resulting in deaf-mustism.

7

Mesodermal Derivatives

THE MESENTERIES AND COELOM

To a degree the middle germ layer resembles entoderm and ectoderm by differentiating some of its organs directly from solid layers of tissue (in this case, *mesoderm*) that are epithelial in character. Such mesodermal derivatives, in the strict sense of that term, are the following: skeletal muscle, from the myotome plates of somites; urinary and reproductive organs, from the nephrotome plates; mesothelium, cardiac muscle, spleen and suprarenal cortex, from the somatic and splanchnic plates of lateral mesoderm. Another group of derivatives arises from the primitive filling-tissue that is known as *mesenchyme*. This tissue consists of cells that gave up their compact epithelioid arrangement in the original mesoderm and became loosely arranged, star-shaped elements. Its derivatives are: connective tissue; cartilage; bone; blood; smooth muscle and endothelium.

The Mesenteries

The Primitive Mesentery: The gut arises when the splanchnopleure is folded into a tube. The splanchnic mesoderm, which is associated with the entoderm, then takes the form of a double-layered partition, extending typically from the roof of the coelom to the midventral body wall (C). This median partition is the *primitive mesentery;* it divides the coelom into halves and contains the gut between its component sheets.

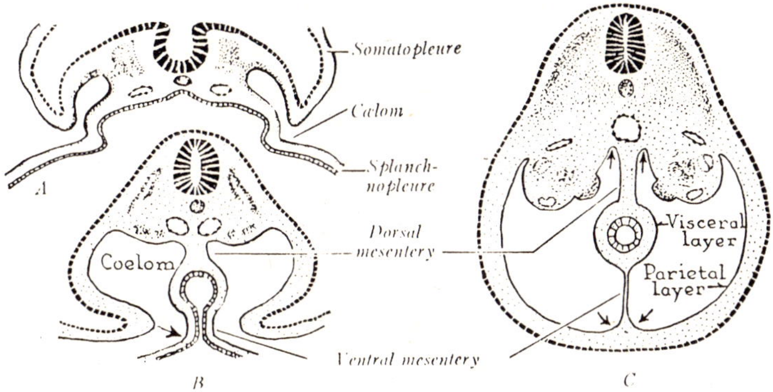

Fig. 7.1. Stages illustrating the formation of the primitive mesentery in human embryos. *A*, at 2 mm. *B*, at 4 mm. *C*, at 8 mm.

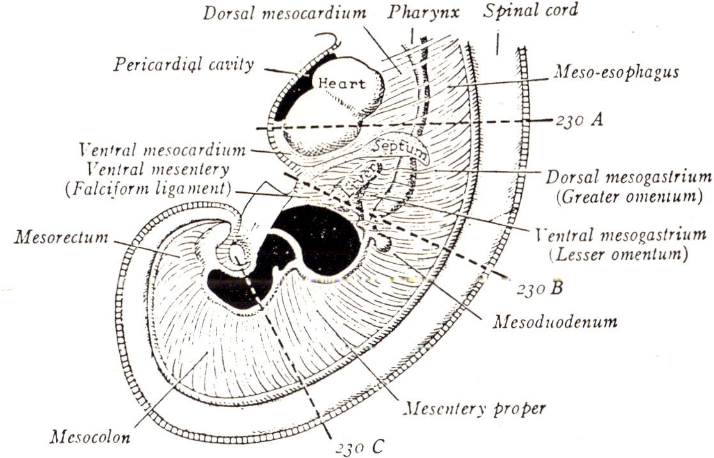

Fig. 7.2. The primitive human mesenteries, shown as a diagram viewed from the left side.

The early, straight gut naturally subdivides the mesentery into an upper and lower portion; for convenience they are identified as the *dorsal and ventral mesentery*. At about the same time the heart, lungs and liver make their appearance, and soon occupy three separate coelomic compartments whose linings are named *pericardium, pleura* and *peritoneum,* respectively. Such a sheet-like layer bounding the coelom is known as a *serous membrane;* it consists of a layer of connective tissue overlaid with simple epithelium that is originally cuboidal but soon becomes flat. The two apposed layers of splanchnic mesoderm that comprise the primitive mesentery enclose the heart, iungs and various abdominal organs; they constitute the *visceral layer* of the pericardium, pleura and peritoneum. The somatic mesoderm furnishes the *parietal layer* of these three sacs; it becomes an innermost part of the body wall.

In addition to the mesenteries of the digestive tube and its associated organs, there are special mesenterial supports for the genital organs.

Specializations of the Dorsal Mesentery: At first the gut is broadly attached dorsally, but presently this region becomes relatively narrower and the gut is then suspended throughout most of its length by a definite *dorsal mesentery* this extends like a curtain in the midplane and supplies the pathway through which blood vessels and nerves reach the gut. Only the pharynx and upper esophagus lack a mesentery, since they lie cephalad in regions where there is no permanent coelom. To the continuous primitive mesentery of the rest of the digestive canal are given distinctive names at its successive, divisional levels. Thus, there are the *meso-esophagus,* the *dorsal mesogastrium* (or *greater omentum)* of the stomach, the *mesoduodenum,* the *mesentery proper* of the jejunum and ileum, the *mesocolon* and the *mesorectum.*

As development advances, parts of the primitive dorsal mesentery become specialized; other regions, following the displacements of the gut, depart from the original midline position and gain secondary attachments, while still other regions are lost by obliteration. Yet most of this mesenterial system persists permanently in some form or other.

The *Meso-Esophagus:* A middle stretch of the esophagus

courses in a mesentery that never thins into a membrane but becomes a thick and specialized median partition known as the *mediastinum*. This partition encloses all the thoracic viscera except the lungs, which expand beyond its confines (Fig. 8 C.). The dorsal meso-esophagus at these levels is short and the esophagus is broadly attached. But below the diaphragm, near its junction with the stomach, the meso-esophagus becomes a typical mesentery.

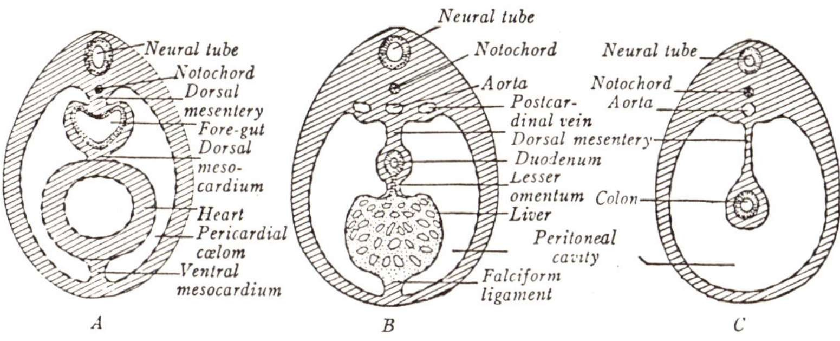

Fig. 7.3. Relations of the human mesenteries, shown in diagrammatic transverse sections through the levels *(A-C)* indicated. on Fig. 7.2.

The Mesogastrium: The history of the dorsal mesogastrium is chiefly concerned with the development of a huge, secondary sacculation, known as the *omental bursa* or *lesser peritoneal sac*. It is so important that its origin and final relations must be traced in some detail.

Although the bursa is often described as a folding of the omentum, brought about by the rotation of the stomach, it actually arises as an independent invagination into the interior of the originally thick mesentery, before rotation begins. The earliest indication of the bursa is in 3 mm. embryos, when a shallow pocket appears on the right surface of the dorsal mesogastrium and straightway proceeds to burrow deeper into the substance of the mesentery. One subdivision of this recess extends cephalad between the esophagus and the right lung bud. Such an extended passage is permanent in reptiles, but in human embryos it is soon interrupted by the developing diaphragm; the pinched-off apex then constitutes a small sac (the *infracardiac bursa*) that frequently persists in the adult *B)*. The other subdivision of the original recess is located more caudally. It enlarges toward the left, behind the stomach *(A)*, and thus creates a blind pocket in the interior of the mesogastrium *(C)*. This is the beginning omental bursa. After the stomach has rotated, the bursa lies dorsal to the stomach; in a sense, the stomach is then carried on the ventral bursal wall *(D, E)*.

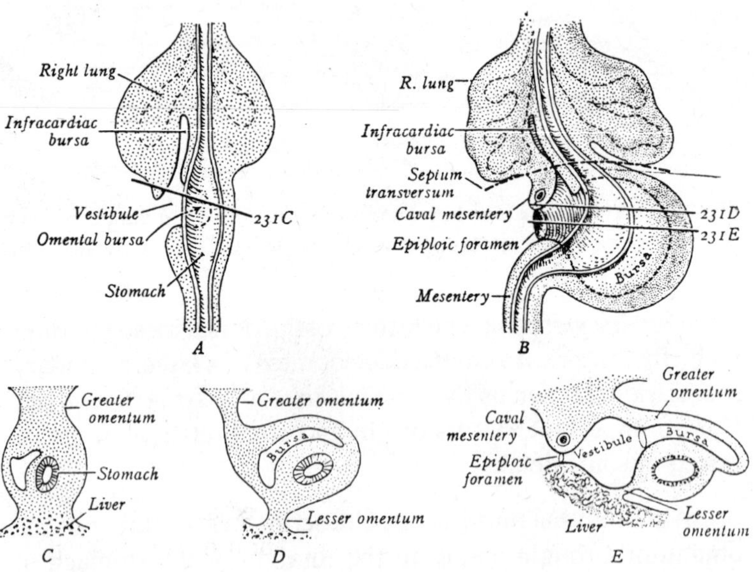

Fig. 7.4.   Early development of the omental bursa in human embryos. A, B, Ventral views, at four and six weeks. C-E, Transverse sections, at the levels indicated on A,B.

The bursa is a progressively growing sac whose expanding walls become thinner as it pushes to the left of the general, medially located mesogastrium. Subsequent sagging of the stomach to a partially transverse position changes the direction of growth of the sac so that it extends caudad (*B*). This flattened, saccular portion of the bursa then overlies the intestines and looks like an apron hanging from the greater curvature of the stomach.

The narrowed mouth of the bursal sac opens into a common *vestibule* which also receives the proximal remnant of the recess that extended lungward. The vestibule, in turn, communicates through an aperture (*epiploic foramen*) with the general peritoneal cavity. It is necessary to emphasize that the epiploic foramen is a wholly different thing from the aperture into the true omental bursa. Also, sections passing through both foramina give the false appearance of a long mesogastrium folded simply upon itself the true nature of a small-mouthed sacculation and its two divisions (Vestibule and bursa proper), cnnected by a narrower passage, is not revealed by such a section.

The *vestibule,* already mentioned, is a peritoneum-lined space captured from the general peritoneal cavity. It is bounded cranially and laterally by a lip-like fold of the dorsal mesentery that continues caudad along the dorsal body wall into the right mesonephric fold; this is the *caval mesentery* in which the upper segment of the inferior vena cava develops. Moreover, as the liver comes to locate within the ventral mesentery, its primitive right lobe both enters into relation with the caval mesentery and grows caudad. In this manner the cavity of the vestibule is extended caudad, to the level of the pyloric stomach, while the caval mesentery and right hepatic lobe form its lateral wall on the right side. The left wall of the vestibule is furnished by the stomach and dorsal mesogastrium. Dorsally the vestibule is limited by the dorsal body wall. As the stomach rotates so that its midventral line becomes the lesser curvature and lies at the right, the position of the lesser omentum (i.e., the ventral mesogastrium between stomach and liver) is necessarily shifted from a sagittal to a frontal plane. This mesentery then makes a ventral floor to the vestibule.

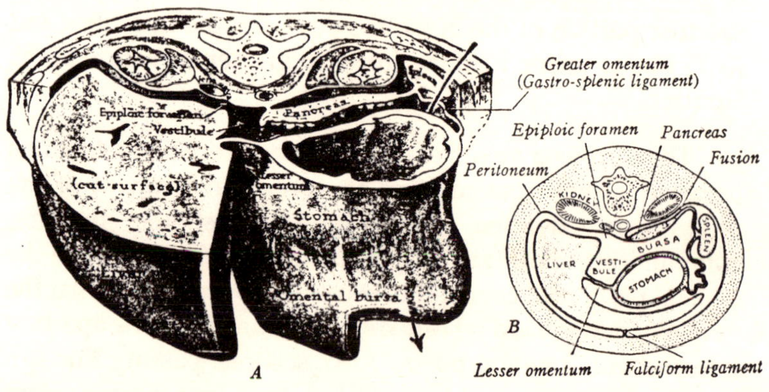

Fig. 7.5.   Relations of the human omenta and general peritoneum, at about four months. *A*, Model, cut transversely; *B* transverse section.

When the changes outlined in the preceding paragraph have been completed, the *epiploic foramen* presents a slit-like opening leading from the peritoneal cavity into the vestibule of the omental bursa. The foramen is bounded ventrally by the free border (originally caudal edge) of the lesser omentum, dorsally by the inferior vena cava, cranially by the caudate process of the liver and caudally by the wall of the upper (transversely directed) duodenum. All these parts are, of course, surfaced with peritoneum. The communication between the vestibule and the true bursal sac is an orifice which is bounded permanently by sickle-shaped *gastro-pancreatic folds*.

In the third and fourth months the omental bursa makes secondary attachments. Its flat, dorsal lamella, into which the pancreas has extended, fuses with the dorsal body wall, thereby fixing the tail of the pancreas and covering the left suprarenal gland and part of the left kidney. This results in the mesogastrium acquiring a new line of origin at the left of the midplane. Where the dorsal lamella of the bursa lies upon the transverse mesocolon and colon, it likewise adheres and fuses. This results in the transverse mesocolon becoming fundamentally a double structure, but, as in all similar fusions, any evidence of compounding soon vanishes. The omental connection between stomach and colon is henceforth designated as the *gastro-colic ligament*. Caudal to this colonic attachment, the walls of the omental bursa unite after birth and obliterate its cavity, the cavity of the adult omental bursa thus may be limited chiefly to a space between the stomach and the dorsal lamella of the greater omentum, which latter layer is largely fused to the peritoneum of the dorsal body wall. The spleen develops in the cranial portion of the greater omentum; that stretch of the omentum between stomach and spleen is known as the *gastro-splenic ligament,* while its continuation beyond the spleen to the left kidney and diaphragm is the *phrenico-splenic ligament.*

The Intestinal Mesentery   As long as the gut remains a straight tube, the dorsal mesentery is a simple sheet whose two attached edges are equal in length. But when the intestine begins to elongate faster than the body wall, the intestinal border of the mesentery grows correspondingly. The result is an elongate, somewhat fan-shaped mesentery, and in this state it is carried out into the umbilical cord between loops of the gut. On the return of the now highly coiled intestine into the abdomen, the characteristic rotation, already begun at the time of herniation into the cord, is completed. It will be remembered that in this process the caecal end of the colon is carried over to the right, whereby the future transverse colon crosses ventral to the duodenum and the small intestine lies mostly at the left of the caecum and future ascending colon. There is thus accomplished a torsion of the mesentery (about the origin of the superior mesenteric artery as an axis), and this rotation is accentuated even more as the limb of the ascending colon elongates and its flexure beneath the liver gains prominence.

From a focal point at the root of the artery the continuous mesentery of the entire intestine spreads out like a funnel.

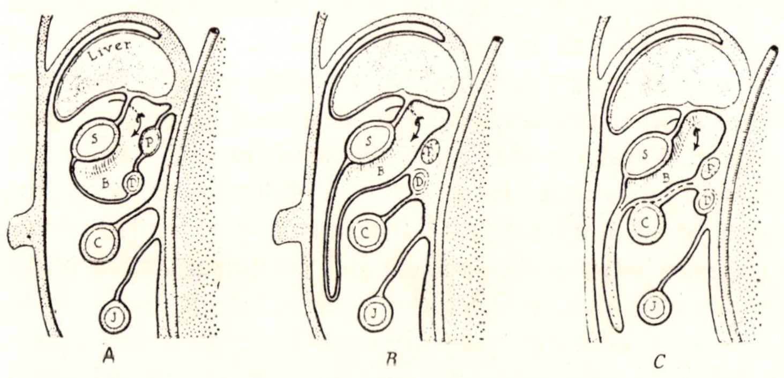

Fig. 7.6.   Secondary fusions of the omental bursa, shown by schematic longitudinal sections of the body. At two months, B, at four months; C, adult. B, Omental bursa; C, transverse colon; D duodenum; J, jejunum, P, pancreas, S, stomach.

Previous to the fourth month the entire intestine is freely movable within the scope of its restraining mesentery, while the latter still retains its primitive line of origin along the mid-dorsal abdominal wall. At this period, however, secondary fusions begin which affix certain portions of the gut and thereby produce new lines of attachment. Rotation of the stomach and the enlarging head of the pancreas cause the duodenum to become curved and laid to the right of the midplane against the body wall. The part of the duodenum nearest the stomach retains its dorsal and ventral mesentery, but the rest of the *mesoduodenum* fuses with the peritoneum of the dorsal body wall at the end of the third month and this portion of the small intestine then becomes permanently fixed. The pancreas, growing dorsad into the mesoduodenum and greater omentum, necessarily shares the fates of these mesenteries and assumes a retroperitoneal position. The *mesentery proper* of the jejuno-ileum is thrown into numerous folds, corresponding to the loops of the intestine, but normally remains entirely free. It does, however, acquire a secondary line of origin where it continues into the presently fixed mesentery of the ascending colon.

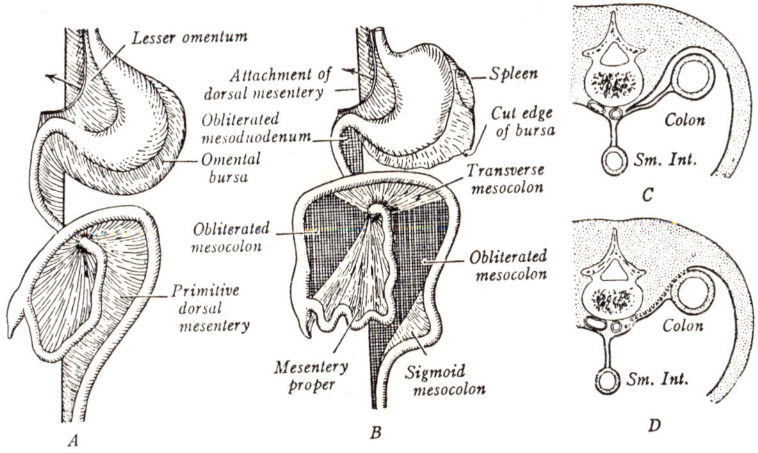

Fig. 7.7.   Secondary fusions of the duodenum and mesocolon. *A*, At three months, before fusions begin. *B*, Later stage; fused surfaces indicated by cross hatching, *C,D*.

The large intestine suffers extensive mesenterial loss. The *ascending* and *descending mesocolons* grow rapidly, carrying the corresponding segments of the colon to the right and left, respectively, far laterad in the abdomen. The mesocolons themselves become pressed against the dorsal body wall and their flat surfaces progressively fuse (mediolaterad) with the adjacent peritoneum. In this manner these two limbs of the colon become permanently anchored by the end of the fifth month, and they themselves attach broadly to the general peritoneum. The *transverse mesocolon* remains largely free, although it does fuse with the cover the duodenum where the colon crosses it this makes the portion of the duodenum become, for the second time, secondarily retroperitoneal in position.

The line of junction of the free, transverse mesocolon with the neighbouring, obliterated, mesocolic sheets gives a new (and transverse) line of origin to the free mesocolon. The fusion between omental bursa and transverse colon has been described in an earlier paragraph. The *sigmoid mesocolon* remains partly free, but the primitive *mesorectum* obliterates as the rectum comes to lie against the sacrum.

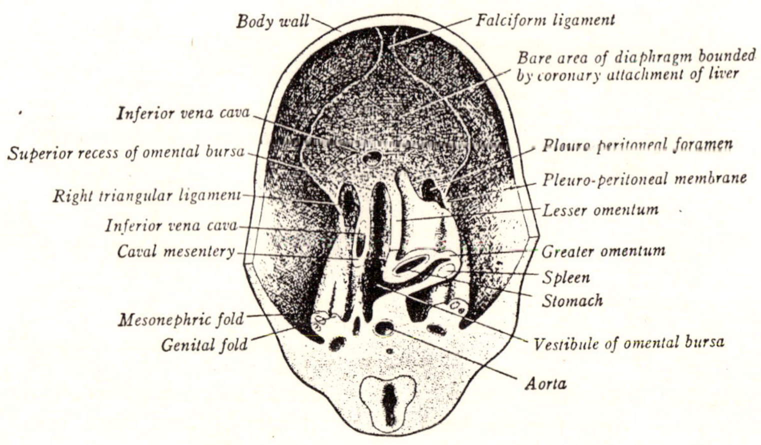

Fig. 7.8.   Mesenterial relations in the region of the diaphragm. shown in a simplified model of a 14 mm. human embryo.

Obliteration of considerable portions of the primitive dorsal mesentery and the resulting fixation of the gut and pancreas are apparently related to upright posture, since these processes especially characterize the anthropoid apes and man. Such mesenterial adhesions, a part of the normal developmental plan, have much in common with those occurring pathologically after inflammation of the peritoneum. It is interesting that the original left side of the dorsal mesentery alone effects fusions with the body wall. This is true of both the omentum and the mesentery proper, except for the short mesoduodenum which by its rotation presents a special case. The fusion of apposed

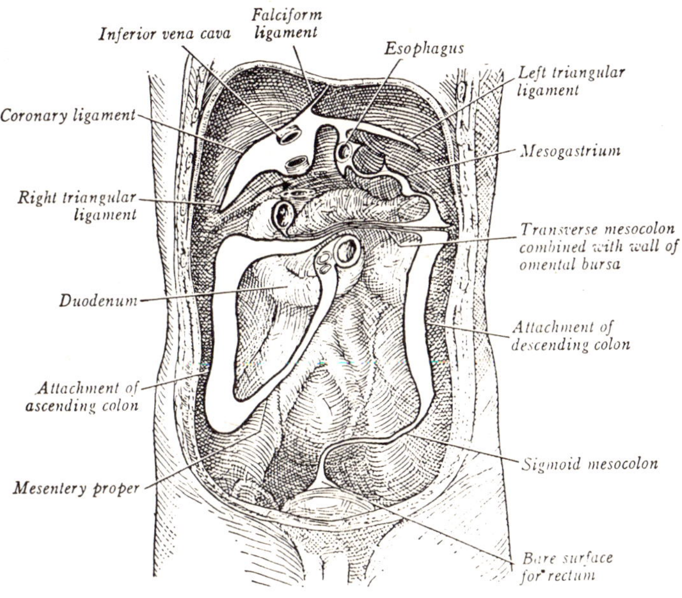

Fig. 7.9.   The final lines of attachment of the human mesenteries to
the dorsal abdominal wall and diaphragm.

pretoneal surfaces is not merely the result of nearness and
pressure; it represents the carrying out of a definite hereditary
pattern.

Specializations of the Ventral Mesentery: The same two layers
of splanch nic mesoderm that comprise the dorsal mesentery
also continue around the gut and recombine beneath it as the
ventral mesentery. In the future thorax the ventral mesentery
is represented by the mediastinal region ventral to the esophagus.
It serves as a mesentery to the lungs (i.e., the pulmonary
roots, or *mesopulmonon*)) and its tissue, carried out by the
enlarging lungs, encloses them *(visceral pleura)*. It is associated
still more intimately with the heart, which becomes an elongate,

single tube by the progressive 'union' of a paired blood vessels, each coursing in a corresponding fold of splanchnic mesoderm. Hence, through its very manner of formation, the heart lies beneath the fore-gut and is fashioned within a specialized region of the ventral mesentery. The dorsal portion of this mesentery constitutes the *dorsal mesocardium*. For a brief period it suspends the heart, but soon disappears. Thereafter the heart lacks any mesenterial support, yet its wall is made of the same substance as the mesentery. The ventral portion, or *ventral mesocardium*, is at best transitory and in mammals is said to have no real existence as such.[5] Secondary shiftings and re-arrangements cause the heart and pericardium to occupy the ventral region of the permanent mediastinum.

The septum transversum, and more particularly the liver which grew within its substance, are permanently related to the typical ventral mesentery of the lower esophagus, stomach and upper duodenum. At first these divisions of the fore-gut directly overlie the septum. When however, the fore-gut soon draws away, its region of attachment with the septum stretches and thins into the definitive ventral mesentery (mostly *ventral mesogastrium*) of this region. At the same period the rapidly enlarging liver begins to project from the surface of the septum. Henceforth the liver can be said to lie within split halves of the ventral mesentery. Caudal to the upper duodenum, no definite ventral mesentery is recognizable even as a transitory feature except in the region of the early cloaca.

Ligaments of the Liver: Since the ventral mesentery encloses the liver, it gives rise to it fibrous capsule and mesenterial supports; the later are designated *ligaments*. Except where the liver impinges on the diaphragm. The enveloping hepatic capsule is covered by mesothelium that is continuous with the general peritoneum. Along its mid-dorsal and midventral lines the liver maintains permanent connections with the ventral mesentery. The portion of the mesentery that extends from the stomach and duodenum to the liver is the *lesser omentum*. For convenience it is more specifically subdivided and given two regional designations; the more cranial part is the *hepato-gastric ligament,* while the more caudal portion is the *hepato-duodenal ligament.* The mesenterial attachment of the liver to the ventral body wall is named the *falciform ligament* because it extends caudad, from diaphragm to umbilicus, in a sickle-shaped fold (Fig. 8.).

The peritoneum does not invade the area of contact where the liver abuts against the septum transversum (later, the diaphragm). Instead it reflects from the diaphragm to the otherwise exposed surfaces of the liver, leaving a *'bare area'* on the liver and diaphragm. This area is continued dorsolaterad by prolongations of the lateral liver lobes known as the coronary appendages. The attachment of the liver to the septum transversum then has the outline of a crown whose name, the *coronary ligament,* is more appropriate at this stage than later. As these illustrations show, the dorsoventral extent of the coronary ligament is relatively reduced during later development and the shape becomes more crescentic. Nevertheless, the coronary ligament is extended caudad somewhat by an attachment established between the right lobe of the liver and the ridge (1caval mesentery') in which the inferior vena cava of this level is developing. The later extensions of the coronary appendages upon the diaphragm give rise to a *triangular ligament* on each side.

In general, the several displacements and secondary fusions of the primitive mesentery, already recorded, cause its line of attachment with the body wall and diaphragm to depart markedly from the original midsagittal position.

The change in position of the lesser omentum to a frontal plane and its participation in creating the vestibule of the omental bursa have already been described. Its caudal free margin, bordering the epiploic foramen, contains the bile duct and the vessels supplying the liver. The original ventral, later right, margin of the omentum attaches to the hilus of the liver and to the groove in which courses the ductus venosus (after birth called the ligamentum venosum). For a time it enclosed the gall bladder and conducted the vitelline veins from yolk sac to duodenum later regression of the omentum exposed the caudal surface of the gall bladder and freed the fused, common vein. The falciform ligament remains in the midplane and carries the umbilical vein (After birth called the ligamentum teres) embedded in its free border. The ligamentum venosum and ligamentum teres are not mesenteries but obliterated blood channels.

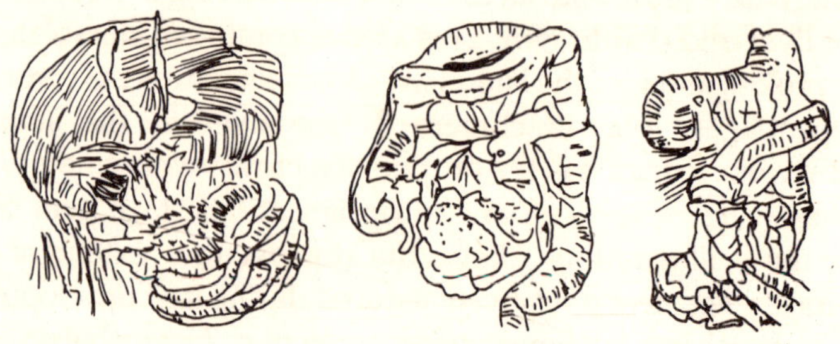

Fig. 7.10.   Anomalies of the human mesenteries (Ladd). *A*, The caecum is fixed in a high position ('undescended caecum'); this condition has obstructed the duodenum. *B*, Free mesenteries, through failure to effect fusions. *C*, Twisting (volvulus) of the unfixed intestine, producing obstruction.

Anomalies: The mesenteries show frequent variations of form and relations. These are commonly due to the persistence of simpler conditions that mark the normal developmental course of the intestinal canal. In about one-fourth of all cases the ascending or descending mesocolon is more or less free, due to faulty fusion with the dorsal peritoneum. At times the ascending colon is anchored at a high position and is termed *'undescended caecum'* (Fig. 8. *A*). Fixation of the entire intestine may fail completely (*B*); in this instance the bowel may twist about the root of its fan-shaped mesentery (*volvulus*) and give rise to obstruction (*C*). The primitive cavity of the omental bursa sometimes falls short of its normal degree of obliteration;

the cavity ('inferior recess') may extend even to the bottom of the sac, as is normal in many mammals. Less common than the preceding condition are those based on changes that exceed normality. For example, reduction in the length and breadth of the sigmoid mesocolon lessens the mobility of that portion of the bowel.

The Coelom

The Primitive Coelom: Originally the coelom of animals was used as a temporary reservoir for excretory wastes, but this function has been superseded in vertebrates so that it now serves as a large bursa to permit frictionless movement of the heart, lungs and abdominal viscera. From the standpoint of development, the coelom permits the visceral organs to grow and shift position without hindrance.

The first occurrence of a body cavity in early human stages is in the extra-embryonic mesoderm which lies between the embryo proper and the primitive chorionic capsule. The earlier appearance here than in the embryo proper is presumably correlated with the precocious development of the extra-embryonic membranes. A cleft appears toward the end of the second week of development; it divides the extra-embryonic mesoderm into a *somatic layer,* which lines the chorion, and a *splanchnic layer* which invests the yolk sac. The space itself is the *coelom,* while the mesodermal cells that bound it flatten into a limiting membrane called *mesothelium.*

About one week later (at the beginning of somite formation) numerous horizontal clefts appear also in the unsegmented mesoderm belonging to the embryo itself; these lie lateral to the midline and begin to split the solid mesodermal sheet of each side into a somatic and a splanchnic layer. Such isolated, coelomic spaces coalesce first in the folding-off head region where they form a canal on each side. The cranial ends of the two coelomic channels are continuous with a similar space located ahead of the embryo, which is destined to be the cardiac region. After this beginning new spaces continue to

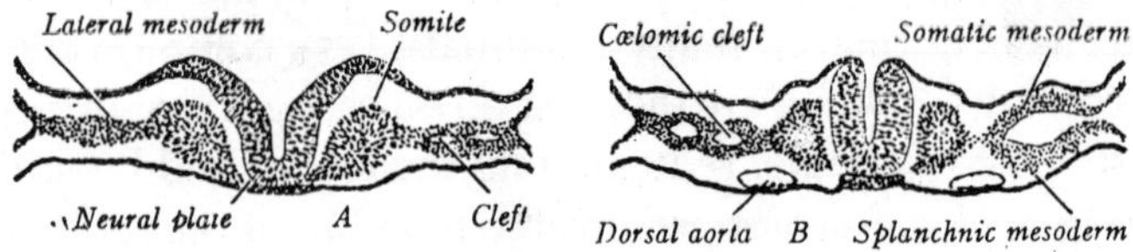

Fig. 7.11. Origin of the human intra-embryonic coelom, shown by transverse sections,. At two somites; *B*, at seven somites. The right half of each section is somewhat more advanced than the left.

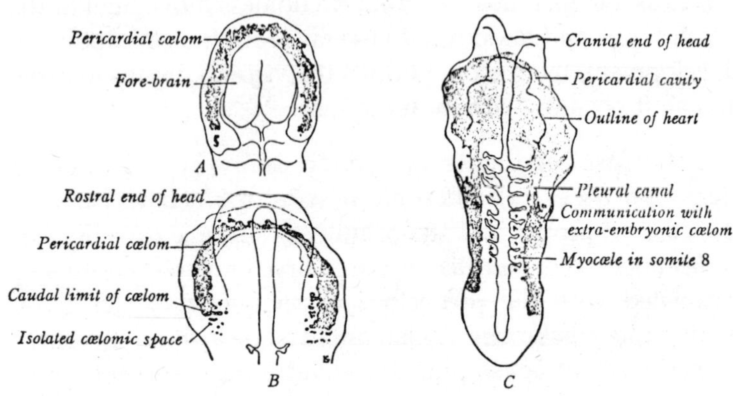

Fig. 7.12. Early coelom of human embryos, in dorsal view (adapted). *A*, At one somite; *B*, at two somites; *C*, at nine somites.

appear, lateral to the somites, as fast as differentiation of the embryo in a tailward direction permits. These then link up progressively to extend the coelomic cavities caudad. In the region where the lungs will develop just caudal to the heart, the coelom remains as two separate canals. At the heart-lung level the head-end of the embryo is separating from the underlying blastoderm, and the body cavity does embryo is separating from the underlying blastoderm, and the body cavity does not connect laterally with the extra-embryonic coelom. Caudad to each prospective lung-region the embryonic coelom communicates freely with the extra-embryonic coelom. Thus, in an embryo about 2.5 mm. long, the coelom of the embryo comprises a -shaped system; the thick bend of then corresponds to the *pericardial cavity*, whereas the right and left limbs may be called *pleural canals* at this stage. The future peritoneal cavity communicates broadly with the extra-embryonic coelom of each side.

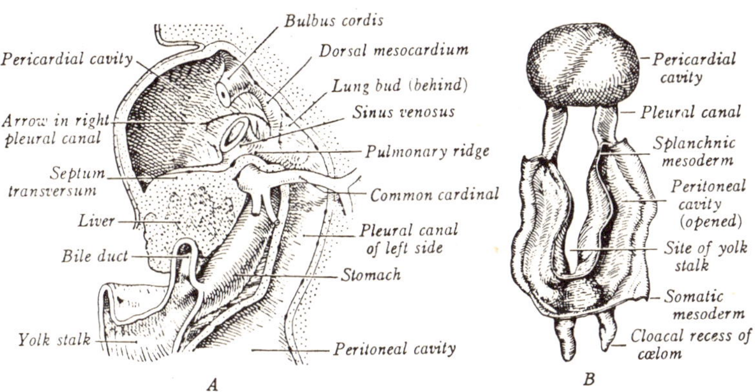

Fig. 7.13. Reconstructions of the body cavities in human embryos of 3 mm. *A*, Section, cut at left of the midplane, viewed from the left side. *B*, The coelomic system, isolated, in ventral view.

The earliest coelom within the embryo occupies a flat, horizontal plane, but the forward growth of the head-end of the embryo and the accompanying reversal of the cardiac region presently swing the pericardial cavity to a more ventral position beneath the embryo. This single, pericardial chamber still communicates at a right-angled bend with the paired pleural canals, which now lie more dorsally. Farther caudad the pleural canals, in turn, connect with the *peritoneal cavity*; as the gut and abdominal wall begin to fold off, this cavity comes to be a single, common chamber. At the end of this early period the coelomic system thus consists of a single pericardial cavity and a single peritoneal cavity, interconnected by a pair of pleural canals. As the embryo continues it folding and elongation, the peritoneal chamber is separated progressively from the extra-embryonic coelom; the last region of closure is at the site of the developing umbilical cord.

Three specialized portions of the intra-embryonic coelom will not be considered in the account that follows: One, the *myocoeles* or tiny cavities of the mesodermal segments, disappears early and has only an historical significance. A second region, the temporary *umbilical coelom* within the umbilical cord at the time of intestinal herniation, has already been discussed. The third portion, the saccular *vaginal processes,* extends from the inguinal region of the abdominal cavity into the scrotum; their development will be described.

The division of the continuous, primitive coelom into separate, permanent cavities is accomplished through the development of the three sets of partitions. They are: (1) the unpaired *septum transversum,* which serves as an early, partial diaphragm; (2) the paired *pleuro-pericardial membranes,* which join the septum and complete the division between pericardial and pleural cavities; and (3) the paired *pleuro-peritoneal membranes,* which also unite with the septum and complete the partition between each pleural cavity and the peritoneal cavity.

The Septum Transversm. When the pericardial region undergoes the reversal of position that brings it beneath the embryo proper, the original cranial margin of the pericardium becomes its definitive caudal wall. This unsplity mass of mesoderm then constitutes a transverse partition occupying the space between the gut, yolk stalk and ventral body wall (Fig. 8. *A*). Standing thus between the pericardial and abdominal cavities, it is called the *septum transversum.* It is, however, an imperfect septum since the paired pleural canals, which connect the pericardial and abdominal portions of the general coelom, course dorsally above the septum on each side. Such permanent communications between the pleural and abdominal cavities characterize amphibia, reptiles and birds. Sharply contrasted are the mammals which supplement this partial septum with additional membranes; these complete the isolation of the pericardial, pleural and peritoneal cavities and, in so doing, produce a true *diaphragm*.

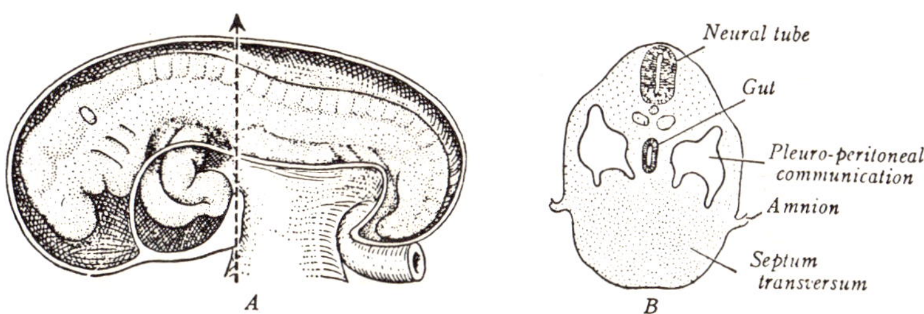

Fig. 7.14. Relation of the septum transversum to the coelom of a human embryo of four weeks. X 25. *A,* Opened coelom, viewed from the left side. *B,* Transverse section, in the plane indicated by the arrow in *A,* showing the position and relations of the septum.

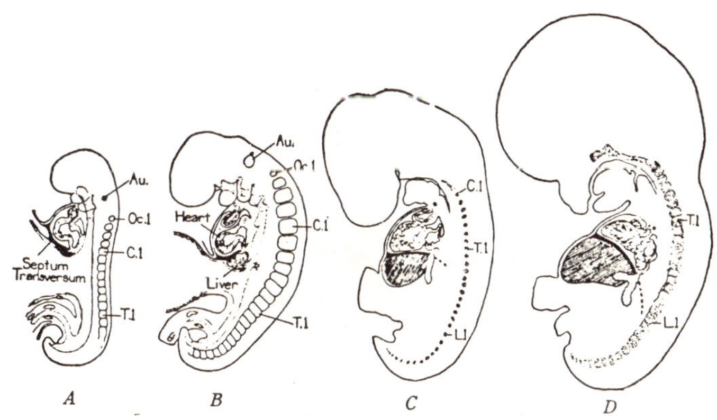

Fig. 7.15. Stages illustrating the 'caudal migration' of the human septum transversum to its final position at two months (Patten). *A,* At 2 mm. *B,* at 3.6 mm. *C,* at 11 mm. *D,* at 25 mm..

Only the cranial part of the original *septum transversum* continues in its role as an actual partition. The liver bud penetrates the more caudal portion of the septum and, as the liver increases in size, this caudal mass draws away, thus producing the ventral mesentery (Containing the liver) as has been described in an earlier chapter. Since both the primitive heart and liver about against the septum, the stems of all the great veins (vitelline, umbilical and cardinal) course through its substance as they join the heart.

The septum transversum of a 2 mm. embryo occupies a position opposite the highest occipital somite. It then enters upon that is usually described as an extensive caudal migration. This displacement, is, however, only relative and is caused chiefly by a faster forward growth of the dorsal body which leaves the more ventral structures behind (in relation to the somites as 'fixed' points of reference). When opposite the fourth cervical segment the septum receives the phrenic nerve, by way of the pleuro-pericardial membrane, and carries it along. The final location of the septum is at the level of the first lumbar segment; this position is attained in an embryo of two months.

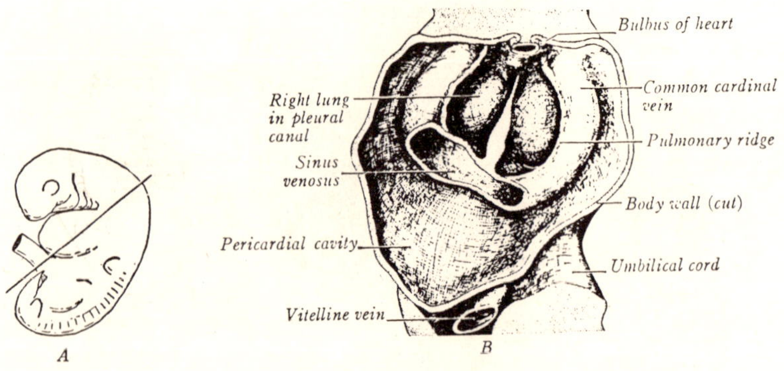

Fig. 7.16. Model of the human pleuro-pericardial cavity, at 5 mm., opened ventrally. The plane of section in *B* is indicated on *A*,

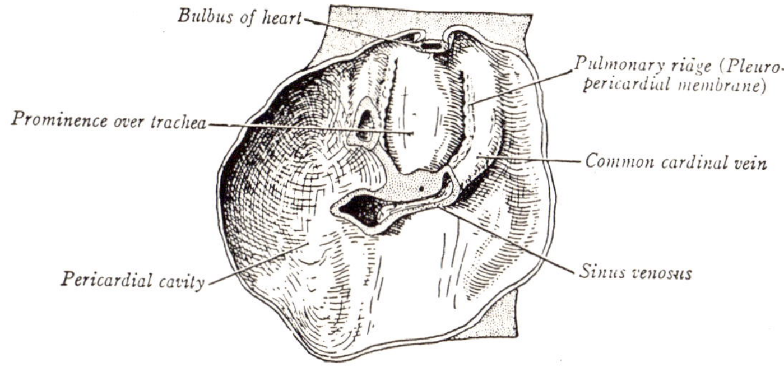

Fig. 7.17.   Model of the human pericardial cavity, at 10 mm., opened ventrally.

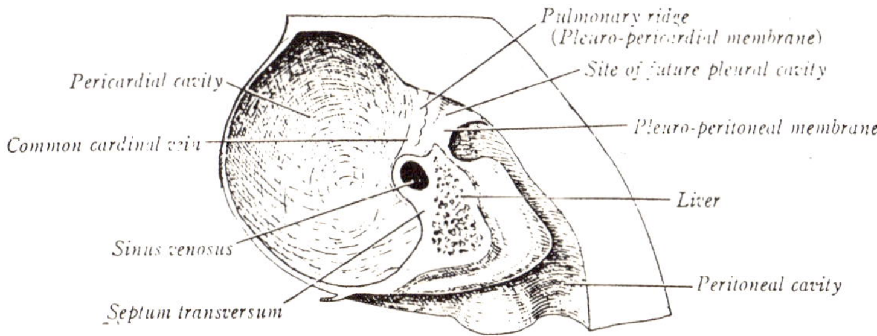

Fig. 7.18.   Model of a right portion of the human coelom, at 5 mm. The cut surface represents a longitudinal section, near the midplane; the heart and lung have been removed.

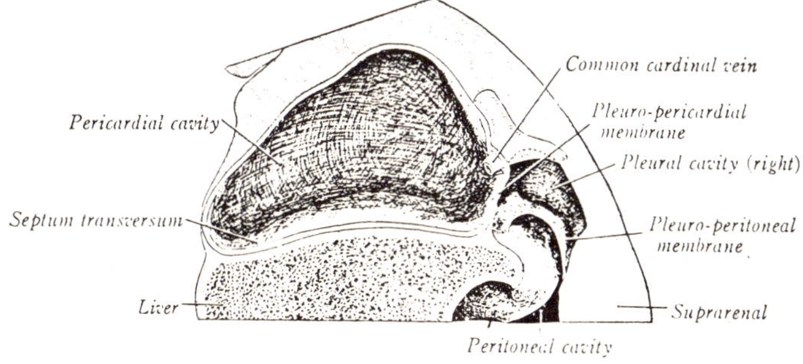

Fig. 7.19.   Model of a right portion of the human coelom at 13 mm. The model is cut longitudinally near the midplane.

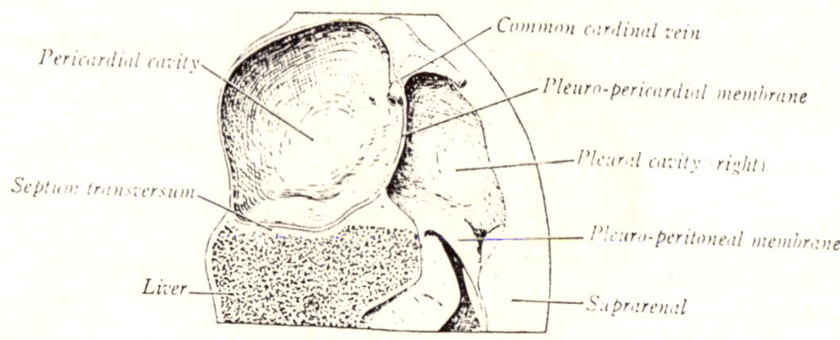

Fig. 7.20. Model of a right portion of the human coelom, at 16 mm. The model is cut longitudinally near the midplane.

The Pleuro-pericardial Membranes: In a 4 mm embryo the lungs begin to develop within the mesial mass of mesenchyme that separates the two pleural canals, and soon bulge into them. The canals thereby become potential *pleural cavities*, and

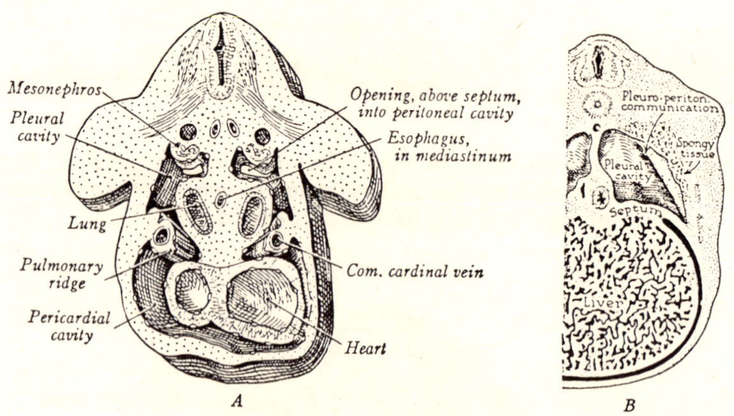

Fig. 7.21. Models of human embryos, cut across the pleural cavities and viewed in a caudal direction. *A,* At 7.5 mm., cut cephalad of the septum *B,* at 16 mm., cut just caudal to the septum.

will be so termed hereafter. At tis period the common cardinal veins (ducts of Cuvier), on their way to the heart, curve around the pleural cavities laterally in the somatic body wall. Each vein courses in a mesodermal ridge that projects mesad into the adjacent pleural canal. This major elevation ends in a projecting, irregular edge known as the *pulmonary ridge* (of Mall). As the common cardinal veins shift toward the midplane they draw out a mesentery-like fold on each side which bears the name of *pleuro-pericardial membrane*. When the membrane of each side presently comes into contact with the median mass of tissue (primitive mediastinum) and fuses with it, the separation of pericardial and pleural cavities is consummated.

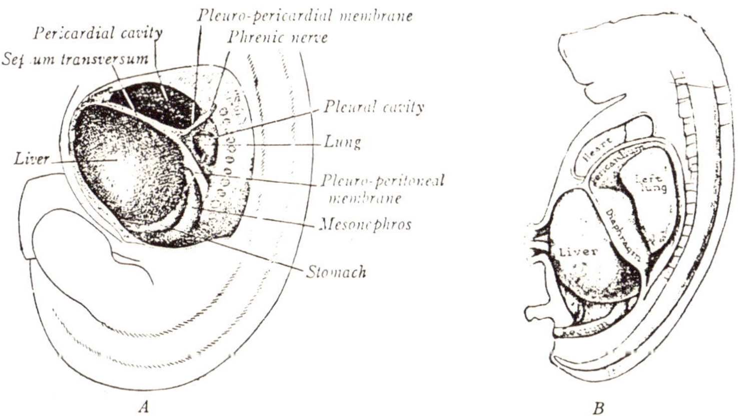

Fig. 7.22. Human coelomic cavities, viewed from the left side after removal of the lateral body wall. *A*, At 11 mm., with incompletely partitioned cavities; note arrow. At 28 mm., After partitioning by the pericardium and diaphragm is complete.

The two stages represent models of the pericardial and pleural cavities after the front half of the body wall had been removed along the plane indicated in. They illustrate the way in which the pulmonary ridges come into relation with the median mass of mediastinal tissue, thereby closing the communication between the pleural and pericardial cavities and masking the lungs from view. In a mere slit still exists, somewhat like the permanent condition in sharks, but this aperture closes in a stage immediately following. Other views of the pleuro-pericardial membrane are displayed in. These illustrations represent models of the body cavities on the right side of the body; they have been exposed by sections cutting a little to the right of the median plane, and the hart and lungs have been removed; the free border of the pulmonary ridge is apparent in the early stage shown in but in an embryo 11 mm. long it joins the median mass of mediastinal tissue. Hence. show the completed (and greatly expanded) pleuro-pericardial membrane.

The Pleuro-peritoneal Membranes: This pair of membranes is largely produced when the lungs can find room for lateral expansion only by invading the adjoining body wall. Representative stages are shown in Figs. At first there is merely a shallow and narrow space between the plumonary ridge, located more cranially, and a quite separate fold now appearing caudally.. The later represents a dorsolateral extension of the caudalmost portion of septum transversum. Soon, however, growth of the lung and shiftings of the liver and common cardinal vein create more room between these pleural boundaries. In such manner a definite pleuro-peritoneal membrane is brought into existence. Continued expansion of the pleural cavity progressively increases the area of this membrane, and of the pleuro-pericardial membrane as well. The opening between pleural and peritoneal cavities becomes reduced during the seventh week and closes by the end of that week (16 mm.). illustrates the relations of the body cavities, at two important stages in their history, as are revealed by lateral dissections.

The Pericardium. The primitive pleural cavities are small. In order to accommodate the rapidly expanding lungs huge extensions are added, so that a major portion of each definitive pleural sac is a new formation brought into existence in the following manner: Since enlargement of the lungs is limited mesially by the mediastinal contents, thy have to grow in other directions. Room is made for the lungs at the expense of the adjacent body wall by the obliteration of its loose, spongy mesenchyme. In this growth the cavities expand, especially in lateral and ventral directions, splitting off additions to the pleuro-pericardial membranes as they advance. Thus the lungs more and more come to flank the heart. The membrane then separating heart from lungs represents not only the original pleuro-pericardial membranes but also the additions to them captured from the body wall. The final, partitioning membrane surrounds the heart like a sac and is named the *pericardium.*

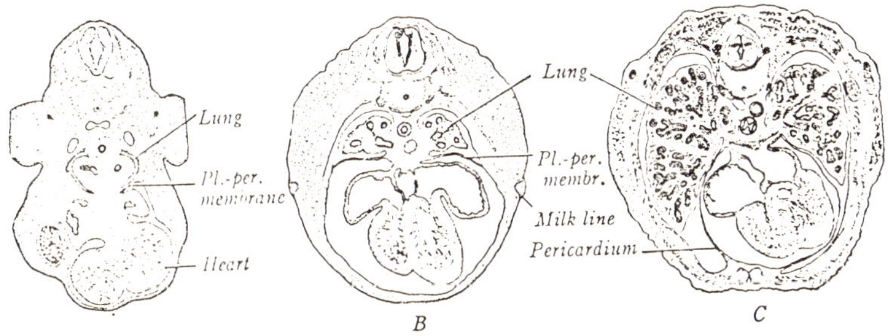

Fig. 7.23. Formation of the definitive human pericardium, illustrated by transverse sections. *A,* At 8.5 mm. *B,* at 16 mm. *C,* at 35 mm.

The Diaphragm: The complete separation of the pleural cavities from the abdomen by a diaphragm is a distinctive mammalian characteristic. It increase greatly the power of inspiration and, in its capacity as a septum, restricts to the thorax the negative pressure produced during inspiration.

The liver grows enormously during the second month, and on both sides some of the adjacent body wall is taken up into the septum transversum and pleuro-peritoneal membranes. The completed *diaphragm* is then derived from four sources. (1) its ventral portion, from the septum transversum; (2) its lateral parts, from the pleuro-peritoneal membranes, plus (3) derivatives from the body wall; (4) lastly, a median dorsal portion is contributed by the meso-esophagus. In addition to these components there is the striated muscle of the diaphragm whose origin is customarily attributed to tissue supposedly derived from the cervical myotomes when the septum transversum still stood at a high level.

Actually the exact source of the muscular component of the diaphragm is not surely established. The early passage of cervical nerve fibres (the phrenic nerve) to the primordial diaphragm is only circumstantial evidence of a concomitant migration of premuscle masses and the retention of their origional nerve supply. On the contrary, the muscle may be derived from the body wall when the burrowing lungs and expanding liver strip off tissue that is added to the periphery of the diaphragm. A central tendinous area of the final diaphragm represents the contribution of the septum transversum.   The mesenterial contribution is easily distinguished. The dorsolateral parts, derived from the pleuro-peritoneal membranes, are not so expansive is man as in some other mammals. It seems probable that the closure of the pleuro-peritoneal communication, as well as that between the pleural and pericardial cavities, is not due primarily to the activity of the membranes themselves but is passive and caused by the growth of adjacent organs. In closing the apertures in the diaphragm it is the liver and suprarenal gland that are chiefly responsible.

Anomalies: The persistence of a dorsolateral opening in the diaphragm, usually on the left side, finds its explanation in the imperfect closure of the pleuro-peritoneal canal. Such a defect leads to the commonest type of *diaphragmatic hernia,* the abdominal viscera projecting to a greater or less extent into the corresponding pleural cavity. *An intact diaphragm, weakened* by being locally deficient in tissue, can also herniate into a pleural cavity, but in this instance the abdominal viscera are contained in a sacculation of the diaphragm. Rarely there is a faulty development of a pleuro-pericardial membrane which permits the cavities containing the heart and lung to communicate. Vaginal sacs of peritoneum which extend into the scrotum may retain their connecting stalks throughout adult life.

THE URINARY SYSTEM

The urinary and reproductive systems are intimately associated in origin development and certain final relations. Both arise from mesoderm that initially takes the form of a common urogenital ridge and both continue to develop in close approximation. Each drains into a common cloaca and, slightly later, into the urogenital sinus which is a subdivision of the cloaca. The history of neither system is simple and direct. Some organs results from the association of structures that were originally quite separate. Other parts arise, only to disappear after a transitory existence during which they may never have functioned. Still other structures, designed for one kind of purpose, abandon their original course and are turned to wholly new use. Certain common primordia transform differently, as will be appropriate to the emerging male or female. Yet because of the interwoven nature of this story, it is far simpler to pursue separate narratives for the urinary and genital systems than to attempt a synchronized, single description of what is often called the *urogenital system.*

Vertebrates have made three distinct experiments in the production of kidneys. Each was an improvement over the preceding type. As might be anticipated, the embryos of the higher vertebrates indicate this progress by repeating the same kidney sequence during development; nowhere can be found a better illustration of the principle of recapitulation. The earliest and simplest excretory organ was the *pronephros*, functional today only in such adult chordates as Amphioxus and myxinoid fishes. The pronephros, nevertheless, does serve as a provisional kidney in larval fishes and amphibians, but it is replaced by the *mesonephros* which remains as the permanent kidney of these animals. The embryos of reptiles, birds and mammals develop first a function less pronephros and then a mesonephros (functional during a part of fetal life), whereas the final kidney is a new organ, the *metanephros*. These three kidneys develop overlappingly, one caudad of the other, in the order indicated by their names.

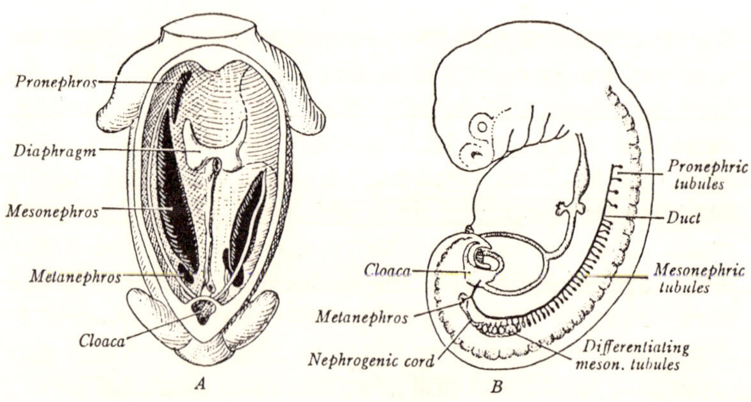

Fig. 7.24.   Locations and relations of the three kidney-types in mammals (semi-diagrammatic). *A*, Ventral dissection, the left side showing a later stage than the right. *B*, Lateral view, with the nephric system and gut showing through.

All three kidney types are organs composed of units known as *uriniferous tubules*, which have a common source of origin and exhibit somewhat the same structural plan. They arise from the mesoderm of the nephrotome; this plate lies just lateral to the somites and connects the latter with the somatic and splanchnic layers of mesoderm which enclose the coelom. In close relation with all three types of secretory tubules (*nephrons*) there is a vascular tuft (*glomerulus*), specialized for separating urinary wastes from out the blood. The collected waste products are then conducted by *collecting tubules* to a common *excretory duct* which discharges them from the body. All three kinds of kidneys, as well as both kinds of gonads, differ from other exocrine glands in that their secretory tissue differentiates from a structureless, cellular blastema. Union with the system of excretory ducts is secondary, whereas other glands owe their origin to a direct budding and branching from the excretory duct.

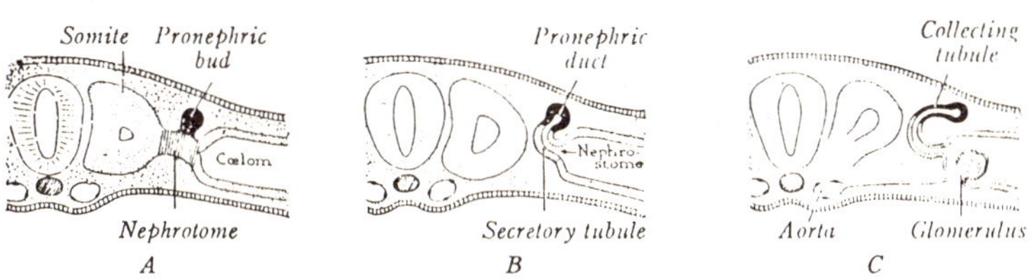

Fig. 7.25. Development of the pronephric tubule, illustrated by transverse sections of early embryos.

The Pronephros

The functional pronephros of lover vertebrates consists of paired *pronephric tubules*, arranged segmentally. One end of each tortuous tubule opens into the coelom, the other into a longitudinal *pronephric duct* which is excretory in function and drains into the cloaca. The ciliated, funnel-shaped communication with the body cavity is the *nephrostome*. Near by, but entirely separate from each tubule, an arterial tuft projects into the coelom. These external *glomeruli*, covered only by thin epithelium, filter wastes from the blood into the coelom. The mixture of urine and coelomic fluid is then taken up by the tubules and carried by means of ciliary currents into the main excretory duct. As implied by its name, the pronephros is located well cephalad in the body; for this reason it has often been called the 'head kidney'.

Although the human pronephros is vestigial, it is as well developed as that of any reptile and better represented than in birds and other mammals. It consists of about seven pairs of rudimentary pronephric 'tubules,' arising as dorsolateral sprouts from those segmentally arranged nephrotomes that border on the region of junction of the future neck and thorax. The distal, or free ends of these solid nodules bend backward, canalize and unite into a longitudinal collecting duct. Caudal

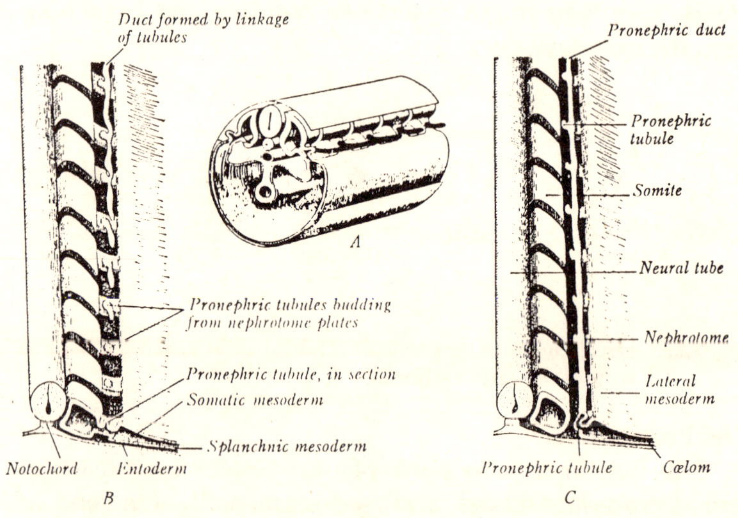

Fig. 7.26.   Development of the pronephric system, illustrated by models *A*, Relation of the pronephric system (ducts in black) to the embryo as a whole. *B*, Younger stage, with tubules still forming and linking together. *C*, Older stage, with tubules and duct completed.

to the lowest tubule (somite 14) the free end of the collecting duct, by a process of terminal growth, pushes between the ectoderm and the nephrotomes until it reaches the lateral wall of the cloaca and perforates it. Thus are formed the paired primary excretory ducts, which at this period bear the name of *pronephric ducts*.

The pronephric tubules develop from the nephrotomes associated with the seventh to fourteenth somites, and sometimes from more cranial nephrotomes as well. The tubules begin to appear in embryos with nine somites and, at the 23-somite stage, all have been formed. Soon afterward, in 4 mm. embryos, the two pronephric ducts reach the wall of the cloaca and promptly communicate with its lumen. The earliest tubules begin to degenerate before the last in the series appear. The degeneration of tubules is virtually complete at the 5 mm stage, but the pronephric ducts persist and acquire new relations which will be described in succeeding paragraphs. At their basal ends the original solid pronephric sprouts may hollow out and open into the coelom as rud' mentary nephrostomes. Nodular representatives of external glomeruli are sometimes seen.

The Mesonephros

The mesonephros, or Wolffian body of vertebrates, is larger than the pronephros; not only does it contain more tubules, but also these are longer and more complicated. It is located farther caudad and is appropriately named the 'middle kidney.' Unlike the pronephros, the primordium of the mesonephros differentiates into tubules only; these drain into the pronephric duct which is retained as the main excretory canal and is henceforth known as the *mesonephric* (or Wolffian) *duct*. Whereas the pronephros is entirely functionless is higher vertebrates, the mesonephros serves these embryos as a temporary excretory organ that overlaps the initial activity of the permanent kidney. In most, but not all, mammals function is attained; even in man, whose mesonephros is not large, this apparently true.

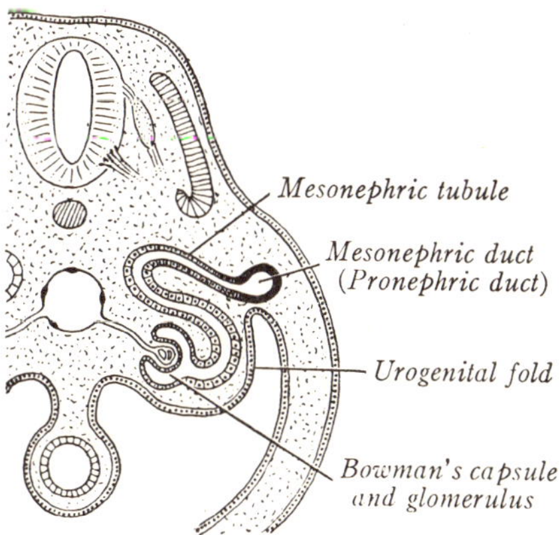

Fig. 7.27.   Form and relation of a human mesonephric tubule, shown in a transverse section at 5 mm.

The mesonephros, like the pronephros, consists essentially of a series of tubules, each of which at one end becomes associated with a knot of blood vessels and at the other end opens into the mesonephric duct. But the mesonephric tubule differs in two important respects : (1) the glomerulus is internal (*i.e.*, it indents the blind end of the tubule, and excreta from the blood pass directly into the lumen of the tubule); and (2) the nephrostome is at best transitory and never serves as a functional coelomic mouth to the tubule proper.

The mesonephric tubules arise caudal to the pronephros and from the same general source, the nephrotome region. In man, however, only a few of the more cranial tubules trace origin to segmentally repeated nephrotomic masses, for caudal to the tenth pair of somites this bridging mesodermal tissue remains unsegmented. Nevertheless, it does retain the same potentialities, and, in preparation for tubule formation, separates into a continuous longitudinal bar; this so-called *nephrogenic cord* extends caudad as far as the twenty-eighth somite. As a whole, the mesonephric tubules bear no significant relation to the body segmentation, and commonly two or three (but even as many as nine) arise alongside a single somite.

Differentiation of the Mesonephros: Tubules take origin from spherical masses of cells, brought into existence by subdivision of the nephrogenic cord. They appear progressively, adding to the series chiefly in a caudal direction. Each spherical mass of mesonephrogenic tissue hollows and the vesicle, so formed, sends out a solid extension which unites with the pronephric (now mesonephric) duct near by *(B)*. To complete a mesonephric tubule there is further canalization, growth with S-shaped bending, and association with a glomerulus *(C,D)*. The free end of the tubule enlarges and becomes thin-walled, as a knot of blood vessels (the *glomerulus*) indents one side. The double-walled vesicle, thus invaginated like a simple gastrula, is the *glomerular* (or Bowman's) *capsule;* the capsule and glomerulus together comprise a unit known as the *mesonephric corpuscle.*

Next in order comes a thicker, lighter-staining *secretory segment* of the tubule and then a thinner, darker-staining *collecting segment* which, in turn, opens into the mesonephric duct *(B)*.

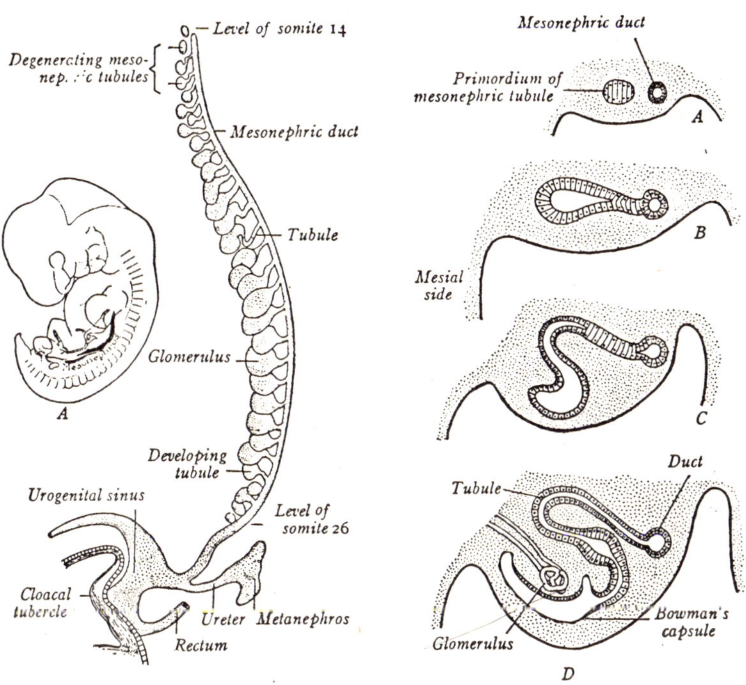

Fig. 7.28.   Location and composition of the human mesonephros. *A*, At 8 mm. *B*, At 10 mm., showing the mesonephric region reconstructed in greater detail.

Fig. 7.29.   Stages in the differentiation of a human mesonephric tubule, shown in simplified sections.

When the developing tubules begin to enlarge, there is not room for them in the body wall and they accordingly bulge ventrad into the coelom. On each side of the dorsal

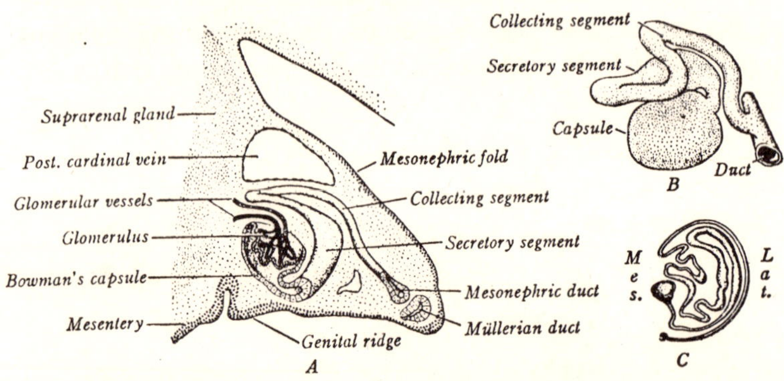

Fig. 7.30.   Models of mature mesonephric tubules. *A*, Human tubule, at
10 mm., partly opened and superimposed on a section of the
left mesonephric ridge *B*, Human tubule, at 13 mm. *C*, Pig's
tubule, at 80 mm.

mesentery there is thus produced a longitudinal *urogenital ridge*, which attains its greatest relative length at about five weeks (7 mm.) by extending a distance of some 15 somite. Soon after its formation this common fold subdivides longitudinally into a more lateral *mesonephric ridge* and a medial *genital ridge (B,C)*. After the fourth week progressive degeneration of the more cranial tubules and continued new formation at the caudal end of the mesonephric ridge effect a wave-like settling of the gland caudad. As a result of this, the upper five-sixths of its extent is lost by the end of the second month. The cranial remnant is reduced to a band known as the *diaphragmatic ligament* of the mesonephros; it presently will serve as a *suspensory ligament* of the gonad. In the remaining one-sixth, new tubules seemingly arise partly by the budding and splitting of those already present. At the end of ten weeks, nevertheless, no tubule remains unbroken at some level. How the male genital system salvages the mesonephric duct and remnants of surviving tubules, and utilizes them for new purposes, will be traced in the following chapter. For the present it will be necessary to consider further the contribution of the mesonephric duct to the permanent urinary system.

The formation of mesonephrogenic spheres commences in embryos with about 18 somites. These appear first opposite the fourteenth somite, whereupon new primordia differentiate chiefly in a caudal direction although some are added above the initial level. Thus in a 5 mm embryo (with nearly all its somites present) the cephalic limit is reached at the ninth somite (definitive sixth cervical), so that the highest tubules overlap those of the pronephros. At 7 mm. the caudal limit is reached at the twenty-sixth somite (fourth lumbar). Embryos four to nine weeks old have a rather constant number of about 30 tubules in each mesonephros, and within these time limits the gland reaches the height of its development. In all, a maximum number of about 80 pairs of tubules is possible, of which some 34 pairs still persist at nine weeks. Half of these are already non-functional, while within another week all become discontinuous; yet the maximum degeneration attained is not complete until the end of the fourth month..

The glomeruli occupy a median column in the gland; the duct is lateral and the tubules are intermediate and dorsal in position. Lateral branches from the aorta supply the glomeruli, while the posterior cardinal veins, dorsally placed, break up into a network of sinusoids about the tubules; these latter channels are continuous in turn with the subcardinal veins and constitute a true renal-portal system, as in lower vertebrates. The mesonephros of man, along with that of the cat and guinea pig, is somewhat small *(B)*. By comparison, the mouse and rat have a tiny, non-functional gland, while that of the sheep is medium-sized and the pig and rabbit have extremely large mesonephric with more completely coiled tubules *(C)*. In a general way these varying degrees of size and differentiation are in inverse relation to the efficiency of the placenta as an early excretory organ.

The Metanephros

The permanent kidney of amniotes (reptiles, birds and mammals) arises far caudad in the body. As in the case of the mesonephros, the final kidney consists of an aggregate of

to the permanent kidney.

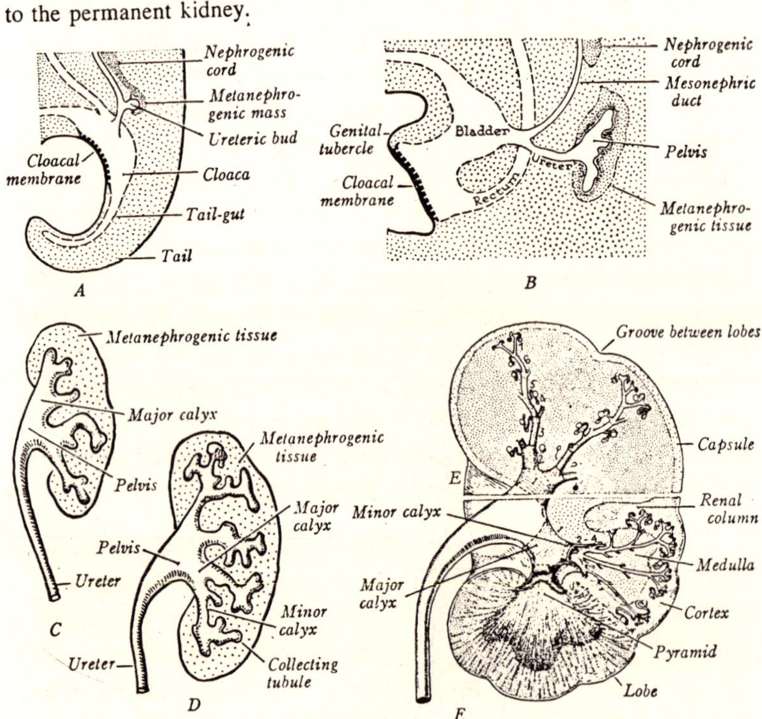

Fig. 7.31. Origin of the human metanephros and the development of its duct system, as shown by reconstructions. *A,B,* Origin and early relations (*A,* at 5 mm., *B,* at 11 mm.,. *C-F,* Further stages in the development of the ureteric bud into the duct system (*C,* at 15 mm. ,*D,* at 20 mm., *E,* at nine weeks, *F,* at birth.

tubules which drain into a common duct. Also like the mesonephros, the metanephros is of double origin; but in this instance the boundary between the two components lies midway of the uriniferous tubules themselves. Thus, the inert system of drainage ducts (ureter; pelvis; calyces; papillary ducts; and straight collecting tubules) is derived from a bud growing off the mesonephric duct. On the other hand, each secretory unit, or *nephron* (Bowman's capsule, both convoluted tubules and Henle's loop), differentiates from the substance of the caudal end of the nephrogenic cord; it thus has an origin similar to that of the entire mesonephric tubule. A collecting and secretory tubule then unite secondarily to complete a continuous *uriniferous tubul.* Yet in structure and function these two components remain as different as was their origin.

The mesonephric duct makes a sharp bend just before joining the cloaca. It is at this angle (level of the twenty-eighth somite, or the future first sacral vertebra) that the so-called *ureteric bud* soon arises, dorsal and somewhat medial in position. The primordium appears in embryos of four weeks (5 mm.), taking the form of a hollow bud which first grows dorsad and then turns cephalad. The proximal, rapidly elongating stalk of this diverticulum is the future *ureter*, while the distal, blind end dilates at once into the primitive *renal pelvis (B).* Shortly after its appearance the ureteric bud pushes into a a mass of condensed tissue which is the caudalmost portion of the nephrogenic cord *(A).* This *metanephrogenic mass* promptly separates from the more cranial mesonephrogenic tissue and then surrounds the pelvic dilatation like a cap *(B).* Such are the early primordia that jointly give rise to the permanent kidney.

Differentiation of the Ureteric Bud: The ureteric stem elongates and, toward the end of the sixth week, the primitive renal pelvis both flattens from side to side and subdivides into the two future *major calyces.* From these calyces secondary branches bud out, which in turn give rise to tertiary branches *(D).* The branching process is repeated still further until, at five months, some twelve generations of collecting tubules have been developed *(E,F)*; this amount of branching increases but little thereafter. The several secondary tubules soon enlarge and absorb into their walls the branches of the third and fourth order; each of the approximately nine composite receptacles so formed is a *minor calyx (E,F).* As a result of the absorption just mentioned, the tubules of the fifth order, some 25 in number, then open into each minor calyx as *papillary ducts*. The remaining higher orders of diverging tubules constitute the straight *collecting tubules*; these are abundant in the *medulla* of the definitive kidney and they also project into the *cortex* where they course in the rays (*pars radiata*).

The aggregate of all the tubular 'trees' (whose trunks are papillary ducts) that drain into any one, minor calyx comprises a renal unit known as a *pyramid*; its base adjoins the overlying cortex and its apex, or *papilla*, projects into the calyx. Later these primary pyramids are subdivided into two or more secondary pyramids in such a way that each papilla comes to serve as a common outlet for the several definitive pyramids (F). The human kidney, with about nine papillae, thus differs from the general primate plan of a single great pyramid and papilla.

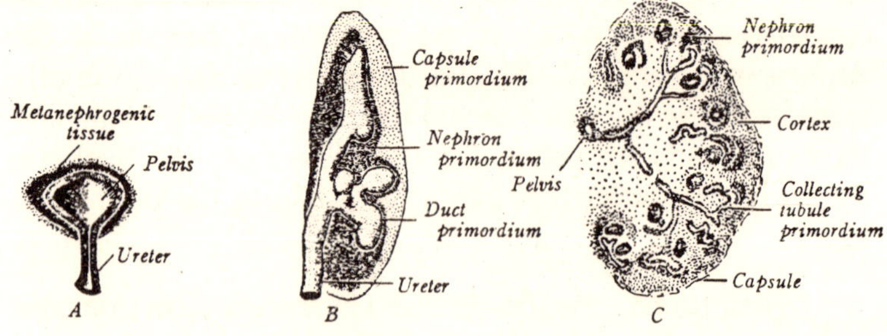

Fig. 7.32.   The early differentiation of human metanephrogenic tissue. *A*, Halved model, at 8 mm. *B*, half model, at 12 mm. *C*, section, at 20 mm.

The simple epithelium of the collecting tubules and papillary ducts elevates to a distinctive, pale columnar type. By contrast the calyces, pelvis and ureter differentiate into a stratified, so-called transitional epithelium; these parts of the urinary tract become invested with coats of smooth muscle and connective tissue produced by specialization of the surrounding mesenchyme.

Differentiation of the Metanephrogenic Tissue: The early cap of tissue that covers the dilated, pelvic portion of the ureteric bud shows two layers of different density. The internal, denser layer differentiates into the secretory tubules, or *nephrons,* whereas the external, looser layer is destined to become the interstitial connective tissue of the kidney and its enveloping *capsule (B,C).*

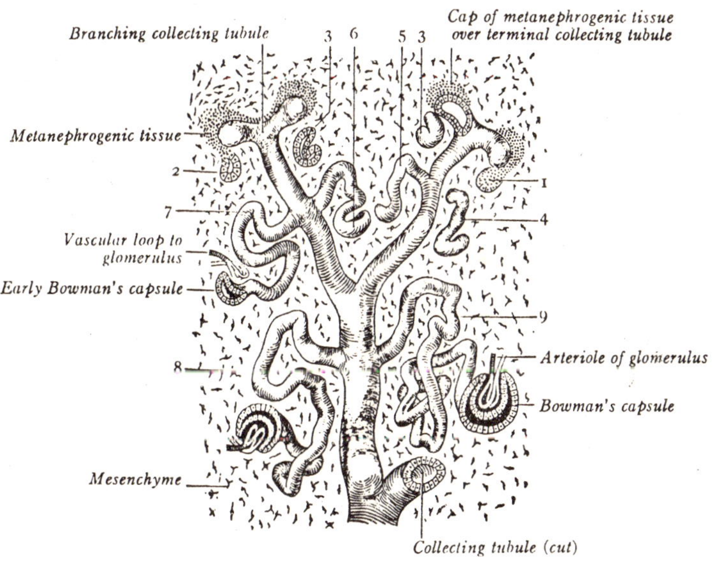

Fig. 7.33. Composite diagram, illustrating stages in the differentiation of nephrons and their linkage with branching collecting tubules. Progressive stages are numbered serially, 1-9.

When the primitive renal pelvis starts on its programme of branching, the internal layer of the metanephrogenic tissue subdivides into a corresponding number of masses. One such lump covers the end of each pelvic subdivision. As new orders of collecting tubules arise progressively, each mass of metanephrogenic tissue not only increases steadily in amount but also subdivides in the same rhythm *(C)*. A small lump is left behind in association with each terminal collecting tubule while other parts of the mass are lifted to higher and higher levels as the stem tubules advance. The shell of tissue that thus comes to overlie the bases of the pyramids is the *cortex* of the kidney; some of its substance fills in the spaces between individual pyramids and is there designated as *renal columns*.

The renal cortex consists of two kinds of territories. The *pars radiata*, or rays, result from the massive extension of radial bundles of collecting tubules, as already explained. The *pars convoluta*, or labyrinth, consists of the aggregate of secretory tubules differentiated from out the metanephrogenic tissue. Each tiny ball of this formative substance is the forerunner of a *secretory tubule*, or nephron, whose developmental course can be followed. A ball hollows into a vesicle which elongates and becomes tortuous. At one end a Bowman's capsule and glomerulus differentiate, while the other end establishes communication with the nearby collecting tubule. Since the newer generations of tubules develop progressively in a capsular direction from the self-perpetuating metanephrogenic tissue, it follows that the oldest tubules are those nearest the medulla. The differentiation of new tubules terminates about one month before birth in the zone just beneath the capsule; at this time above one million have been produced in each kidney. All later increase in renal size results from the enlargement of tubules already present.

The details of the tubule-differentiation are as follows: During the seventh week some of the nephrogenic tissue about the ends of the collecting tubules condenses into spherical masses; these hang down in the angles between the end-buds of collecting tubules and their parent stems. One such metanephric sphere is the fore-runner of each secretory tubule. The formation of new spheres and their transformation into

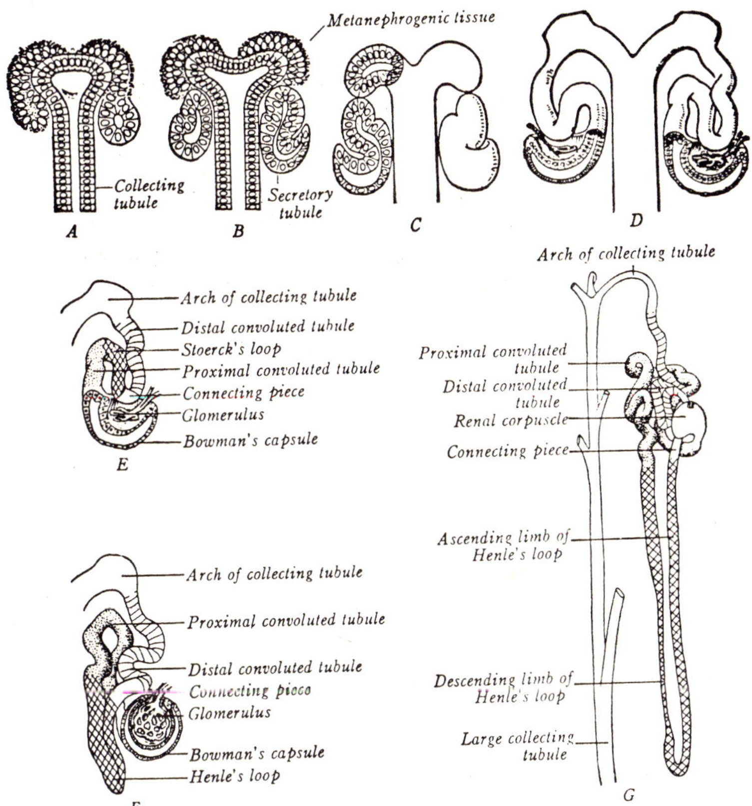

Fig. 7.34.   The differentiation of human uriniferous tubules (after Huber). *A-D*, stages so arranged that the left tubule in each drawing illustrates an earlier condition than the right tubule. *E-F*, Reconstructions differentially marked to show the changing relations during the growth and specialization of a tubule.

tubules continue at progressively higher levels as the cortex thickens and the stem tubules continue to branch. The stage of a solid sphere is soon converted into a vesicle with an eccentrically placed cavity (*A,B*). The vesicle then elongates, thereby producing an S-shaped *secretory tubule (C)* which unites at one end with the adjacent terminal *collecting tubule (D)*. The thinner-walled, blind end of the tubule becomes the *glomerular capsule* (Bowman's) of a renal corpuscle *(D,E)*. The stage of the S-shaped tubule is followed by marked elongation and twisting *(F,G)*.

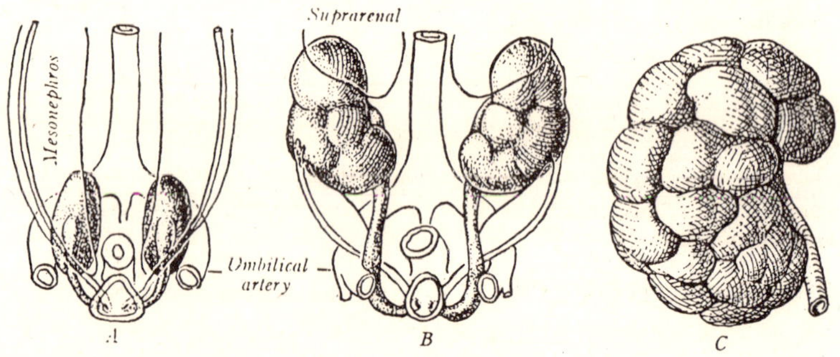

Fig. 7.35.   Ascent and lobation of the human kidney, *A,B,* Shift in position between six and nine weeks, and early lobation. C, External lobation at birth.

The fully formed uriniferous tubule is arranged in a definite and orderly manner. Beginning with *Bowman's capsule* each tubule consists of a *proximal convoluted* portion, a U-shaped *loop* (of Henle) with descending and ascending limbs, a *connecting piece* which lies close to the renal corpuscle and a *distal convoluted* portion continuous over into the collecting tubule. These parts are derived from the S-shaped primordium in a manner more easily traced by the differential markings in Fig. 8.39 *E-G* than through a written description. It should, however, be noted that the primitive loop (of Stoerck) includes not only the definitive Henle's loop but a portion of the proximal convoluted tubule as well. Into the concavity of Bowman's capsule first grows the afferent arteriole, then the capillary loops of the *glomerulus* differentiate, and from them the efferent arteriole finally buds outward. The concavity of the capsule is at first shallow (*E*). Later, the walls of the capsule grow about and enclose the vascular knot except at the point where the arterioles enter and emerge (*F*). The first few generations of secretory tubules are temporary, or provisional, and ultimately degenerate. There is considerable specialization of the original epithelial lining of the secretory tubules to produce the characteristic modifications of the simple layer that are encountered at the various levels of a functional tubule.

Other Features: The kidneys are unusual among the organs of the body by coming to lie at a higher level than their site of origin. In this shift the ureters do elongate, but it is perhaps a straightening of the body curvature at this time that chiefly brings about a displacement in which the kidney primordium seems actually to move cephalad. This displacement is through the distance of four somites, and at eight weeks the center of the kidney is already at its permanent level (opposite the future second lumbar vertebra). A good deal of the apparent migration of the kidneys and the corresponding elongation of the ureters is due to the marked growth of the lumbar and sacral regions of the body. As the kidneys rise out of the pelvis they undergo a rotation of 90 degrees, whereby the original dorsal border becomes the convex, lateral border and the hilus faces mesad rather than ventrad. At all times the kidney is retroperitoneal in position.

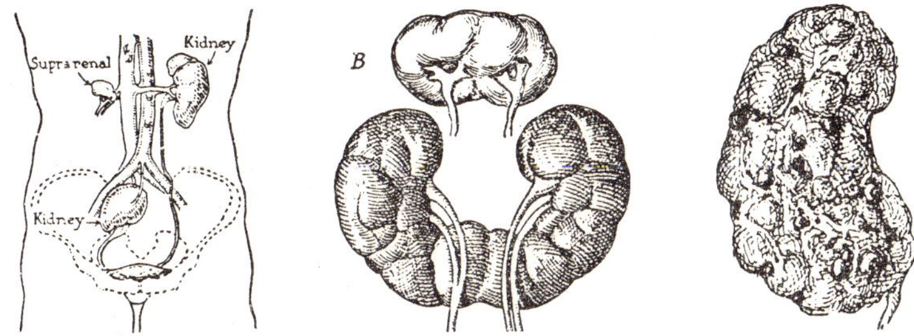

Fig. 7.36. Anomalies of the human kidney and ureter. *A*, Unascended right kidney. *B*, Fused, unascended kidneys forming a 'cake.' *C*, Horse-shoe kidney, with fetal lobation retained; the right ureter is cleft, the left double. *D*, Congenital cystic kidney.

As the Cortex organizes, the region over a primary pyramid becomes demarcated from similar territories by a deep groove; later the pyramids subdivide and this is accompanied by a corresponding increase in the number of cortical areas *(F)*. These *lobes* may reach a maximum of twenty in the fetus, but this external lobation disappears progressively in infancy and early childhood as the grooves fill in.

The human kidney is capable of secretion early in the third fetal month. Since, however, excretion is adequately performed by the placenta, renal function is not necessary before birth. But even through the physiological conditions are unfavourable to efficient renal activity, urine is produced slowly. Not only does the bladder filled in the early months, but also some urine voids into the amniotic sac; this, in turn, is drunk along with the amniotic fluid proper.

Causal Relations: The pronephros seems to originate through an inductive influence emanating from the notochord. The differentiation of the mesonephric tubules is induced by the nearby pronephric duct. In the absence of a definite ureteric bud, growing out of the mesonephric (pronephric) duct and into the nearby metanephrogenic tissue, the latter is unable to differentiate and hence no kidney is formed.

Anomalies: One or both kidneys may be lacking. A primary cause of such agenesis lies in the failure of the mesonephric ducts to develop ureteric buds. Partial developmental failure is shown in a kidney that appears but remains stunted and poorly differentiated. Another kind of failure is *non-ascent*, or the retention of the primary pelvic position of the organ. This may involve one or both kidneys; when both remain they commonly fuse into a *'cake' (B)*. A joined, common kidney may lie on one side of the body and drain into one normal and one crossed-over ureter. Most commonly kidneys join only by their lower ends, forming a *horse-shoe* kidney, and these may or may not ascend properly *(C)*. Such union presumably results from early fusion due to a converging course taken by the ureteric buds. An arrest by which the external lobation persists *(C)* duplicates the normal adult condition in reptiles, birds and some mammals *(e.g.,* whale; bear; ox).

Opposite in nature is developmental excess. Supernumerary kidneys are rare; they are due either to an extra primordium or to the subdivision of an ordinary one. Double ureters and renal pelves that supply a single kidney result from two ureteric buds growing out of the same mesonephric duct; a branched ureter and double pelvis trace origin to the forking of a single ureteric bud *(C)*. Congenital *cystic kidney* is usually bilateral. It is characterized by the presence of blind secretory tubules that become dilated cysts through the retention of fluid *(D)*. The cause has been attributed to the primary non-union of secretory and collecting tubules, to the cystic degeneration of secondarily detached tubules, and to secondary obstructions.

The Cloaca

The Primitive Cloaca: Vertebrates below the placental mammals retain a common entodermal chamber into which fecal, urinary and reproductive products all pass, and from which they are expelled to the exterior. Higher mammals have subdivided this *cloaca* into a dorsal rectum and a ventral bladder, urethra and urogenital sinus. In such manner two separate outlets were gained for fecal and urogenital discharge. These changes were consequent on the evolution of an external penis in higher mammals; cloacal subdivision has also brought into existence a *perineum,* separating the rectal orifice from the urogenital vent. The developmental course of the human cloaca, before complete division is attained, recapitulates several stages permanent in lower mammals.

In human embryos with six somites the future cloaca is merely a blind, caudal expansion of the hind-gut which already stands in contact ventrally with the ectoderm. This area of union between ectoderm and entoderm constitutes the *cloacal membrane,* a region originally caudal to the primitive streak that has been turned under by the tail fold. At first the cloacal membrane extends from the tail bud to the body stalk, but later this expanse is diminished relatively by the ingrowth of mesoderm to produce the infra-umbilical belly wall *(B)*. At its cephalic end the cloaca gives off the ventrally directed allantoic stalk; laterally the cloaca receives the mesonephric ducts, while it is prolonged caudad as the transitory tail-gut.

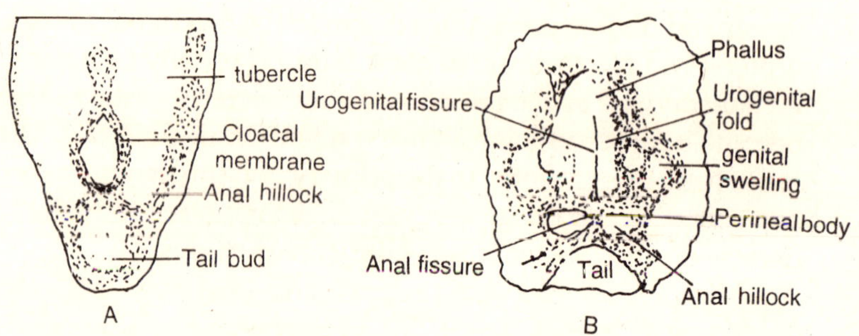

Fig. 7.37. Region of the human cloacal membrane, in ventral view. *A,* at 3 mm.; *B,* at 21 mm.

Subdivision of the Cloaca: The facing walls of the hind-gut and allantois meet in a saddle-shaped notch, or fold, whose apex points caudad. The wedge of mesenchyme filling this interval is the so-called *cloacal septum;* it is also known as the *urorectal septum.* This mesenchymal mass pushes caudad as the fold advances, thereby dividing the cloaca into a dorsal *rectum* and a ventral *bladder* and *urogenital sinus.* Division is completed during the seventh week.

Even at the end of the sixth week (11 mm.), before the cloacal division is wholly finished, certain future regions can be recognized in the ventral half. The *bladder* is continuous with the allantois and, at its caudal end, receives the two common stems of the paired mesonephric ducts and ureters. These stems also mark the approximate upper end of the primitive urogenital sinus, which shows two emerging regions. Nearest the bladder there is a *pelvic portion;* more distad is the *phallic portion*, which extends into the genital tubercle. Presently these two parts become clearly defined in both sexes; their fates will be explained in a subsequent paragraph.

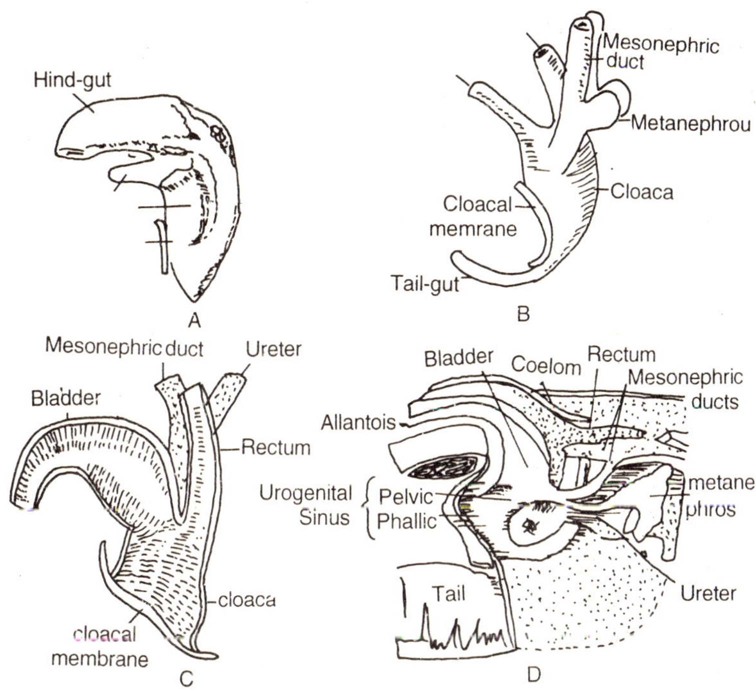

Fig. 7.38. Partial division of the human cloca, illustrated by models viewed from the left side. *A, B,* At 3.5 mm. and 4 mm., respectively; *C,* at 8 mm.; *D,* at 11 mm. An asterisk indicates the position of the cloacal septum in *A, B.*

The Perineum: When the cloacal septum extends to the level of the cloacal membrane, rupture of this entodermal-ectodermal plate follows promptly (7 weeks). For this reason, separate *anal* and *urogenital membranes* have no real existence as such. The rupture exposes the caudal edge of the septum, which is, naturally, surfaced with the entoderm of the advancing fold. This projecting wedge, interposed between anus and phallus, is the *perineal body*. The external fissure, resulting from the disappearance of the cloacal membrane, is closed again in its middle region by the merger of the perineal body with lateral folds flanking the fissure. The area so produced, covered finally by ectoderm and marked by a median raphe, is the primary *perineum*. Hillocks, located behind the anus, encircle its orifice and create a definite proctodeum or *anal canal*. This canal is lined with ectoderm as far as the level of the now ruptured anal portion of the cloacal membrane.

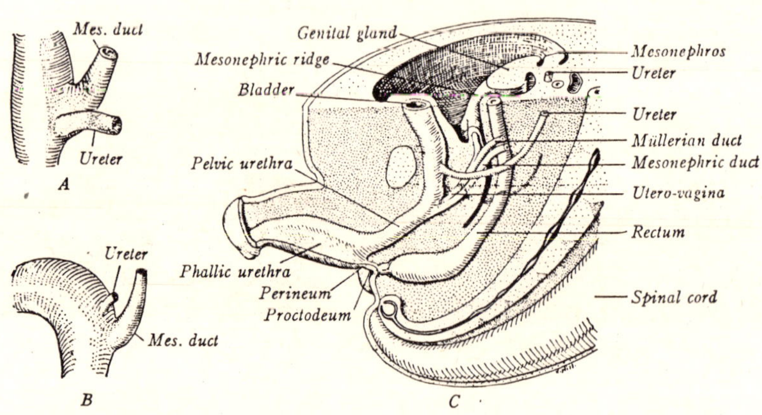

Fig. 7.39. Completed division of the human cloaca, and associated changes, shown by models viewed from the left side. *A, B,* At six week and seven weeks, respectively, illustrating the absorption of the left ureter and mesonephric duct into the wall of the bladder. C, At nine weeks.

The Bladder: At the time of its emergence as a separate entity, the *bladder* still receives on each side the common stem of a mesonephric duct and ureter. Growth processes quickly lead to the absorption of these stems, so that the four ducts acquire separate openings. A somewhat complicated shifting then displaces the mesonephric ducts farther caudad (B,C). The two ureters come to lie well apart from each other, but the mesonephric ducts open close together at an elevation in the future urethra known as *Muller's tubercle*. The triangular area on the dorsal wall of the bladder, and its continuation along the dorsal wall of the urethra to Muller's tubercle, as marked off by these four ducts, is the *trigone* of adult anatomy. Temporarily it is an island of mesodermal epithelium amid the general entodermal lining of the bladder, because of the process of absorption already described. However, this original mesodermal island is replaced by encroaching sinus epithelium, so that the trigone is finally entodermal also.

After the second month the bladder proper expands to an epithelial sac whose apex tapers into an elongate tube; this latter portion is named the *urachus*. The early urachus, in turn, is continuous at the umbilicus with the remnant of the allantoic stalk, but it is believed that the latter contributes nothing to either urachus or bladder. The bladder and urachus elongate proportionately as the infra-umbilical body wall is progressively brought into existence. The general organization of the bladder, both as regards its stratified epithelial lining and muscular wall, is attained during the third month. The urachus persists throughout life as a cord into which a patent canal extends for one-third of the distance to the umbilicus. After birth it is known as the *middle umbilical ligament*.

The Urethra and Urogenital Sinus: The caudal region of the cloaca that separates away from the rectum becomes the primitive urogenital sinus; its pelvic and phallic segments have already been mentioned.

The Female: The originally short neck between the bladder and the urogenital sinus elongates into the permanent *urethra*. The pelvic and phallic portions of the sinus merge to create the shallow slit-like *vestibule* into which the urinary and genital tracts open separately. The female urethra does not extend into the clitoris, which otherwise is homologous to the penis of the male.

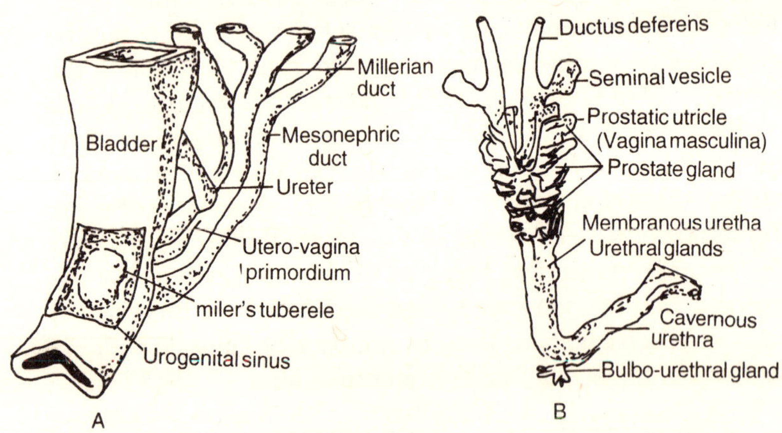

Fig. 7.40.   Differentiation of the human urogenital sinus, illustrated by models. *A*, Female fetus of nine weeks, in left front view. *B*, Male fetus of four months, in right rear view.

The Male: The urethra is more complicated than that of the female. The counterpart of the entire female urethra is a short tube extending from the bladder to Muller's tubercle, which presently becomes the permanent seminal colliculus. This canal comprises most of the *prostatic urethra.* Next below the level of Muller's tubercle is the pelvic portion of the urogenital sinus; this becomes the rest of the *prostatic* and all of the *membranous urethra (B).* Finally, the phallic portion of the sinus adds the *cavernous urethra* which extends through the penis. Since the mesonephric ducts are utilized by the male as the chief genital ducts, all the permanent urethra distal to their outlets on Muller's tubercle serves as a true urogenital canal. This includes both the membranous and the cavernous urethra *(B).*

Accessory Genital Glands: Several glands, associated with the genital system, develop from the region of the urogenital sinus and can logically be included in the present chapter.

The prostate Gland: This organ develops as multiple outgrowths of the entodermal urethral epithelium, both above and below the entrance of the male ducts. The tubules arise at eleven weeks in five distinct groups and total an average number of 63. The groups correspond to the future lobes of the final gland. The surrounding mesenchyme differentiates both connective-tissue and smooth-muscle fibers, into which the prostatic buds grow. The prostate of the new-born shows some evidence of temporary activation by maternal hormones, but it remains relatively undeveloped until puberty.

In the female the urethral glands are presumably homologous, a group of which on each side is drained by *para-urethral ducts.*

The Bulbo-Urethral Glands: These glands (of Cowper) arise in male embryos of nine weeks as a pair of solid buds from the entodermal epithelium of the urogenital sinus at the beginning of the future cavernous urethra. The outgrowths extend backward, almost paralleling the sinus, and penetrate through the investing mesenchyme of the primitive corpus cavernosum urethrae. At four months the epithelium becomes glandular.

The *major vestibular glands* (of Bartholin) are the female homologues. They appear at the same age as the male glands, grow through puberty and involute after the menopause. These glands open into the vestibule, near the hymen.

The Urethral Glands: The cavernous urethra begins to bud off numerous small glands (of Littre) at three months. Homologous structures, the *minor vestibular glands,* open into the vestibule of the female.

The Seminal vesicles: Although of somewhat different origin, these saccular glands belong functionally in the present group. They are exclusively male organs which outpouch from the lower ends of the mesonephric (now deferent) ducts in fetuses of thirteen weeks and gain a muscular wall from the adjacent mesenchyme. By the seventh month the seminal vesicles have attained their adult form; until puberty they grow but slowly.

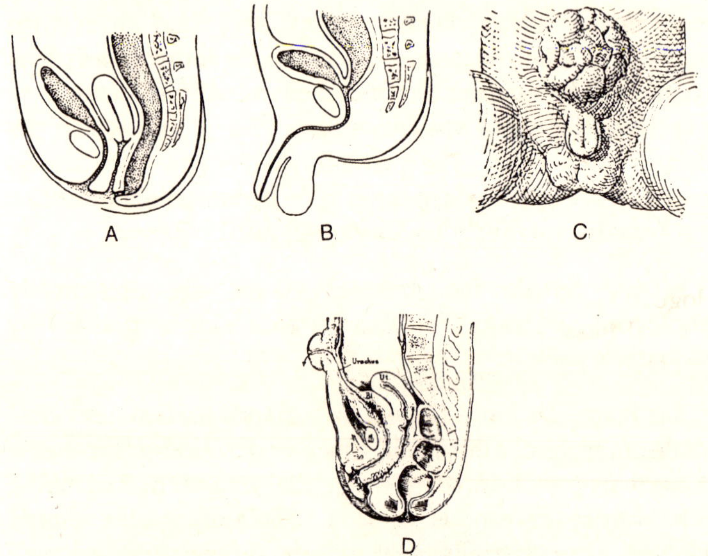

Fig. 7.41. Anomalies resulting from faulty differentiation of the human cloaca. *A*, Persistent cloaca (recto-vestibular fistula) in the female, shown in sagittal section. *B*, Similar condition (recto-urethral fistula) in the male. *C*, Exstrophy of the bladder in a newborn, combined with epispadias of the penis and undescended testes. *D*, Patent urachus in a female, shown in sagittal section; the urachus opens on the abdomen as a urachal fistula.

Anomalies: Imperforate anus results from a retention of the anal portion of the cloacal membrane; an actual anal canal (proctodeum) is shallow. A conspicuous malformation is a *persistent cloaca*, as occurs normally in most vertebrates. A *recto-vestibular fistula* in the female or a *recto-vesical* or *recto-urethrai fistula* in the male *(B)* is the usual outcome. These conditions are due to the failure of the rectum and urogenital sinus to separate completely. Only rarely is the bladder duplicated or divided into two chambers. It sometimes opens and everts broadly onto the ventral body wall; this condition is known as *exstrophy of the bladde*. In severe cases this is associated with spread public bones and either a grooved or bifid penis or clitoris. Failure of mesoderm to invade the infra-umbilical region and reduce the expansive cranial extent of the early cloacal membrane would predispose to this condition rupture of this membrane would then expose the bladder. Because of the primary relation of the mesonephric ducts to the ureter, and their normal absorption into the differentiating cloaca, variations in the ureteric openings occur; they may terminate in the seminal vesicles, urethra, rectum, uterus or vagina. At times the urachus remains patent even to the umbilicus and establishes a *urachal fistula* there through which urine escapes. Less complete remnants of the urachus are blind sinuses, leading from the bladder, or isolated epithelial cysts. Anomalies of the urethra and accessory genital glands are not common.

THE GENTIAL SYSTEM

The intimate relations of the urinary and genital organs as components of the developing urogenital system have already been enumerated. As with other hollow viscera, the epithelial constituents of all these organs are primarily important and it is their development that naturally receives major attention. The accessory, investing coats of muscle and connective tissue organize in both systems during the third month from condensed, neighbouring mose chyme.

The Indifferent Stage

During the fifth and sixth weeks (5-12 mm.) the genital system makes its apperance. This has been named the 'indifferent period' because the sex of the embryo cannot be determined at that time, either by gross or microscopic insecation of the internal and external genitalia. In addition to such a pair of generalized sex glands, all vertebrate embryos are equipped at an early stage with a double set of sex ducts (male and female). Both are held in readiness for the time when sexuality is declared, but only the one appropriate duct system will then advance significantly beyond its primitive state; the complementary duct suffers regression.

To be sure, the sex-determining mechanism is present from the moment of fertilization, but diagnosis of sex on the basis of chromosome counts or identification of the sex chromosomes cannot be accomplished reliably as a routine procedure. Not until the eighth week, at the earliest, does sex recognition become fairly practicable by inspection of the external genitalia.

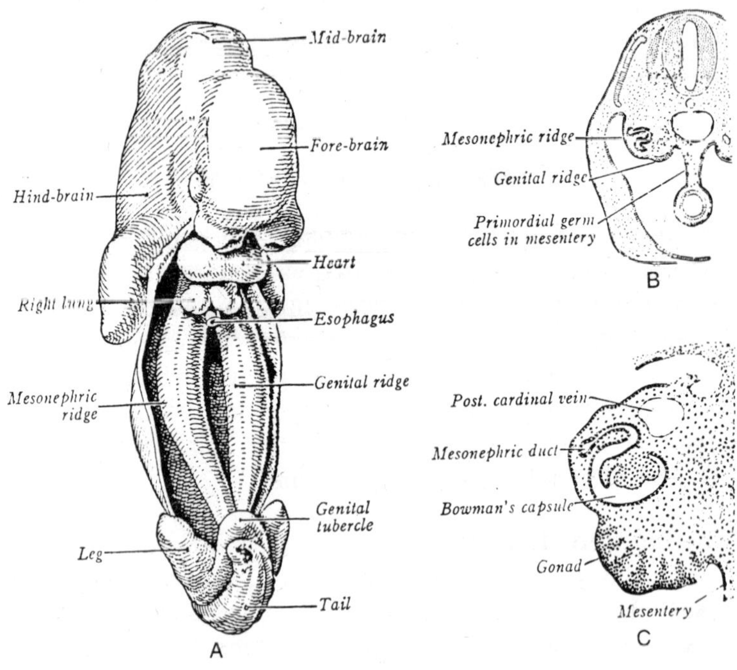

Fig. 7.42.   Urogenital ridge of the human embryo. *A*, Dissection, at 9 mm., in ventral view *B*,*C*, Transverse section, at 7 mm. and 10 mm.

The Gonads: As long as the prospective testis and ovary are structurally indistinguishable they are given the non-committal name, *gonad.* The primitive sex gland makes its appearance within a localized region of the thickening that has already been described as the *urogenital ridge;* this folded ridge is appropriately named since it contains both the nephric and genital primordia.

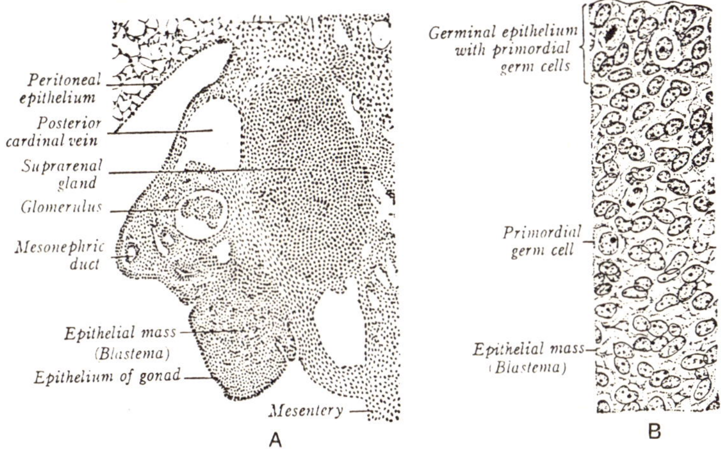

Fig. 7.43. Indifferent stage of the human gonad, illustrated by transverse sections. *A,* At 12 mm., including associated regions. *B,* At 12 mm., showing structural details in a sample area.

On the ventromedian surface of the urogenital ridge the peritoneal epithelium begins to thicken (6 mm. embryos) and rapidly becomes several layers thick. Proliferation soon causes this region to bulge into the coelom as the *genital ridge.* This thickened strip extends longitudinally and thus parallels the mesonephric ridge, but lies mesial to it. At six weeks the resulting, 'sexless' gonad consists of a superficial *germinal epithelium* and an internal *blastema,* somewhat loosely arranged. The blastemal mass is mostly derived by proliferative ingrowth from the epithelium which early loses its basement membrane and shows continuity with ill-defined cords in the blastema. Longitudinal furrows separate the indifferent sex gland from the mesonephros, laterally, and from the mesentery of the gut, medially. During the next week the gonad begins to assume characteristics that identify it as testis or ovary.

Even in presomite embryos certain large, distinctive cells can be recognized caudal to the embryonic disc in the yolk-sac entoderm. Soon afterward they lie in the cloacal entoderm, and in 4 mm embryos they are migrating cephalad, by way of the entodermal gut and dorsal mesentery, into the region of the future genital ridge. Such cells are called *primordial germ cell*; in a 4 mm embryo nearly 1400 were counted. Some claim that all definitive sex cells of the genital glands are descended from them. This contention has been challenged, as already discussed, and it is uncertain whether these cells are the ones actually used or whether some or all of the definitive sex cells originate locally from the germinal epithelium.

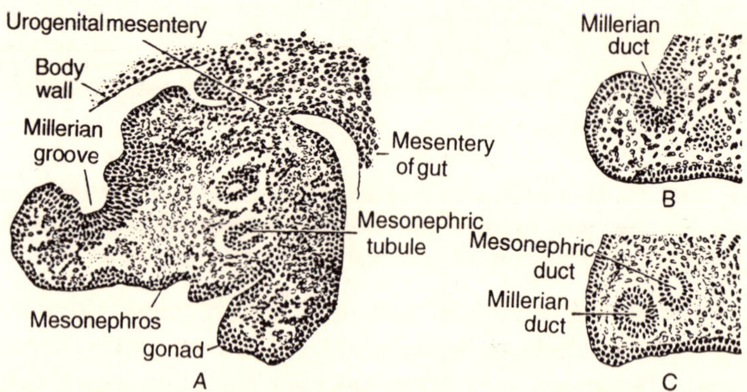

Fig. 7.44.   Origin of the human Mullerian duct, illustrated by transverse sections of the urogenital ridge at 12 mm. *A,* Through open groove. *B,* Slightly lower level, showing closure. *C,* Still lower level, showing free tube.

The Primitive Genital Ducts: The male does not elaborate any ducts intended primarily for the service of the testis.

Instead, with the degeneration of the mesonephros, it merely appropriates the abandoned mesonephric ducts and some of the mesonephric tubules and converts them into genital canals. The origin and early history of these parts have been adequately described in previous paragraphs. The duct is complete at four weeks and new tubules cease differentiating at five weeks. Use will also be made of urethra as a terminal sexual passage.

Both sexes also develop somewhat more tardily a pair of female ducts (of Muller). Embryos of nearly six weeks (10 mm.) first indicate the future *Mullerian ducts* by a groove in the thickened epithelium of each urogenital ridge; this furrow is located laterally on the mesonephros, near its cephalic pole. The extreme cranial end of the groove remains open, while more caudally the lips of the groove close into a tube (*B,C*). Starting thus as an epithelial inrolling, the Mullerian duct continues to advance in a caudal direction by the progressive growth of its solid, blind end. The female duct courses just beneath the surface epithelium and lateral to the mesonephric (male) duct, with which its tip is intimately related. It is, nevertheless, generally held that the mesonephric duct does not contribute directly to the growth of the Mullerian duct. This is quite different from the condition in sharks where the Mullerian ducts arise from the direct longitudinal splitting of the mesonephric ducts.

Near the cloaca the two urogenital ridges swing mesad to the midplane and fuse into the so-called *genital cord*. In this maneuver the progressively elongating Mullerian ducts, originally lateral in position, necessarily are brought side-by-side in the midplane, whereas the mesonephric ducts assume a more lateral position. In embryos of nine weeks the portions of the Mullerian ducts coursing within the genital cord have fused and end blindly at *Muller's tubercle*. This tubercle is a median protuberance into the dorsal wall of the urogenital sinus which was originally brought into existence by the earlier arrival of the mesonephric ducts. The fused, common Mullerian tube is the first indication of a *uterus* and *vagina*, whereas the more cranial portions of the ducts remain separate and will serve as the *uterine tubes.*

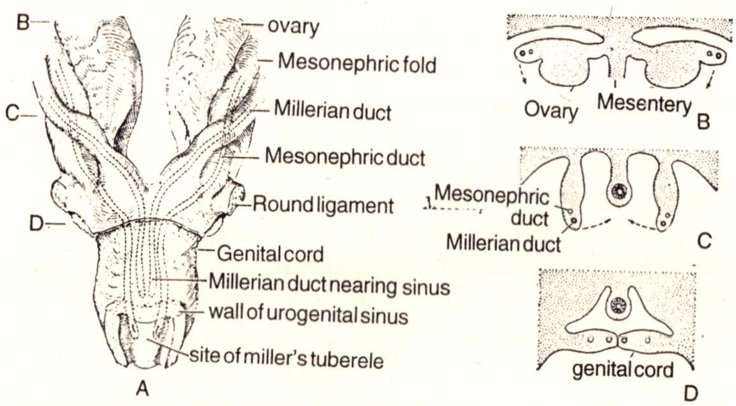

Fig. 7.45. Course of the human urogenital ducts and formation of the genital cord. *A,* Model, at two months. *B-D,* Transverse sections, at the three levels indicated alongside *A.*

The double set of ducts, male and female, is similar in all embryos throughout the second month. After the sex of the embryo is well-established, the provisional ducts of the opposite sex regress and largely disappear.

The External Genitalia: Embryos at the end of the fifth week (8 mm.)shows conical *genital tubercle* in the midline of the ventral body, between the umbilical cord and tail. Its caudal bears a shallow *urogenital groove,* whose floor is the thin *urogenital membrane* and whose side walls are slightly elevated *urogenital folds.* During the seventh week the genital tubercle elongates into a somewhat cylinderical *phallus,* with its tip rounding into the *glans.* Lateral to the base of the phallus, a rounded ridge then makes its appearance on each side; they are the *genital swellings.* presently to be designated labial or scrotal swellings, Rupture of the urogenital membrane in a region near the base of the phallus provides an external opening for the urogenital sinus at seven weeks. From this generalized set of primordia, the external genital organs of the male or female will be modeled in an appropriate and distinctive manner during the ensuing weeks.

Causal Relations: Experiments on amphibians show that the early nephrotome region, when transplanted into the body wall of a larval host, develops a gonad. The fate of the prospective sex cells in the graft is, therefore, irreversibly determined before the genital ridge is visible as such. This determinative control is exerted by the dorsocaudal entoderm, and probably by the mesonephros as well. The growth caudad of the Mullerian ducts of amphibians and birds can be proved to be dependent on the presence of the mesonephric duct alongside. Causal development in the mesonephric system is summarized.

INTERNAL SEXUAL TRANSFORMATIONS

The Gonads

Differentiation of the Testis: As the male genital glands increase in size, they shorten relatively into more compact organs located farther caudad. At the same time the originally broad attachment to the Mesonephros is converted into a gonadial mesentery known as the *mesorchium.* In embryos, about 15 mm long, destined to be males the gonads begin to show two characteristics that mark them as testes : (1) the appearance of prominent, branched and anastomosing strands of cells, the *testis cords;* and (2) the occurrence, between the covering (germinal) epithelium and the centrally located testis cords, of a layer of tissue that foreshadows the *tunica albuginea,* or fibrous capsule of the gland.

This testis cords of human embryos perhaps arose in the indifferent period as direct extensions from the germinal epithelium, but this relation is not plain. At seven weeks they seem to organize suddenly within the blastemal mass. Primordial germ cells are incorporated into these cords at this time, but they soon become unrecognizable as such. The radially arranged

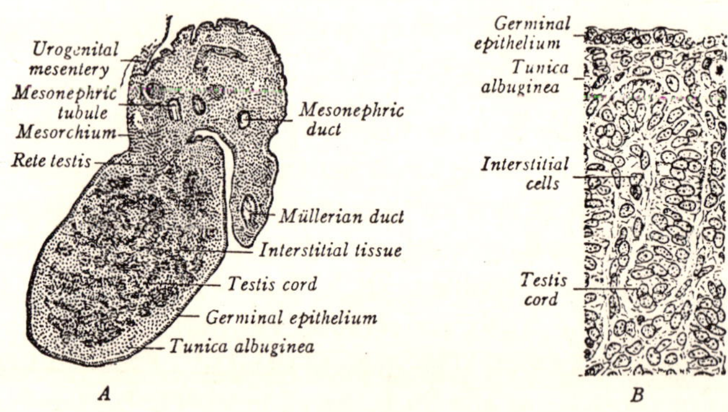

Fig. 7.46.   Early differentiation of the human testis, illustrated by transverse sections at nearly eight weeks. *A*, General organization and relations of the urogenital ridge. *B*, Structural details of the testis.

testis cords converge toward the mesorchium where a dense portion of the blastemal mass is also emerging as the primordium of the *retetestis*. Soon the cell clusters of the *rete* primordium become a network of strands which are continuous with the testis cords. Each of the latter splits into three to four daughter cords—the forerunners of the *seminiferous tubules*. Their peripheral ends join in looping arches (*B*), while the main portions of the cords soon elongate into twisted *tubuli contorti*. Nearer the rete testis, however, they remain straight, as the *tubuli recti*. The rete testis unites the tubuli recti with the mesonephric components of the duct system in a manner to be described presently. Actually the testis cords do not canalize into permanent tubules until the time of puberty (*B, C*). Their central cavities then unite with the cavities of the rete cords which were completed before birth. Thus the originally solid cords of both kinds end their development as a continuous system of tubules, lined with epithelium.

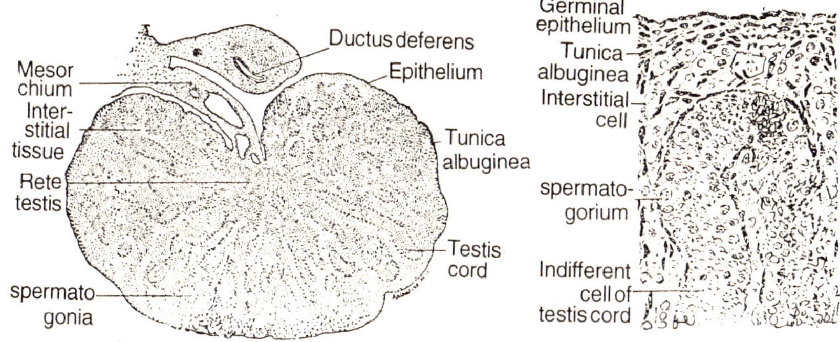

Fig. 7.47. Advancing differentiation of the human testis, illustrated by transverse sections at fourteen weeks. *A,* General plan and relations. *B,* Structural details.

The early testis cords are composed chiefly of so-called indifferent cells. Some of the primordial germ cells, enclosed

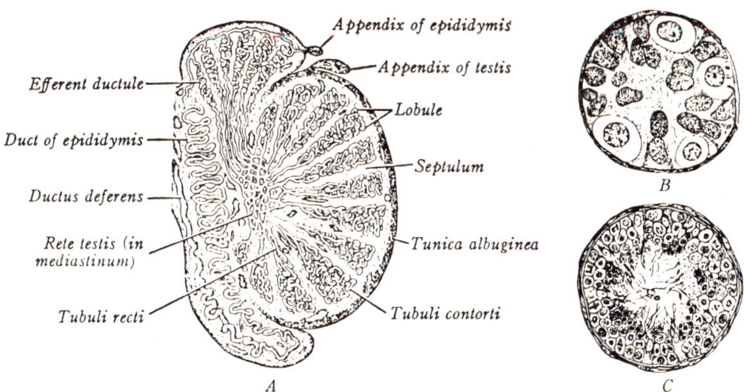

Fig. 7.48. Late differentiation of the human testis. *A,* Hemisection, showing the plan of organization of the testis and its ducts in a newborn. *B,* Section of a seminiferous tubule of a newborn. *C,* Section of a seminiferous tubule of an adolescent.

within the cords, perhaps become early *spermatogonia*, but it is wholly possible that the later generations of sex cells differentiate from certain of the 'indifferent' elements. Other indifferent cells of the cords transform into the *sustentacular cells* (of Sertoli). The full course of development of spermatogonia into spermatozoa, which first begins at puberty.

The general bed of mesenchymal tissue, in which the tubules of the testis lie, organizes into the connective-tissue framework of the organ. Thus the 250 *lobules* of the testis, each containing three or four seminiferous tubules derived from a primitive testis cord, become isolated by partitions. In one direction these *septula* converge to the *mediastinum testis* (where the rete tubules lie); in the opposite (peripheral) direction they extend to the *tunica albuginea* which, after the second month, becomes a definite encapsulating layer. Coincidently with the differentiation of a tunica, the germinal epithelium reverts to the ordinary type of peritoneal mesothelium and does not bodily accompany the testis on its scrotal journey. Certain cells of the mesenchymal stroma transform into large, pale elements which lie in the unspecialized connective tissue between the seminiferous tubules and hence are designated *interstitial cells*. They are very abundant in the fourth to sixth months and again increase in number after puberty. They are generally believed to be responsible of the endocrine secretion of the testis.

Differentiation of the Ovary: Like the testis, the ovary gains a mesentery (*mesovarium*) and settles to a more caudal position. Yet this gland does not exhibit any distinctive ovarian features until several weeks after the gonad of the male has declared itself as a testis. However, gonads that do not differentiate epithelial cords during the seventh week can be diagnosed negatively as ovaries. In the eighth week the blastemal mass of the indifferent period begins to show clusters composed of small, indifferent cells and one or more primordial germ cells. Soon there may be distinguished a denser *primary cortex* beneath the germinal epithelium and a looser *primary medulla* internally.

In addition, a compact cellular mass bulges from the medulla into the mesovarium and establishes there the primitive *rete ovarii,* or homologue of the rete testis. Neither epithelial cords nor tunica albuginea are developed at this stage, as in the testis.

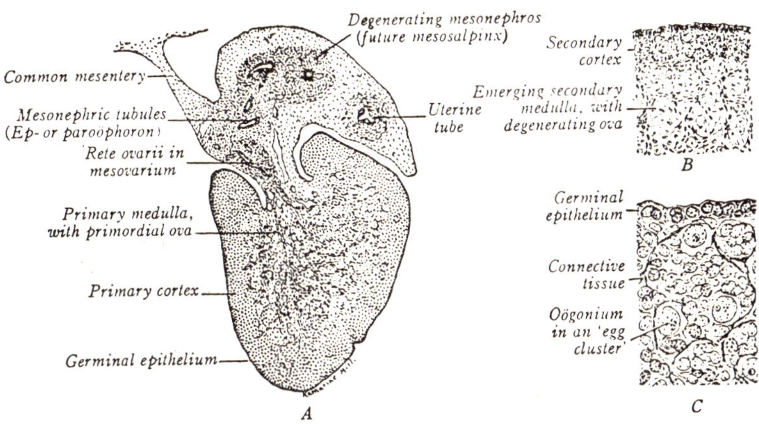

Fig. 7.49.   Early differentiation of the human ovary, illustrated by transverse sections. *A,* General organization and relations, at three months (after Prentiss. *B,* Structural details, at fourteen weeks. *C,* Secondary cortex, at four months.

In fetuses three to four months old three important changes are taking place: (1) Most of the cells comprising the original internal cell mass transform into young ova, the conversion spreading from the region of the rete peripherad (*A*). (2) The ovary enlarges rapidly, due to the deposition of a new, definitive *cortex* upon the original blastemal mass (*B, C*). This secondary cortex arises partly by the division of cells of the blastema, already present, and also through a renewal of proliferation by the germinal epithelium. In the human ovary this new stratum is largely a homogeneous mass; distinct, cellular cords ('Pfluger's tubes') do not grow in from the germinal epithelium as conspicuously as in other mammals. (3) Ingrowth of connective tissue (accompanied by blood vessels) from the region of the rete ovarii produces supporting structures similar to the mediastinum and septula of the testis. The latter tissue now isolates the cortical substance into cord-like masses. C). At the periphery of the ovary the septula expand during the sixth month into a loose, connective-tissue layer known as the *tunica albuginea*; its appearance marks the end of the period of deposition of the new cortex, but it never becomes such a firm layer as in the testis.

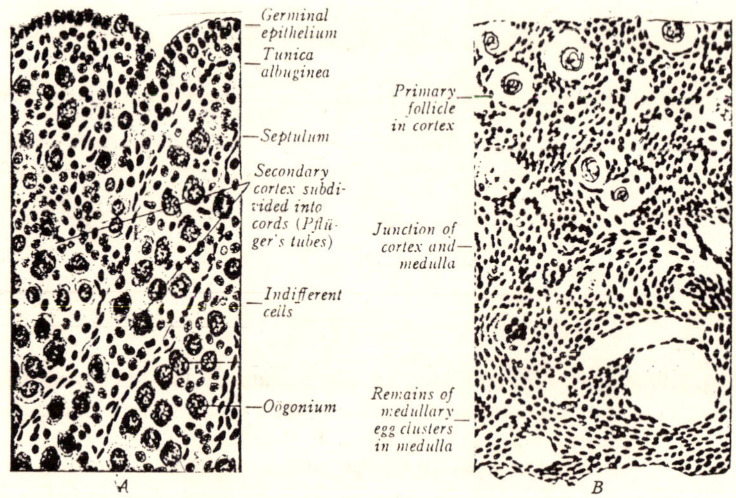

Fig. 7.50. Later differentiation of the human ovary, shown in vertical sections. *A*, Cortex, at six months. Junction of cortex and medulla, at eight months.

Coincidental with the addition of new cells (secondary cortex) at the periphery of the ovary goes the decline of the earlier ova which were growing in the primary medulla and cortex. Such clusters of germ cells, separated by invading connective tissue, regress and are replaced by a vascular, fibrous stroma, thus arises the permanent *medulla*. In the secondary cortex the epithelial cords fragment, and egg clusters and single eggs are similarly isolated by connective tissue, but they do not succumb. Instead, indifferent epithelial cells surround the young cortical ova in the later fetal months, and even after birth, and thereby produce the *primary follicles*. Although some of these advance further during fetal life and after birth, the development of *vesicular* (Graafian) *follicles* is mostly characteristic of the active sexual years.

Bipotentiality of the Gonad: The gonads of birds and mamals show a definite tendency toward bisexual organization. The first set of sex cords, which constitutes the primitive cortex and medulla of the ovary, is the equivalent of the definitive seminiferous tubules of the male; on the other hand, the functional cortex of the ovary is a distinctive, female characteristic. Correspondingly, the testes of some birds and mammals (including man) exhibit for a short time the structural equivalent of an ovarian cortex in addition to the medullary, or male component. This double potentiality is the basis of sex reversal or of a combined ovary and testis (ovotestis). That is, further stimulation of the germinal epithelium of a prospective male gonad adds an "ovarian' cortex, while inhibition of the cortical addition to a prospective female gonad leads to testis formation. At the same time, the appropriate sex ducts undergo progressive development and those of the opposite sex are suppressed. These shifts in sex direction occur sometimes in nature and are producible in a certain degree by experimental hormone administration.

Causal Relations: Experiments on the newt and chick indicate that only sterile gonads develop when the normal migration of primordial germ cells into the gonad is prevented. This might mean either that primordial cells are the only source of definitive germ cells, or that the primordial cells furnish a stimulus without which the gonad cannot complete its differentiation. The germ cells of a gonad cannot self-differentiate completely, but depend on the surrounding tissues for the specific influences that bring about their differentiation. Moreover, studies on sex reversal imply that the particular chromosome-complex within a germ cell does not determine its sex differentiation; rather, it is the influence of the cortex (in the female) and the medulla (in the male) that directs differentiation into male or female sex cells.

Anomalies: Congenital absence of duplication of the testes and ovaries, is very rare. Fused testes and lobed ovaries are recorded. A combined *ovotestis* is a common accompaniment of true hermaphroditism. The tumor-like growths, known as *dermoid cysts* and *teratomas,* which occur most frequently in the ovary, have been discussed in a previous chapter.

Transformation of the Mesonephros

The mesonephric system of male amphibians performs a double function. Some of the more cranial tubules unite with the testis, while the caudal ones continue to excrete urine. Hence the mesonephric duct conveys both urine and spermatozoa to the cloaca. In higher vertebrates the same potential arrangement is laid down, but the replacement of the mesonephros by the permanent kidney makes available individual ducts for the sexual and urinary products of the male. In the female the two kinds of ducts are separate from the start.

The growth of the gonad soon surpasses that of the mesonephros, which thereafter appears as an adjunct alongside. Nevertheless, both in male and female embryos of nine weeks there still remain some thirty mesonephric tubules; of these, half are intact and the rest more or less fragmented. All the tubules that escape complete degeneration can be divided into a cranial and a caudal group on the basis of their subsequent history. The cranial group soon consists of but 8 to 15 tubules; these project against the adjacent primordium of the rete testis or rete ovarii, as the case may be. Union of the rete cords and mesonephric tubules begins in fetuses of three or more months. The point of union is usually at the junction of the secretory and collecting limbs of each tubule, but may occur at Bowman's capsule. The caudal group of tubules does not make such unions.

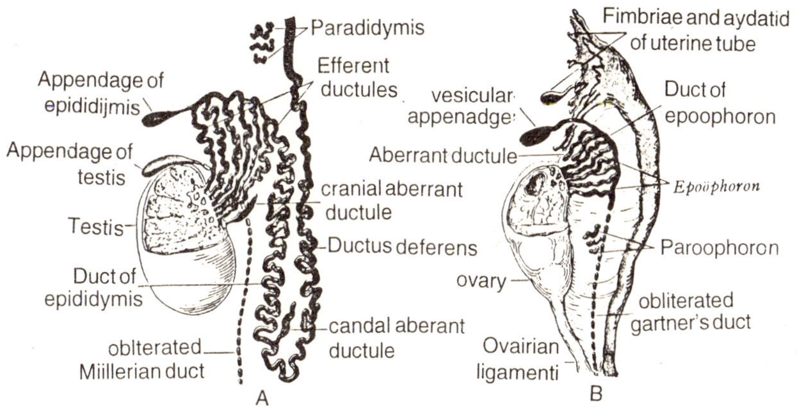

Fig. 7.51. Diagrams illustrating the diverse fates of the mesonephric tubules and the mesonephric and Mullerian ducts in the two sexes. *A*, Male; *B*, female.

The Male: Portions of the mesonephric system are salvaged by the male and used as functional sexual ducts. The fates of these parts can be followed in. The lumina of the rete tubules and cranial mesonephric tubules become continuous by the end of the sixth months, whereupon the mesonephric tubules are given a new name—the *efferent ductules* of the epididymis. Each coiled ductule makes a conical mass known as a *lobule of the epididymis.*

The efferent ductules are destined to convey spermatozoa from the rete testis into the mesonephric duct. The latter, accordingly, undergoes certain regional specializations which transform it into the chief genital duct. In completing these changes the cranial end of the mesonephric duct becomes highly convoluted and is named the *duct of the epididymis;* the caudal portion remains straight and, as the *ductus deferens* and terminal *ejaculatory duct,* extends from epididymis to urethra. Near its opening into the latter canal the male duct dilates to form the *ampulla,* from the wall of which is evaginated the saccular *seminal vesicle* in fetuses of 13 weeks.

Vestiges. One of the cranial group of mesonephric tubules ends blindly; it is the *cranial aberrant ductule.* The entire caudal group of tubules is vestigial, yet it persists as the *paradidymnis* and the blindly ending *cavdal aberrant ductule.* The cranial tip of the mesonephric duct is generally believed to become the cystic *appendage of the epididymis.* Some have claimed that this appendage is formed by certain cranial mesonephric tubules and others attribute its origin to an accessory Mullerian funnel.

The Female: The entire mesonephric system undergoes loss or atrophy in the female. The *rete ovarii* is vestigial, though retained in the adult. Some time before birth it canalizes and often unites with the persisting cranial group of mesonephric collecting tubules, thus duplicating the functional connections in the male. Nevertheless, the cranial group of mesonephric tubules always remains a functionless vestige, located within the broad ligament. Most of its components are tiny, blind canals attached to a short, persistent segment of the mesonephric duct; the whole complex is the *epoophoron.* A few of the most cranial tubules of the cranial group may become cystic *aberrant ductules.* The caudal group of mesonephric tubules constitutes the smaller *paroophoron;* it usually disappears before adult life is attained.

The greater part of each mesonephric duct atrophies and disappears in the female, the process beginning early in the third month. A cranial portion persists as the *duct of the epoophoron* and its tip becomes the cystic *vesicular appendage.* More caudal portions, known as *Gartner's duct,* may occur as vestigial structures at any level between the epophoron and hymen. Representatives are to be found in about one-fourth of all adult females; usually they are located in the broad ligament or in the wall of the uterus or vagina, and sometimes they give rise to cysts.

Transformation of the Mullerian Ducts

All vertebrates below marsupials retain separate female ducts which open into the permanent cloaca. In placental mammals, on the other hand, there is fusion to varying degrees at the caudal ends. In primates, complete caudal union of these ducts produces a common uterus as well as a common vagina, but only in the highest primates does the union reach its fullest development.

A previous page has described how the female ducts develop in the urogenital ridges, enter the genital cord, fuse there, and end at Muller's tubercle. When the urogenital ridges are crowded laterad by the enlarging suprarenal glands and permanent. kidneys. Kidneys, the Mullerian duct naturally participate in this displacement. As a result, each duct in its course makes two bends which roughly establish three regions, different in future potentialities: (1) a cranial, longitudinal portion (uterine tube); (2) a middle more or less transverse portion (uterine fundus and corpus); and (3) a caudal, longitudinal portion which fuses with its fellow to produce a common tube (uterine cervix and, at least, a provisional vagina).

The Female: The cranial segment of each Mullerian duct retains its separate existence and is presently called a *uterine tube.* Its open, upper end early gains a fringe, or fimbriae, by adding a series of minor pits (accessory Mullerian funnels) to the rim of the tube. After a time, at the level of the transverse limbs of the Mullerian ducts, the cranial walls of both tubes bulge in a cephalic direction, so that their original angular junction becomes a convex dome. In this manner a considerable extent is added to the *uterus;* it comprises the definitive fundus and corpus of the original fusion between the Mullerian ducts; until puberty it is the longest segment of the uterus. The more caudal portion of the fused ducts represents the *vagina.* Its permanent lining was formerly believed to be the original epithelium, but it is now known that the entodermal epithelium of the urogenital sinus invades this level of the genital cord and replaces the Mullerian epithelium wholly or in part. The

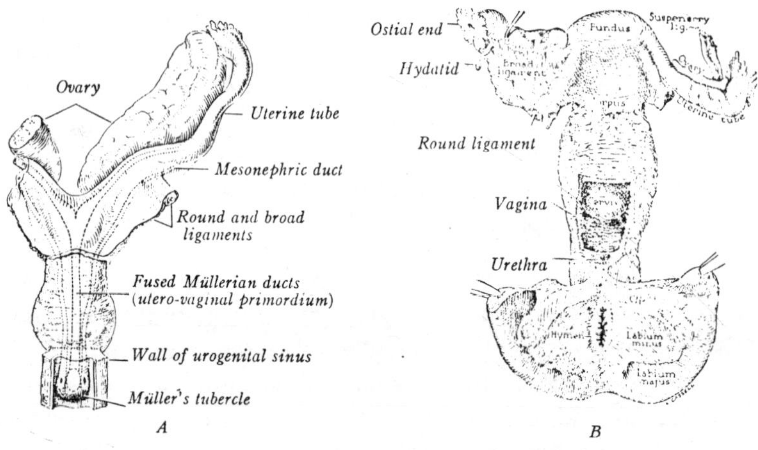

Fig. 7.52.   Genital tract of the human female, in ventral view. *A*, At ten weeks; *B*, at birth.

hymen arises at the site of Muller's tubercle as a ring-shaped fold between the future vagina and the urogenital sinus. when the vagina acquires a lumen, the hymen serves as a perforate membrane guarding the entrance to the vagina.

The uterine tube and uterus are lined with a simple epithelium. Only the uterus develops *glands;* these invaginate by the seventh month, yet remain small until puberty. A distinction between uterus and vagina is not evident until the middle of the fourth month when the fornices appear. For a time the vaginal epithelium is a solid column; when the lumen reappears as a central cleft in fetuses of about five months, the epithelium continues to be stratified. The muscular wall of the entire genital tract is foreshadowed at three months by mesenchyme condensing about the epithelial lining. This investment is especially thick in the genital cord where the uterus develops. The uterus grows rapidly in the last fetal months, loses one-third of its length shortly after birth, and does not recoup this loss until just before puberty. This strong prenatal and pubertal growth is directed by the female hormone, estrogen, supplied first by the mother and later by the maturing girl. The vagina, including its epithelial lining, shares in this interrupted response.

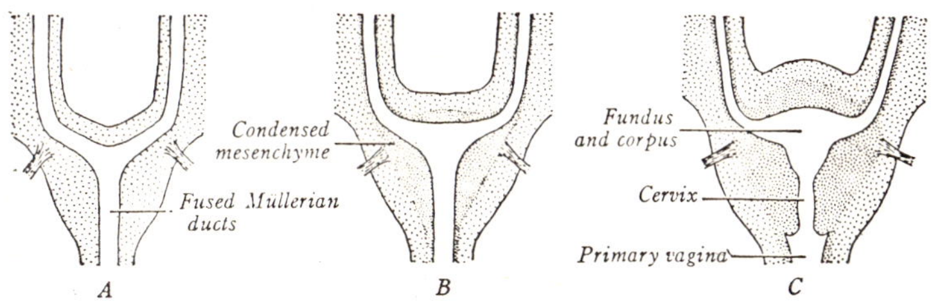

Fig. 7.53. Diagrams illustrating the later history of the transverse limbs of the Mullerian ducts and the fused portions of the ducts within the genital cord.

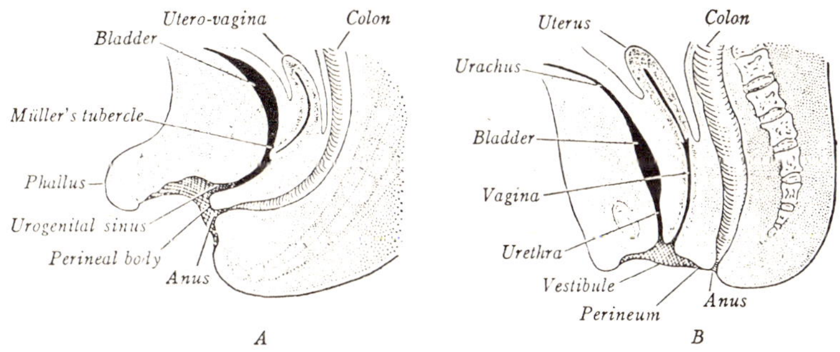

Fig. 7.54. Sagittal selections of female fetuses, demonstrating the relative shortening through which the tubular urogenital sinus becomes a shallow vestibule. *A,* At ten weeks; *B,* at five months.

The young uterin tubes fail to match the elongation of the trunk as a whole, and their trumpet-like upper ends finally lie opposite the fourth lumbar vertebra, thirteen segments below their level of origin. The vagina is originally some distance above the outlet of the urogenital sinus, similar to the permanent condition in many mamals. The intervening stretch of sinus thereafter undergoes a great relative shortening to become the shallow vaginal *vestibule* into which both urethra and vagina open independently (*B*). From the standpoint of specialization of the primitive cloaca, this arrangement is an advance over the condition found in the male since a common urogenital sinus has been practically eliminated.

The Male: Similar primordia also develop in the male, but remain rudimentary. Degeneration of the Mullerian ducts occurs in the third month and only the extreme cranial end is finally spared; this vestige is called the *appendix testis.* The vaginal primordium persists as a tiny pouch on the dorsal wall of the urethra to which has been given the name *prostatic utricle* or *vagina masculina.* Like the vagina of the female, its original Mullerian epithelium is replaced by invading epithelium from the urogenital sinus. The Mullerian tubercle is represented by the elevated *seminal colliculus,* from whose summit leads off the prostatic utricle, unlike the female, the male retains its primitive urogenital sinus as a tube (The permanent membranous and cavernous urethra).

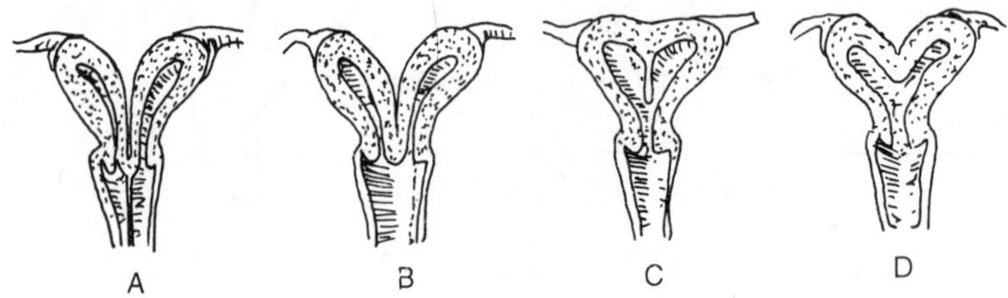

Fig. 7.55.   Anomalies of the human uterus. *A, Double uterus and vagina;* B, double uterus; *C*, bipartite uterus; *D,* bicornuate uterus.

Anomalies: The more common anomalous conditions are based on the failure of the Mullerian ducts to approximate, fuse or form a fundus: (1) Duplication of the uterus and vagina (as in monotremes and lower marsupials; *A*). (2) Duplication of the uterus, but not the vagina (as in many rodents and bats; *B*). (3) Bipartite uterus, more or less subdivided by a median septum (as in carnivores, some bats and the sow; *C*). (4) Bicornuate uterus, in which the uterine horns fail to elevate into a domed fundus (as in ungulates, cetaceans and most bats; D). (5) Retention of the fetal or infantile condition. (6) Congenital absence of one or both uterine tubes, of one uterine horn, or of the uterus or vagina occurs rarely but may be associated with hermaphroditism of the external genitalia. (7) The vagina sometimes remains solid; the hymen may retain its primary imperforate condition which, at puberty, prevents the discharge of menstrual products. (8) Accessory Mullerian funnels may give rise to an accessory tube, a branched tube or an accessory mouth; an occasionally seen cyst (hydatid) has a similar origin.

THE GENITAL LIGAMENTS AND DESCENSUS

The Genital Ligaments

At six weeks the urogenital ridge is attached broadly near the root of the intestinal mesentery but soon a common urogenital mesentery suspends the gonadic and mesonephric regions. Toward the end of the second month there develop additional mesenterial and ligamentous supports for the internal genitalia. These are comparable in both sexes, but only in the female do they become structures of permanent importance.

The Female: The ovary is prmiarily suspended by a short mesentery, named the *mesovarium*, which comes into prominence as the gonad outgrows the mesonephros. The remains of the atrophic urogenital ridge at more cephalic levels connects with the cranial pole of the ovary and persists as the *suspensory ligament*. Similarly, the terminal portion of the genital ridge unites the caudal end of the ovary first to the transverse bend of the urogenital ridge and then to the uterus which develops in it. This connection becomes fibromuscular and is known as the *proper ligament of the ovary.*

With the degeneration of the mesonephric system, the uterine tube lies in a mesenterial fold, the *mesosalpinx.* Somewhat earlier the mutual fusion of the caudal portions of the two urogenital ridges has produced the *genital cord*. This is a mesenchymal shelf that bridges in the frontal plane between the two lateral body walls and contains the uterus in its center. The shelf itself persists as the sheet-like *broad ligaments* on each side of the uterus. The pelvic coelom is thereby subdivided into two blind bays, the *recto-uterine pouch* (of Douglas) and the *vesico-uterine pouch*. After the ovary and uterine tube 'descend' to a lower position, the mesovarium and mesosalpinx are intimately associated with the broad ligament.

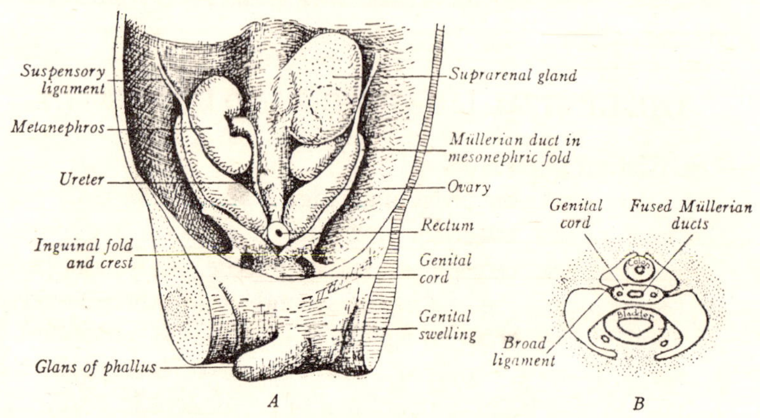

Fig. 7.56.   Urogenital organs of the female fetus, and especially the early genital ligaments. *A*, Ventral dissection, at two months. *B*, Transverse section through the lower trunk, at the level of the genital cord, at three months.

During the seventh week a pair of uterine ligaments of another sort is begun. At the level where each urogenital ridge bends horizontally toward the midplane in forming the genital cord, a continuation of the common urogenital mesentery (inguinal fold) bridges across to a prominence (*inguinal crest*) on the adjoining abdominal wall. Within these parts the mesenchyme condenses as a primitive ligament. A direct continuation of this band extends to the subcutaneous tissue of the genital (future labial) swelling of the same side. Thus, by the beginning of the third month, a continuous cord of dense mesenchyme extends from the region of the uterus to the labium majus. This cord soon becomes fibro-muscular and persists as the *round ligament* of the uterus. Where the ligament passes through the abdominal wall, muscles organize about it and constitute the slanting, tubular *inguinal canal*.

The Male: The primitive mesentery of each testis is named the *mesorchium*. It loses prominence in later fetal life and is represented in the adult merely by the fold between the testis and epididymis. A *suspensory ligament*, equivalent to that of the ovary, for a time extends down to the cranial pole of the testis, but later it disappears. The *ligamentum testis*, comparable to the proper ligament of the ovary, develops in a caudal continuation of the genital ridge; it extends from the caudal pole of the testis to the transverse bend in the urogenital ridge. From the caudal surface of this bend a primitive ligament soon bridges across to the adjacent body wall and continues into the scrotal swelling of the same side of the body, similar to the formation of the round ligament in the female. Since a uterus does not develop in the male, at the beginning of the third month there thus exists a continuous mesenchymal cable extending from the caudal pole of each testis through the inguinal canal to the scrotal swelling below. This cable is named the *gubernaculum testis*. It is composed of. (1) the ligamentum testis; (2) a connecting cord in the urogenital ridge (region of the regressive mesonephros and uterine primordium); and (3) a ligament extending from this ridge into the scrotal swellings. In terms of female counterparts it is the homologue of the proper ligament of the ovary plus the round ligament of the uterus, between which the uterus intervenes.

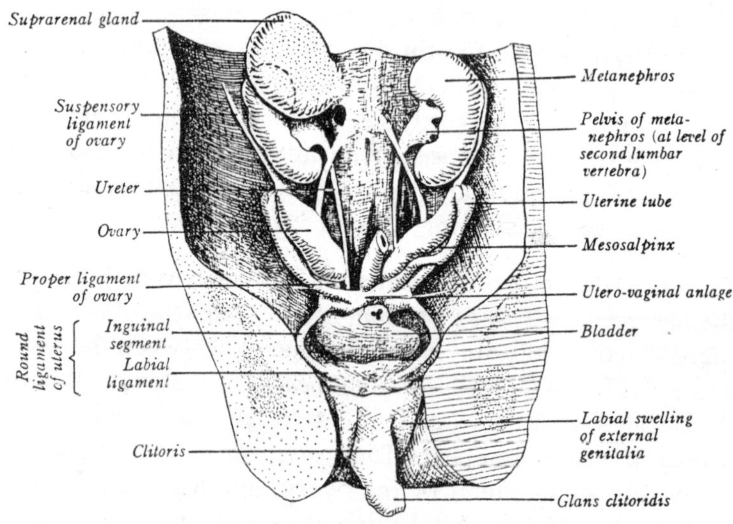

Fig. 7.57. Urogenital organs of a female fetus, at ten weeks (Prentiss). x 5. The ventral dissection displays especially the genital ligaments.

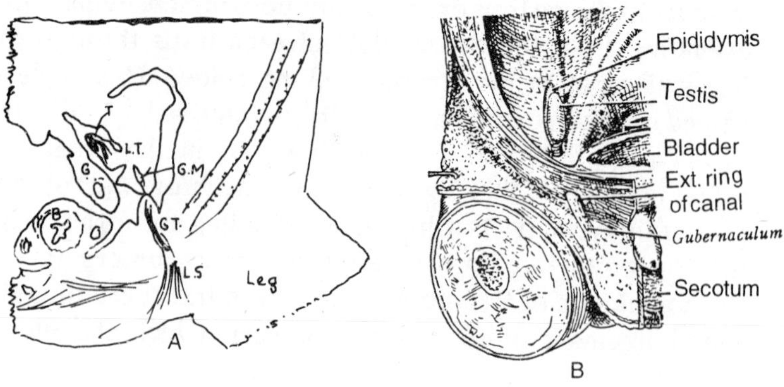

Fig. 7.58   The composition and relations of the human gubernaculum

cont....

testis. *A*, The several components, demonstrated by a schematic section through a male embryo of two months. *B*, Relations at the start of descent (seventh month.

B, Bladder; *G*, gut; *T*, testis; *L.T.*, ligamentum testis; *G.M.*, mesonephric segment of gubernaculum; *G.I.*, inguinal segment; *L.S.*, ligamentum scroti.

Descent of the Gonads

The Male: The Original positions of the testis and ovary change during development. At first they are slender structures, extending caudad from the diaphragm. A faster elongation of the trunk cephalad, in contrast to the slower growing gonad, produces a relative shift of the latter in a caudal direction until the sex gland lies ten segments below its level of origin. When this process of growth and shifting is complete (10 weeks), the caudal end of the gonad lies at the boundary between abdomen and pelvis, and close to the groin.

In addition to its early 'migration' caudad (internal descent), just mentioned, the testis later leaves the abdominal cavity and descends bodily into the scrotum (external descent). At the beginning of the third month, sac-like pockets of the peritoneum protrude into each side of the ventral abdominal wall. These are the beginnings of the so-called *vaginal processes*, and until the end of the sixth fetal month the lower poles of the testes continue to lie near them (at the site of the future internal abdominal ring of the inguinal canal) without change of position. Each process, or sac, evaginates (herniates) through the ventral abdominal wall, thus producing a slanting *inguinal canal*; it then passes over the pubis, and so into the scrotum which it invades from the seventh month on. In doing this the muscular and fascial layers of the abdominal wall are also carried down and added to what now becomes a scrotal sac.

During the seventh to ninth months the testes descend along the same path. The mechanics of this process, including the role of the gubernaculum testis, is poorly understood. The actual time spent in passing through the inguinal canal is very short, whereas the continued journey into the scrotum proper is leisurely. The testes are usually found in the scrotum early in the ninth lunar month, or at least before birth.

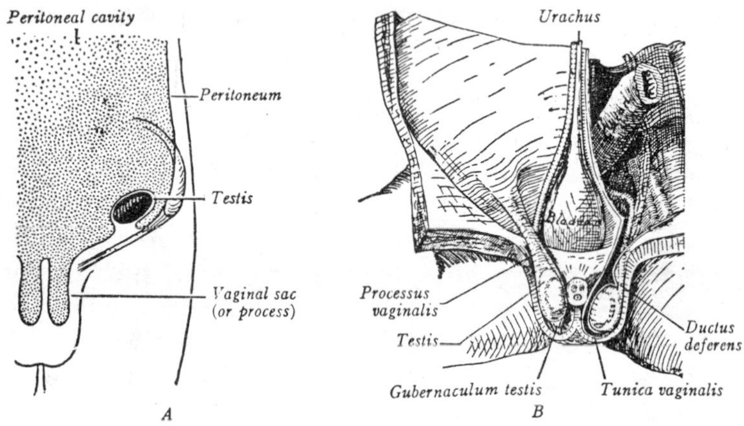

Fig. 7.59. Perioneal relations before and after the descent of the human testis. *A,* Diagrammatic phantom at the onset of external descent (seventh month). *B,* DIssection of a newborn. the left vaginal process has been opened and the testis rotated 90°.

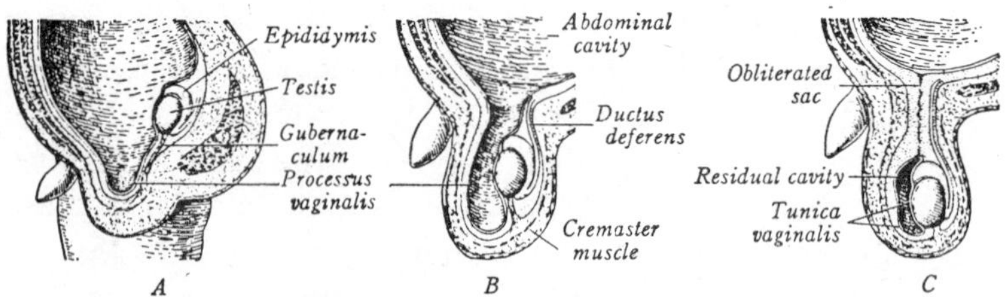

Fig. 7.60. Descent of the human testis and its subsequent relations, shown in diagrammatic hemisections.

It must be understood that the testis and gubernaculum are covered by peritoneum before the descent begins; consequently, the testis follows the gubernaculum along the inguinal canal, behind (i.e., dorsal to) the peritoneal tube. On reaching the scrotum, the testis continues to be covered by a reflected fold of the processus vaginalis but lies entirely outside its cavity (*C*). Even before birth the narrow peritoneal canal, which connects the main sac of the processus vaginalis with the abdominal cavity, frequently begins to become solid, whereupon its epithelium disappears. The vaginal sac, now isolated, represents the *tunica vaginalis*. Its visceral layer is closely warpped about the protruding testis and epididymis, whereas the parietal layer forms a lining to the scrotal sac. Thus, the scrotum proves to be a specialized pouch of skin into which an extension of the muscular and fascial body wall accompanies the evaginating sac of peritoneum. The ductus deferens and the spermatic vessels and nerves are carried down into the scrotum along with the testis and epididymis. They are embedded in connective tissue and constitute the *spermatic cord*. Owing to the path taken by the testis in the scrotal migration, the ductus deferens loops over the ureter.

The hypothysis is suspected of activating the process of descent through the agency of male hormones. The role of the gubernaculum testis is obscure and disputed. From the caudal pole of each testis the corresponding gubernaculum extends through the inguinal canal into the scrotum. During the seventh month the gubernaculum not only ceases growth but also actually shortens one-half. This shortening, both relative and actual, is commonly said to draw the testes into the scrotum. Others deny that the guber-naculum exerts any kind of traction since its lower end is only loosely attached; on the contrary, it is said to convert into mucoid tissue, to dilate the inguinal canal and to become incapable of sustaining the testis which perches on its crater-like upper end. As a result, the testis both sinks downward by a process of normal herniation (possibly driven by increasing intra-abdominal pressure) and carries the processus vaginalis with it. The gubernaculum of a newborn infant is only one-fourth of its length when descensus began; after birth it atrophies almost beyond recognition.

A permanent scrotal location of the testes is normal for primates and various other mammals. This is advantageous because of a lower temperature there, since spermato genesis does not occur at the higher temperature of the abdomen. In other mammals (armadillo; whale; elephant) an abdominal position of the testes is normal, but the temperature in this group is well below that found in mammals with scrotal testes. Still other mammals (rodents; insectivores; bats) maintain open inguinal canals; their testes remain in the abdomen except during the mating season when they descend to the cooler scrotum. Some hibernating animals have a periodic descent of the testes that follows the sharp rise in temperature on awakening.

The Female: The internal descent, already mentioned, brings the ovaries to the pelvic brim during the third month. Here they still lie in a newborn female. Afterwards the ovary and uterus attain their final positions in the minor (true) pelvis as the result of marked growth by the pelvis. Each ovary rotates into a transverse position and also revolves about the uterine tube until it comes to rest dorsal to the tube.

Shallow peritoneal pockets, frequently persistent as the *diverticula of Nuck,* Correspond to the vaginal processes of the male. There is no external descent of the ovaries; one reason lies in the mechanical block presented by the uterus which is interposed between the ovarian and round ligaments.

Anomalies: Testicular descent may be arrested at any point along its normal path, but most often at the external inguinal ring A on right). This condition, known as *crypotorchism* (i.e., concealed testis), is due to developmental anomalies, mechanical obstruction or, perhaps, to hormone deficiency. Prior to puberty, 4 per cent of boys still have imperfectly descended testes . In permanent cryptorchism of any significant degree the testis fails to produce competent sperms or makes none at all. After transversing the inguinal canal descent into an abnormal location, such as the pubes, thigh or perineum, results in an *ectopic testis A*, on left). Rarely, in instances of faulty development of the internal genitalia, more or less complete descent of the ovary into the lubium majus occurs.

When the stalk of a vaginal process does not obliterate, wholly or in part, conditions are favourable for congenital *inguinal hernia* of the instestine into the inguinal canal or even into the socotum.

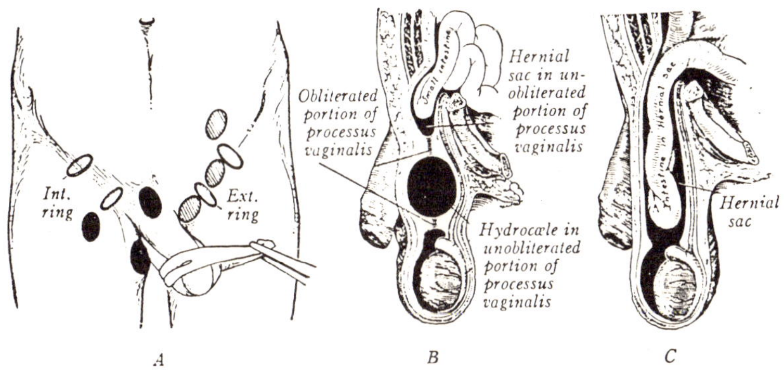

Fig. 7.61. Anomalies of testicular descent. *A*, Diagram showing final positions of the testis in arrested descent (hatched black) and in ectopic descent (solid black). *B, C*, Dissections of partial and total inguinal hernia, correlated with different degrees of failure of the vaginal process to obliterate (Callander).

The External Genitalia

The acquisition of an external penis, with a penile urethra, in the male of higher mammals has paralleled the evolution of a vagina, uterus and the intra-uterine development of the young in the female. The relation of the external penis to the subdivision of the cloaca and the production of a perineum has already been mentioned. Progressive stages in these several changes are illustrated in reptiles, monotremes, marsupials and placental mammals.

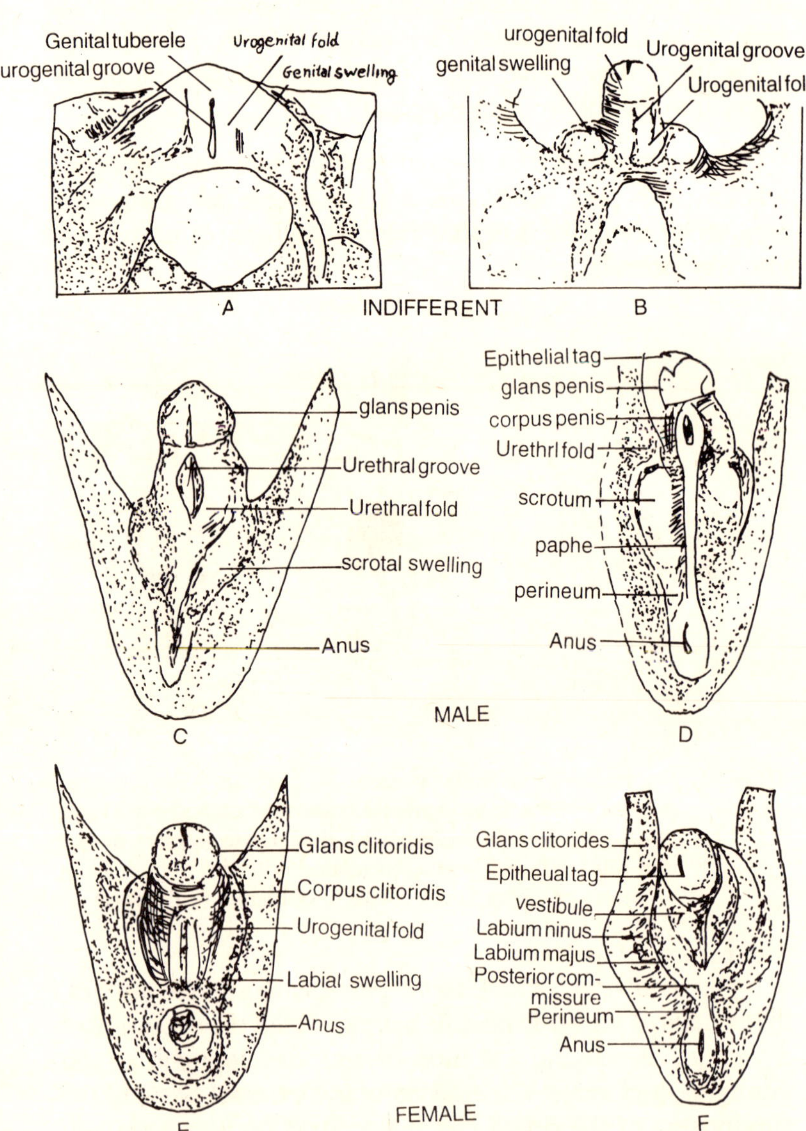

Fig. 7.62. Differentiation of the human external genitalia. *A, B,* The indifferent period, at nearly seven and nearly eight weeks. *C,D,* The male, at ten and twelve weeks. *E,F,* The female, at ten and twelve weeks.

For a week or more after the external genitalia are first indicated, they are indifferent, or sexless, in appearance. During the eighth week sex begins to be indicated grossly through certain external characteristics, chief among which are the erectness of the phallus, the length of the urethral groove and the relations of the urogenital folds to the genital swellings (*C, E*). Yet for a time these criteria are only fairly reliable; especially is there liability to error in diagnosing retarded males for females. At three months the progressive modeling of the external genitalia has attained characteristics that are recognizable as distinctively male or female.

The Male: Fetuses of ten weeks are at the beginning of the definitive stage. In the male, the phallus becomes the *penis.* The edges of the urogenital groove fold together progressively in the direction of the galns; this fusion transforms an open urogenital sinus, bordered by urethral folds, into the tubular *cavernous urethra* within the shaft of the penis while the fused edges constitute a raphe. The genital swellings shift caudad and are recognizable as *scrotal swellings;* each becomes a half of the *scrotum,* separated from its mate by the *scrotal septum* and superficial *scrotal raphe.* In the meantime the shaft of the penis elongates, and by the fourteenth week the urethra has closed as far as the glans. The urethra is then continued along an epithelial plate, which represents a solid part of the original urethral primordium, now incompletely partitioning the glans. By splitting, the plate is first converted into a through; this promptly recloses into a tube that continues the urethra to its permanent opening at the tip of the glans (*B*). During the third month a fold of skin at the base of the glans begins growing distad and two months later surrounds the naked, spheroidal glans. This is the tubular *prepuce,* or fore-skin (*A*). Fusion occurs between the epithelial lining of the prepuce and covering of the glans (*B*), but clefts appear later in this combined membrane and free the prepuce once more; this separation, however, is still incomplete at birth. A region of incomplete prepuce-formation on the under surface of the glans produces

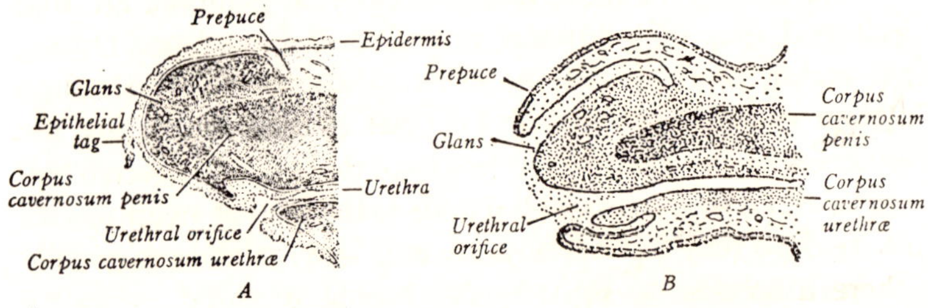

Fig. 7.63. The tip of the human penis, in longitudinal section. *A*, At four months; *B*, at five months.

the fold known as the *frenulum.* The *Corporacavernosa penis* are indirected in the seventh week as paired mesenchymal columns within the shaft of the penis. The unpaired *corpus cavernosum urethrae* results from the linking of similar mesenchymal masses, one in the glans and the other in the shaft.

The Female: Changes in the female are less profound , yet slower. The phallus lags in development and becomes the *clitoris,* with its homologous *galns clitoris* and *prepuce.* The urogenital groove never extends onto the glans, as in the male; it remains open as the *vestibule.* The urogenital folds, which flank the original groove, constitute the *labia minora;* the junction of ectoderm and entoderm is assumed to be at the margin of the lip. The primitive genital swellings grow caudad and fuse in front of the anus as the *posterior commissure,* while the swellings as a whole enlarge into the *labia majora;* these parts now form a horse-shoe shaped rim, open toward the umbilicus. The cephalically located *mons pubis* arises later to complete the gap in the ring; it develops independently of the primitive swellings.

Anomalies: Absence or doubling of the penis or clitoris is very rare; doubling, really division, is usually associated with severe exstrophy of the bladder. The penis may remain rudimentary or the clitoris may hypertrophy; both conditions are common in hermaphroditism. If the lips of the slit-like

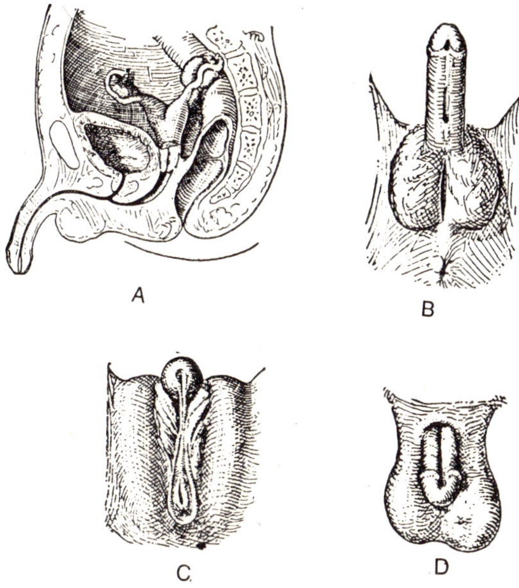

Fig. 7.64. Anomalies of the human genitalia. *A,* True adult hermaphrodite, in sagittal section; the external genitalia are typically male, except for the empty scrotum; the internal genitalia are female and include a bicornuate uterus; an imperfect testis and ovary occur on each side. *B,* Hypospadias, showing in one darwing a composite of the common locations. *C,* Hypospadias, of a severe degree, in a false hermaphrodite. *D,* Epispadias.

urogenital opening on the under surface of the penis fail to fuse anywhere along their extent, *hypospadias* results (*B*); a type involving the scrotal region is a common occurrence in false hermaphroditism that stimulates the female type (*C*). Rarely the urethra opens on the upper surface of the penis, an abnormality known as *epispandias (D)*. This defect commonly accompanies severe exstrophy of the bladder, just above, caudal to the cloacal membrane.

The name hermaphroditism (i.e., Hermes plus Aphrodite) has been given to the condition that actually or apparently combines both sexes in one person. *Ture hermaphroditism* is a

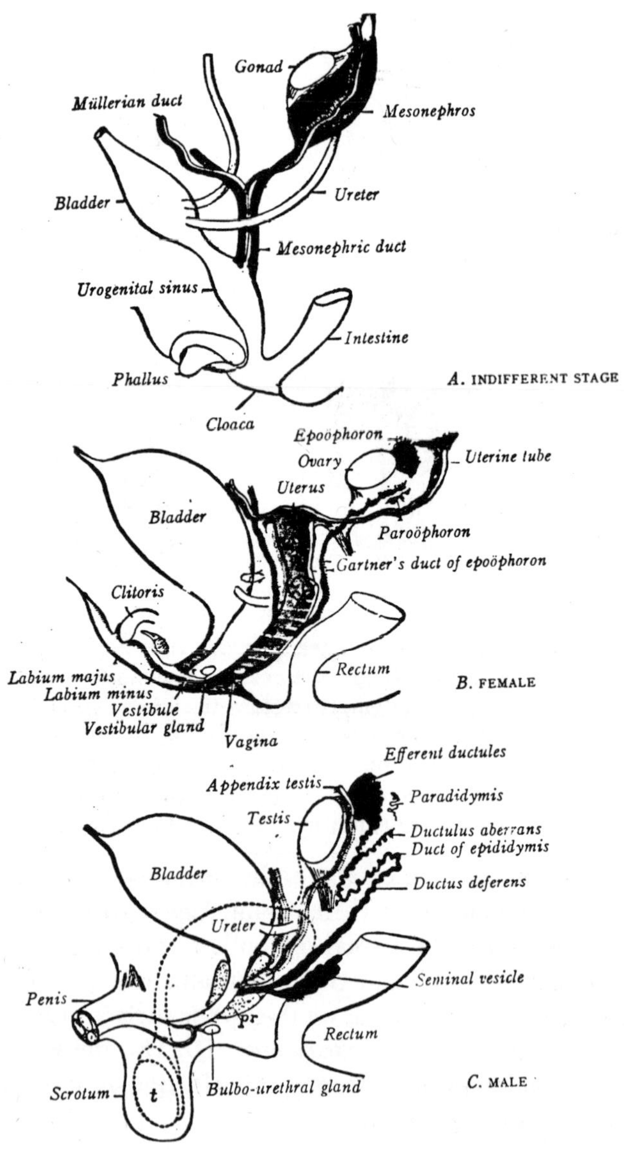

Fig. 7.65. Diagrams illustrating the transformation of an indifferent, primitive genital system into the definitive males and female types.

condition of intersex in which testis and ovary are both present in the same individual. It occurs rarely in birds and mammals, is not uncommon in the lower vertebrates, and is the normal condition in hag fishes and many invertebrates (worms; molluscs). Forty-two human cases have been described that can be accepted as authentic. These include persons with separate ovary and testis, with a pair of combined ovotestes, and with an ovary or testis paired with an ovotestis. No such inter-sex is known to have been fertile either as a male or as a female. The internal genitalia are faultily bisexual, although double ducts and gonads may be represented on one or both sides. When a female gonad and duct occur on one side and a male gonad and duct on the other, this combination approximates the condition of *gynandromorphism*. The external genitalia of hermaphrodites often show features intermediate between male and female characteristics. The secondary sexual characters (beard; breasts; voice; etc.) are usually mixed, tending now one way, now the other.

Relatively common is *false hermaphroditism*. This condition is characterize; ably the presence of the genital glands of one sex in an individual whose secondary sexual characters and external genitalia tend to be indeterminate or to approximate those of the opposite sex. The correct diagnosis of sex is often not established until puberty, and even then it may require the microscopical examination of a sample bit of the gonads obtained by biopsy (or a simpler, recent technique on a skin sample). The internal sexual tract can be that of either sex, or it may be double or mixed; it is commonly atrophic in some of its parts. In *masculine hermaphroditism* an individual possesses testes, often undescended, but the external genitals (by retarded development and severe hypospadias) and secondary sexual characters are like those of the female. In the rarer *feminine hermaphroditism* ovaries are present and sometimes descended, but the other sexual characters, such as enlarged clitoris or perfectly fused labiae, tend to simulate the male.

No one theory accounts satisfactorily for all hermaphroditic conditions. In general, an abnormal genic balance between the sex and ordinary chromosomes, which controls the relative activity of the cortical (female) and medullary (male) components of the primitive bisexual gonad, is seemingly responsible for many of the conditions observed. This would offer a possible explanation for the separate testis and ovary or the combined ovotestis of a true hermaphrodite. Also significant in this regard are the masculinized pseudohermaphrodites (with ovaries) which have shown tumors of the ovarian medulla. On the other hand, the feminization of a congenital male pseudo hermaphrodite might trace origin to undue influence of the maternal hormones during pregnancy.

Homologies of the Urogenital System

In the appended table are summarized the equivalent permanent derivatives of the indifferent reproductive system; persisting vestigial parts are printed in italics.

THE VASCULAR SYSTEM

As the embryo becomes larger and more complex, it develops a more efficient delivery method than that provided by the fluid field coelomic spaces. A distinct circulatory blood vascular system brings fluids into the body and delivers item to the tissues where they give up oxygen, numents and some water in exchange for carbon dioxide and other waste products.

The formation of the system begins in the wall of the yolk sac which was the source of nutrients in ancestral forms, as it is in the chick. Groups of mesoderm cells from blood islands in which the cells around the outside become the lining cells—*endothelium*—of blood vessels and the cells on the inside become *primitive blood cells*. The islands link up to form a vascular network in the yolk sac wall. Meanwhile, empty vessels (containing no blood cells) from independently in the mesoderm of the connecting stalk and of the choriom including its villi.

Isolated vessels also form within the embryo and the join to form continuous vessels on each side of the body.

The chorionic, connecting stalk, intraembryonic and yolk sac vessels then join to complete a single vascular system, and primitive blood cells from the yolk sac spread throughout it. An effective means of transport between the villi of the placenta and the embryonic tissues is thus established, and the fluid within it contains primitive red blood cells specialized for the transport of oxygen. Cells from the yolk sac also settle in other sites where blood formation later occurs.

Within the embryo, the blood vessels of each side meet in the wall of the pericardial cavity and form the heart tube which pumps blood around the circulation. Blood returns to the heart via the *cardinal* (intraembryonic), *umbilical* (from the placenta via the connecting stalk)and *vitelline* (Yolks sac) *veins*. From the heart, blood passes via *aortic arches* into the *aortae* which give off *intersegmental* (passing between somites) *vitelline* and *umbilical* arteries.

In an animal such as the shark, the heart receives deoxygenated blood from the tissues and delivers it, via symmetrical ortic arches, to the gills whence blood vessels deliver oxygenated blood directly to the tissues. In mammals. However, the heart is partitioned in such a way that its right side receives deoxygenated blood from the tissues and delivers it to the lungs whereas its left side receives oxygenated blood from the lungs and delivers it to the tissues. Although a single organ, the mammalian heart consists of twin pumps placed in series between the pulmonary and the systemic circulations.

In the fetus the lungs are unexpanded and blood flow through them is limited because oxygenation occurs in the placenta. Nevertheless, the left side of the heart must be prepared to assume its full load after birth. The fetal circulation is thus balanced by two right-to-left shunts; one—the *foramen ovale*— from the right atrium to the left atrium, and the other—the *ductus arteriosus*—between a pulmonary artery and the aorta.

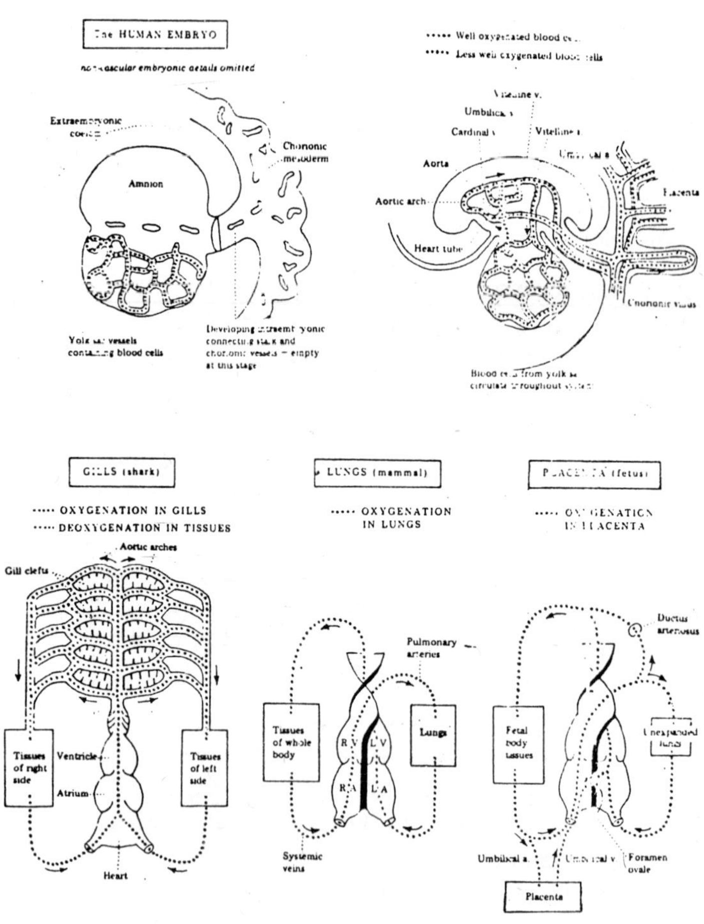

Fig. 7.66.   Early Vascular System and Gills, Lungs or Placenta

After birth the blood flow through the lungs increases as they inflate, circulation through the placenta ceases and the foramen ovale and ductus arteriosus are closed.

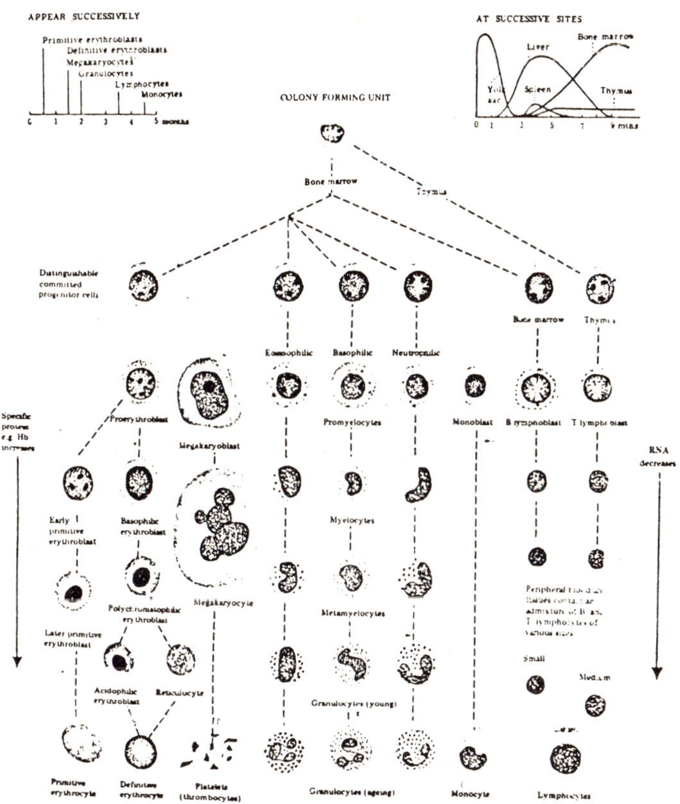

Fig. 7.67. Colony Forming Unit.

Blood Formation

In the yolk sac wall certain mesenchyme cells form primitive blood vessels. Other within them differentiate into *colony-*

forming units (CFUs) which are the stem cells for blood cells of all types. The CFUs undergo proliferative and quantal mitoses, the first clones accumulating haemoglobin and entering the large-celled primitive erythrocyte series. This is superseded by the smaller-celled definitive erythrocyte series. Circulating red cells of both series are at first nucleated and contain only fetal haemoglobin (Hb-F). Which takes up oxygen and gives up carbon dioxide, more readily than does adult haemoglobin (Hb-A). However, at term, erythrocytes contain about four times as much Hb-F as Hb-A. In subsequent generations the proportion falls rapidly.

Other cells types appear in sequence in CFL's seed a succession of other sites where they produce clones. Some CFUs remain pluripotent but others become the progenitors of a progressively more limited range cell types. In particular, cells seeding the thymus become the progenitors of T-lymphocytes. The progenitors of other cell types, including B lymphocytes, of cells are initially found at each of the other sites but later become limited to the marrow.

Progenitor cells are not sufficiently differentiated to be recognized by microscopy but can be distinguished by the regulatory factors (glycoproteins) to which they respond and the clones they produce. All bone marrow is of the red haemopoietic type throughout infancy and early childhood.

The early circulation

The earliest blood vessels are simple endothelial tubes which differentiate in situ in the somatopleuric mesoderm of the chorion, the allantoic mesoderm of the connecting stalk, the splanchnopleuric mesoderm of the yolk sac and, somewhat later, in intraembryonic mesenchyme. From these centres, *chorionic, umbilical, vitelline* and *intraembryonic* systems extend by differentiation in situ of further vessels until they join and a primitive, bilaterally symmetrical, circulation is established. Thereafter, new vessels form as outgrowths of existing vessels whenever and wherever there are functional demands. *Diffuse*

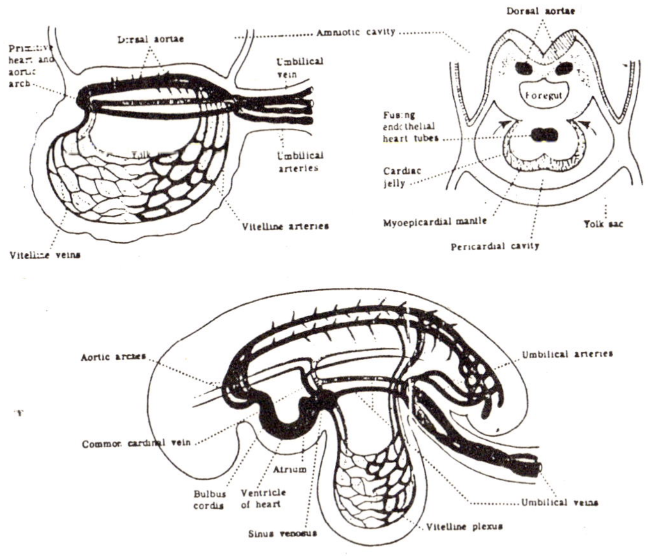

Fig. 7.68. The earliest blood vessels.

capillary plexuses precede defined channels in any region. Such channels arise by the enlargement of some vessels at the expense of others in response to genetic and local haemodynamic influences. All vessels are initially endothelial, but are named arteries and veins in anticipation of the development of medial and adventitial coats from surrounding mesenchyme.

Similarly, the *heart* is first represented by a pair of endothelial tubes which differentiate in the cardiogenic area. With head-fold formation the tubes fuse ventral to the foregut and bulge into the primitive pericardial cavity. The cavity extends dorsally and them medially so that the hearts is suspended within it by a *dorsal mesocardium*. Outside the endothelial tubes is a loose *gelatinoreticulum*. This is invaded by myoblasts which are derived from the epipcardium and, with it, form the *myoepicardial mantle*. Caudal to the fusing heart tubes is the mesoderm of the septum transversum. In it the vitelline, imbilical and intraembryonic—*common cardinal*—*venis* empty via a short venous trunk—the *sinus venous*—into the *primitive atrium of each* side. The atria lead to the single *primitive ventricle*. A single *bulbus cordis* then leads to the *aortic arches* and *aorsal aortae*.

The Heart

Growing within the confined space of the pericardial cavity, the primitive heart tube folds on itself and forms a *bulboventricular loop* which falls the right, probably under the influence of flow and of miliary action of the coelomic epithelium. The notch between the two limbs of this loop becomes shallower as the proximal part of the bulbus merges with the primitive ventricle into a common chamber. *Interventrcular sulci* soon appear and demarcate the definitive right and left ventricles. Meanwhile, the primitive atria merge into a *common atrium* which rises out of the septum transversum into the pericardial cavity to lie dorsal to the ventricles. On each side the sinus venosus also rises to lie dorsal to the atrium.

At first the *sinuatrial junction* is ill-defined, but it soon becomes invaginated on each side. Of the resulting folds, that on the left is more extensive and forms a septum which displaces the sinuatrial orifice to the right, creating a transverse chamber—*the body of the sinus venosus*—connecting the two sides (horns). The free edges of the folds are now termed *venous valves*.

Meanwhile, the left vitelline and left and right umbilical veins have lost their connections with the sinus venosus and an evagination—the future *common pulmonary vein*— has grown from the sinuatrial junction into the dorsal mesocardium and has tapped the pulmonary venous plexus. The terminal part of the right vitelline vein now connects the liver and the right horn of the sinus venosus and is called the *right hepatocardiac*

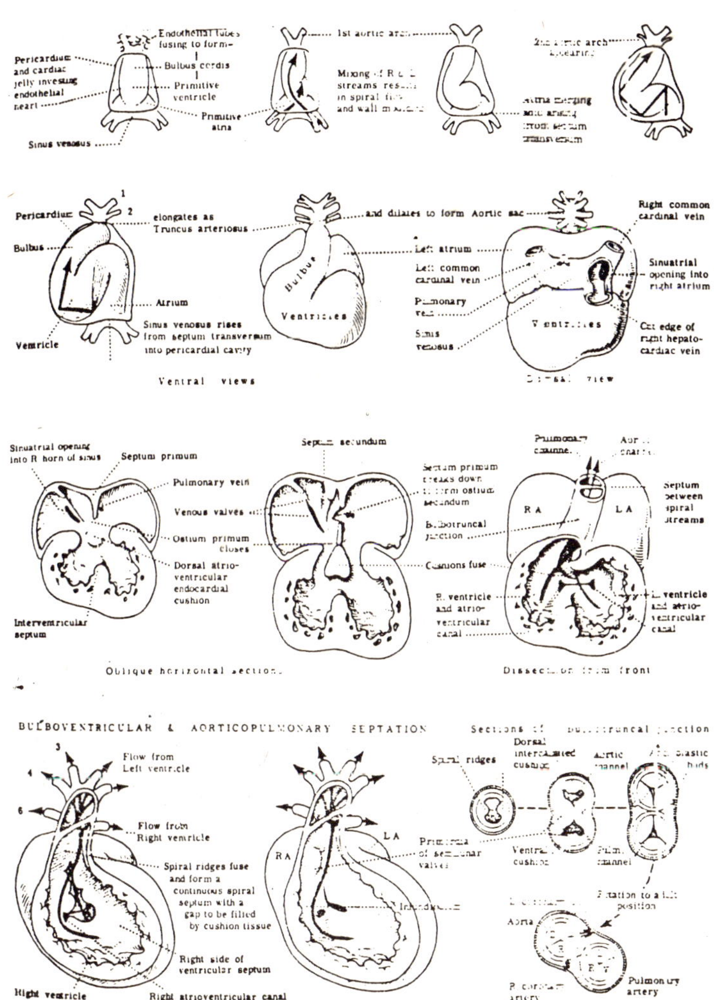

Fig. 7.69. Bulboventricular and Aorticopulmonary Septation

vein. The common pulmonary vein, and hence the pulmonary venous return, is now relegated to the left atrium as the sinuatrial septum has grown across caudal to it and the interatrial septum primum has appeared on its right.

The *septum primum* is a sickle-shaped projection, produced by proliferation of myoblasts, which grows down towards the narrowing atrioventricular function in which persisting gelatinoreticulum forms *endocardial cushions.* At first these translate the contractions of the myocardium to the endocardial heart and form a primitive valve mechanism. Later, they meet and fuse and separate *left and right atrioventricular canals.* The advancing septum primum and fused cushions bound a narrowing *ostium primum.* This permits the right-to-left shunt which is necessary for the normal development of the left side of the heart. The ostium primum eventually closes but the shunt is maintained because the cranial part of the septum primum breaks down and an *ostium secundum* is formed.

The bloodstream within the bulboventricular loop is directed first caudally, then to the right and then cranially. Where the stream impinges on the wall and is reflected and changes direction, loops and strands grow in from the myoepicardial mantle and interdigitate with outpouchings from the endocardium to form the trabeculae of the ventricles. The ventricles enlarge by deepening on either side of a zone of stasis, which thus forms the *ventricular septum* and the interventricular sulci. The horns of the crescentic septum reach the corresponding atrioventricular cushions.

As the bulboventricular notch shallows, the right side of the atrioventricular canal comes to lie dorsal to the bulbar orifice. At this stage the ostium primum, atrioventricular canal and interventricular foramen form a continuous hiatus between the right and left sides of the heart. After the atrioventricular cushions have fused, the stream from the left atrioventricular canal enters the left ventricle and is ejected via the interventricular foramen into the dorsal part of the bulbar orifice. The right ventricular outflow passes into the ventral part of the bulbar orifice.

The intersection of left dorsal and right ventral streams produces clockwise spiralling as the two streams traverse the bulbus cordis and a short arterial trunk—the *truncus arteriosus*--to reach the origin of the aortic arches—the *aortic sac*. Beginning between the fourth and sixth aortic arches, an *aorticopulmonary septum* grows back between the spiralling streams. In the truncus arteriosus it spirals through 180° so that the aortic channel which is ventral in the sac is dorsal at the bulbotruncal junction—the site of semilunar valve formation.

The bulbus cordis is lined with deformable *cushion tissue* which is pressed aside by the spiralling streams and fills in between them as *bulbar ridges*: they begin at the bulbotruncial junction where they are continuous with the aorticopulmonary septum and lie left and right; traced back in the bulbus they rotate through 90° and end at the ventral and dorsal horns of the ventricular septum respectively. The *right bulbar ridge* reaches the dorsal horn of the ventricular septum by crossing the fused atrioventricular cushions.

Beginning at the bulbotruncal junction, the bulbar ridges meet, fuse and are reinforced by the ingrowth of muscle tissue. The resulting septum divides the bulbus into a major part—the *infundibulum* of the right ventricle—and a minor part—the *aortic vestibule*. The remaining gap between the bulbar septum and the ventricular septum is closed by cushion tissue from the right side of the fused atrioventricular cushions. This grows along the crest of the ventricular septum, fusing with it and the bulbar septum, and eventually forms the *membranous part of the interventricular septum*. At the bulbotruncal junction, bisected *right and left septal cushions*, with an additional (*nonseptal*) cushion in each channel, form the primordia of the semilunar valves. Flow conditions produce excavation of their downstream aspects to form cusps and sinuses.

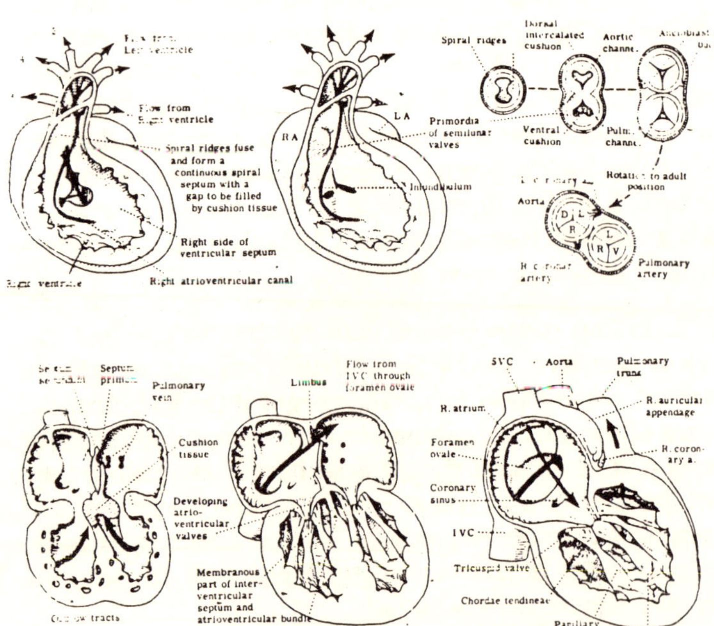

Fig. 7.70. Interatrial and Atrioventricular Valve Formation.

The bulbar septum separates the *outflow tract of the right ventricle*—the infundibulum—from the *outflow tract of the left ventricle*—the aortic vestibule. The outflow tracts lie ventral to the atrioventricular canals, which have accumulations of cushion tissue projecting into them. In the *left canal, septal and lateral cushions* are the primordia of the anterior and posterior leaflets of the mitral valve. In the *right canal*, there is a single *septal cushions*—the primordium of the septal leaflet of the tricuspid valve—but there are two *lateral cushions*. These are derived largely from the right bulbar ridge and are the primordia of the anterior and posterior leaflets. The shafts of muscle trabeculae

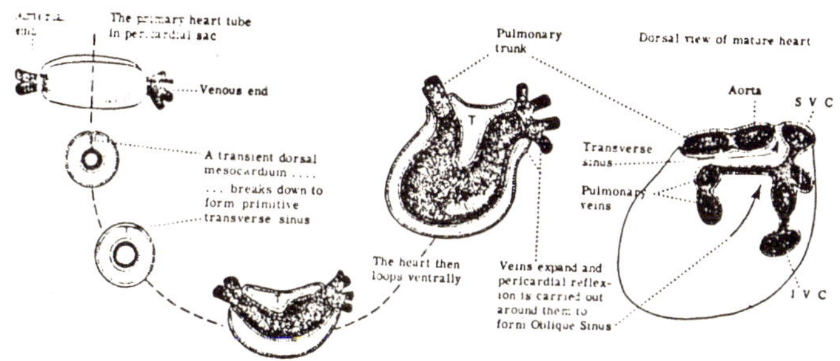

Fig. 7.71. The Pericardium.

which extend into the bases of the cushions, split off from the ventricular wall, while the cushions themselves become excavated on their downstream aspects to form valve cusps. The mural ends of the trabeculae remain muscular and form papillary muscles. Collagen fibres are deposited in their cushion ends which form the chordae tendineae attached to the valve cusps.

Meanwhile, a sickle-shaped *septum secundum* grows down from the atrial roof and obtrudes on the ostium secundum. The part of the septum primum attached to the endocardial cushions becomes a flap slung from the left side of the septum secundum. The resulting *atrial septal complex* consists of a *foramen ovale* bounded by a *limbus* (septum secundum) and a flap-like *valve* (septum primum). It is a competent valve system which permits a right-to-left shunt but prevents a left-to-right shunt. The common pulmonary vein and its right and left tributaries have dilated and have been incorporated into the *left atrium* as the smooth-walled vestibule. The primitive left atrium with its ridged lining is represented only by the auricular appendage. The right horn of the sinus venosus has also dilated and has been incorporated into the *right atrium* as the smooth-walled sinus venarum. The vessels which formerly opened into the right horn now drain directly into this part of the right atrium. These vessels are the right common cardinal vein, the right hepatocardiac vein and the body of the sinus venosus. They form the proximal parts of the superior vena cava, the inferior vena cava and the coronary sinus respectively. The intermediate part of the coronary sinus is derived from the left horn of the sinus venosus, and the distal part from the left common cardinal vein. The venous valves, which formerly bounded the sinuatrial opening, are now much reduced. The left venous valve fuses with the atrial septal complex. The right venous valve, though reduced by resorption, continues to separate the ridged primitive right atrium from the derivatives of the sinus venosus and forms the crista terminalis, most of the valve of the inferior vena cava and the valve of the coronary sinus. The valve of the inferior vena cava is completed by the spur of tissue which separates the orifices of the inferior vena cava and coronary sinus.

The cardiac myoblasts form a non-syncytial network. They develop myofibrils which increase in number and size, align with those of adjoining cells and develop cross-striations. At first atrial and ventricular myocardium are continuous across the atrioventricular junction. The *atrioventricular bundle* is then seen as a specialized fascicle with densely straining nuclei which extends from the right side of the dorsal wall of the common atrium, behind the dorsal atrioventricular cushion, on to the crest of the primitive ventricular septum. Later, the atrioventricular node is defined at its atrial end, and left and right branches appear successively at its ventricular end. Elsewhere, continuity across the atrioventricular junction is lost as the fibrous skeleton of the heart develops. The sinuatrial node develops in relation to the base of the right venous valve.

The Arteries

The *aortic arch arteries* appear in the mesenchyme of the pharyngeal arches in craniocaudal sequence and connect the *aortic sac* with the two *dorsal aortae*. These fuse between the fourth lumbar and fourth thoracic segments and form the single descending aorta. As new arches appear, the flow pattern within the system changes and blood is diverted from preceding arches. The first two arches thus become redundant and do not persist as complete arches. With the development of the neck and descent of the heart from its primitive cervical position, the third and fourth arches draw out the aortic sac on each side to form *right* and *left horns*. The part related to the third arch further elongates and forms the common carotid artery. From this the external carotid artery sprouts as a new formation, though it may incorporate segments of the first two arches. The proximal part of the internal carotid artery is derived from the third arch itself. The distal part is derived from cranial extensions of the dorsal aorta, including segments of the first two arches.. The fourth arches persists as channels connecting the horns of the aortic sac, via the corresponding dorsal aorta, to the upper limb artery. The right horn thus form the brachiocephalic trunk, and the right fourth arch the proximal part of the subclavian artery. The aortic sac and its left horn and fourth arch form and arch of the aorta. The fifth pair of arches is transient or absent in man.

The sixth arches are atypical. They supply primitive pulmonary plexuses and only secondarily become connected with the dorsal aortae. The ventral stems become the pulmonary arteries. On the right side connection with the dorsal aorta is lost, but on the left it persists as the *ductus arteriosus*. The dorsal aortae between the third and fourth arches and the unfused part of the right dorsal aorta distal to the limb artery then disappear. Meanwhile, the spiral ridges have fused and separated the pulmonary trunk (leading from the right ventricle to sixth arch derivatives) from the ascending aorta (leading via the arch of the aorta and the brachiocephalic trunk to fourth arch derivatives).

Specialized pressor-receptor areas (e.g., carotid sinus) develop in persisting arches and are supplied by the appropriate arch nerves. The recurrent laryngeal nerves reach the larynx by passing caudal to the sixth arch on each side. On the left, this arch (ductus arteriosus) descends into the thorax, dragging a loop of nerve with it. On the right, the sixth arch loses

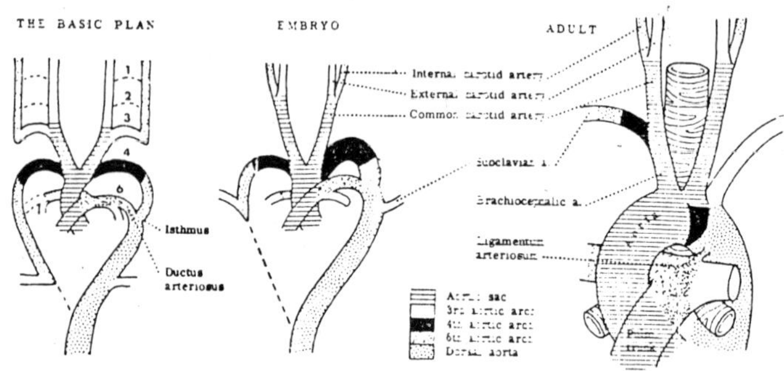

Fig. 7.72.   The Aortic Arches

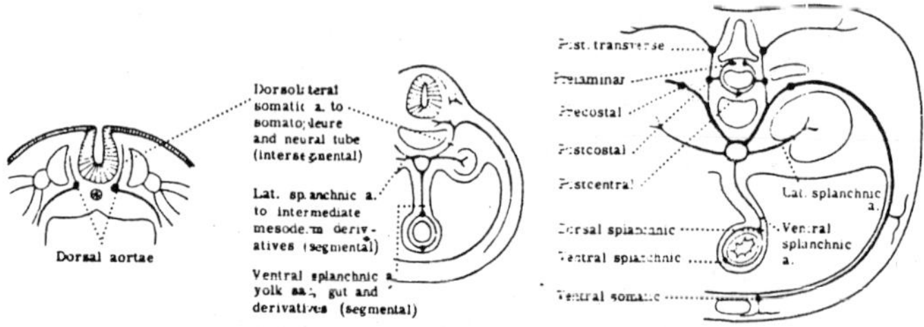

Fig. 7.73.   Segmental Arteries and Longitudinal Anastomoses

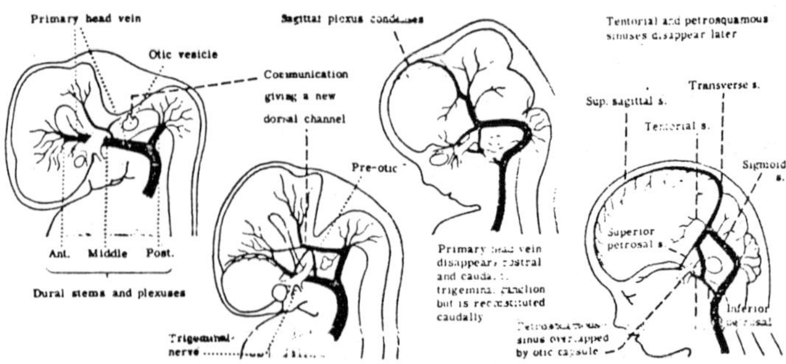

Fig. 7.74.   Dural Venous Sinuses

connection with the dorsal aorta, and the fifth is transient so the nerve slips these arches and hooks around the fourth arch (subclavian artery), thus descending only to the root of the neck.

The primitive dorsal aortae lead directly to the umbilical arteries and develop, in turn, ventral segmental (vitelline) branches, dorsolaortae intersegmental (somatic) branches, lateral segmental (intermediate) branches and longitudinal anastomoses which connect the different levels. The caudal loops of the umbilical arteries are bypassed by anastomotic connections with the fifth lumbar intersegmental arteries which pass dorsal and lateral to the Wolffian duct. The paired *ventral segmental branches* fuse and are reduced to the midline arteries of the foregut (oesophageal and coeliac trunk), midgut (superior mesenteric) and hindgut (inferior mesenteric). Dorsal splanchnic anastomoses form the gastroepiploic and pancreaticoduodenal arteries and the mesenteric arcades. Ventral anastomoses form the gastric and hepatic arteries. The *dorsolateral intersegmental branches* form the posterior intercostal, subcostal and lumbar arteries and the stems of the limb arteries. The *longitudinal somatic anastomoses* and their derivatives are post-transverse (deep cervical), postcostal (most of the vertebral), prelaminar and postcentral (spinal), precostal (ascending cervical; supreme intercostal) and ventral (internal thoracic; superior and inferior epigastric). The *lateral segmental branches* form the phrenic, suprarenal, renal and gonadal arteries.

The developing *brain* is covered by a capillary plexus fed by the branches of the internal carotid artery. Its primitive maxillary branch supplies the telencephalon but is succeeded by a terminal branch which gives rise to the anterior and middle cerebral and choroidal arteries. The diencephalon and mesencephalon are supplied by another terminal branch which gives rise to the posterior communicating and cerebral and the anterior cerebellar arteries. The arteries of the hindbrain and cord differentiate from longitudinal plexuses fed by presegmental branches of the internal carotid artery and by spinal branches of intersegmental arteries. On each side the capillary plexuses drain into the precardinal vein   via a *primary head vein:* this lies medial to the trigeminal ganglion, where it persists as the cavernous sinus. Secondary connections between the plexuses develop dorsal to the primary head veins and supersede them as main channels. The deeper vessels form the cerebral veins whilst the most superficial vessels form dural venous sinuses.

The Veins and Lymphatics

The intrinsic veins of the somite embryo are the *vrecardinal* and *postcardinal* veins which join to form a *common cardinal vein* on each side.

The right *precardinal vein* forms the right internal regular and brachiocephalic veins and that part of the superior vena cava cranial to the azygos vein. Caudal to the azygos vein the superior vena cava is derived from the right common cardinal vein. An oblique cross-connection between the precardinal veins forms part of the left brachiocephalic vein. As venous return is thus diverted to the right, the left common cardinal vein retrogresses and forms the distal part of the coronary sinus, the oblique vein of the left atrium and a fibrous connection between it and the proximal parts of the left pre-and postcardinal veins. These parts form the left superior intercostal vein which drains upwards into the left brachiocephalic vein. The left precardinal vein forms the rest of the left brachiocephalic vein and the left internal jugular vein.

The *postcardinal veins* drain the lower limb buds, the body wall and the mesonephric ridge and lie dorso-lateral to the last-named. They are soon supplemented and largely superseded by other longitudinal channels and plexiform connections. The venous drainage of the mesonephric ridges is taken over by *subcardinal veins* lying medial to them and ventral to the aorta. That of the body wall is taken over by *supracardinal* and *azygos line veins* which lie dorsal to the aorta and lateral and medial to the sympathetic trunk. With the diversion of venous

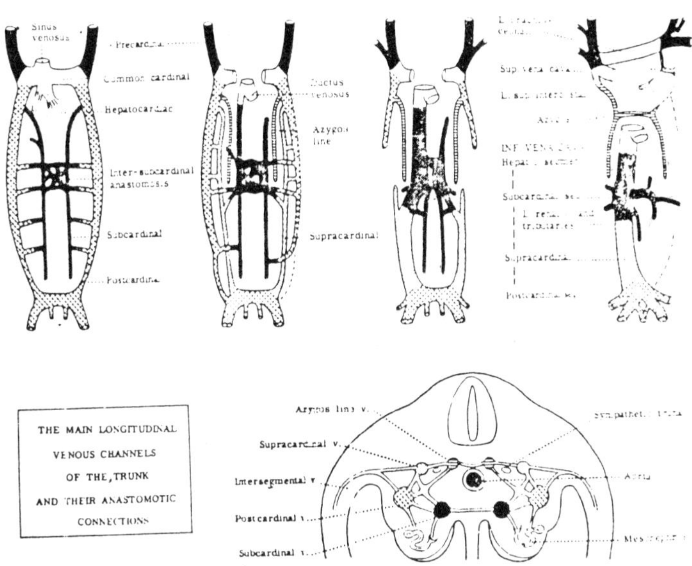

Fig. 7.75.   The Somatic Veins.

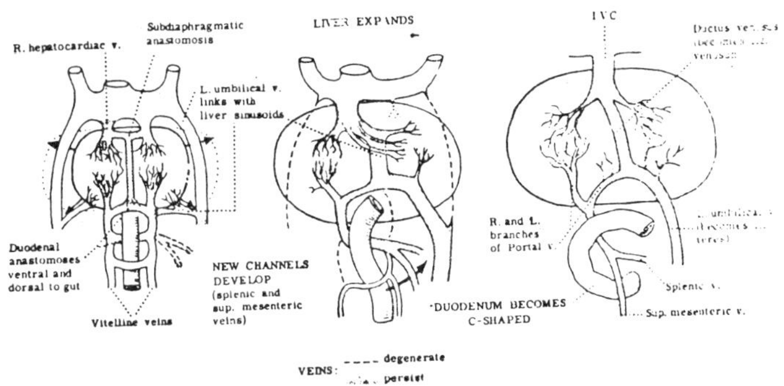

Fig. 7.76.   The Visceral Veins.

return to the right, most of the channels on the left disappear. Those that persist drain to the right by cross-anastomoses. Beginning caudally, the inferior vena cava is formed from the right postcardinal vein (draining lower limb and pelvis), the right supracardinal vein (draining lower abdominal wall), the right subcardinal vein (draining kidneys, suprarenals and gonads), the right hepatocardiac vein (representing hepatic sinusoids and the terminal part of the right vitelline vein draining the liver) and anastomotic segments between them.

The further development of the *visceral veins* is linked with that of the liver. As the vitelline venis pass cranially, a *dorsal anastomoses* connect them and around the primitive duodenum. Beyond this they break up into hepatic sinusoids. Their cranial ends still drain into the sinus venosus as hepatocardiac venis and are connected by a *subdiaphragmatic anastomosis*. A midline channel— the *ductus venosus*— connects the subdiaphragmatic and the cranial duodenal anastomoses. With expansion of the liver and diversion of venous return to the right side of the heart, the right umbilical and left hepatocardiac veins disappear completely and he left umbilical vein, after being tapped by the liver sinusoids and cranial duodenal anastomosis, disappears in its cranial part. The whole of the placental venous return then traverses the liver, but whereas some traverses the sinusoids, the rest traverses which effectively connects the left umbilical and right hepatocardiac veins.

Parts of the vitelline veins also disappear in such a way that the persisting parts of the figure 8 system may be represented as a letter S. However, since the developing duodenum is meanwhile assuming its characteristic C shape, the persisting parts pursue a progressively straighter course to the liver. The superior mesenteric vein and, later, the splenic vein join the left end of the original dorsal anastomosis which thus, with the extrahepatic part of the right vitelline vein, forms the portal vein. The more distal parts of he vitelline system disappear with the vitelloinestinal duct.

The earliest *lymph vessels* may be derived from clefts in mesenchyme which become secondarily connected with the venous system, or they may be derived directly by capillary offshoots from the venous system. They form a bilaterally symmetrical system with transverse anastomoses across the midline. Some confluesce and form lymph sacs. Paired sacs are found in relation to the veins a the roots of the limbs, and midline sacs in the root of the mesentery. New lymph vessels but off from the sacs or arises in situ and connect with them. The lymph sacs at the roots of the upper limbs have, or develop, connection with the internal jugular veins and, via bilateral anastomotic connections with the midline sacs, form bilateral primitive thoracic ducts. The definitive thoracic duct is derived from the caudal part of the right vessel, a transverse anastomosis and the cranial part of the left vessel. The right llymphatic duct is derived from the cranial part of the right vessel. The lymph sacs mainly retrogress but a midline sac persists as the cisterna chyli. Mesenchyme aggregates around the others and around more peripheral lymphatics, and is seeded by lymphocytes to form lymph nodes.

Fetal and Neonatal Circulation

Cardiovascular development before birth has led to a system which is adequate for prenatal needs and yet capable of undergoing changes at birth which fit it for postnatal needs. Fairly well-oxygenated blood returns from the placenta via the unpaired vein in the umbilical cord and via the persisting left umbilical vein in the body. Some of the blood discharges into hepatic sinusoids but the rest by passes them, is joined by blood from the portal vein and traverses the ductus venosus.

veins. The mixture passes into the left ventricle and is ejected into aorta, whence most of this *relatively well-oxygenated blood reaches the arteries of the heart, head, neck and arms.* The residuum traverses the *isthmus aortae* to join the stream from the *ductus arteriosus.* Meanwhile, the remainder of the inferior caval stream joins the blood from the superior vena cava and coronary sinus in the right atrium proper and enters the right ventricle. Some of the right ventricular blood goes to the lungs. The remainder bypasses them via the ductus arteriosus and joins the stream from the isthmus aortae in the descending aorta for distribution partly to the lower parts of the body but mainly via the umbilical arteries to the placenta.

Both ventricles are thus working in parallel, and at birth their walls are of similar thickness. However, because they work in parallel, their outputs need not be, and probably are not, equal. Similarly, the placenta and lungs are in parallel with the systemic body tissues. The placenta constitutes a relatively low-resistance circuit and has a greater volume of blood flow than the systemic tissues. The unexpanded fetal lungs constitute a relatively high-resistance circuit and have a smaller volume of blood flow than the systemic tissues.

At term, the principal cardiovascular and respiratory reflexes are believed to be active. During delivery the oxygen-saturation in the umbilical arteries falls below 30 per cent. This is probably due to compression of the umbilical cord, to partial separation of the placenta and to a reduction in uterine blood flow during contractions. Respiration is then stimulated reflexly by the effect of anoxia on chemoreceptors, and probably by exteroceptive stimuli. With the onset or respiration, intrathoracic pressure falls and the lungs are inflated. Pulmonary vascular, resistance falls rapidly and the pulmonary blood flow increases up to tenfold in as many minutes. This response is partly due to gaseous expansion of the lungs, a local mechanical phenomenon, and partly due to changes in alveolar and arterial oxygen and carbon dioxide tensions.

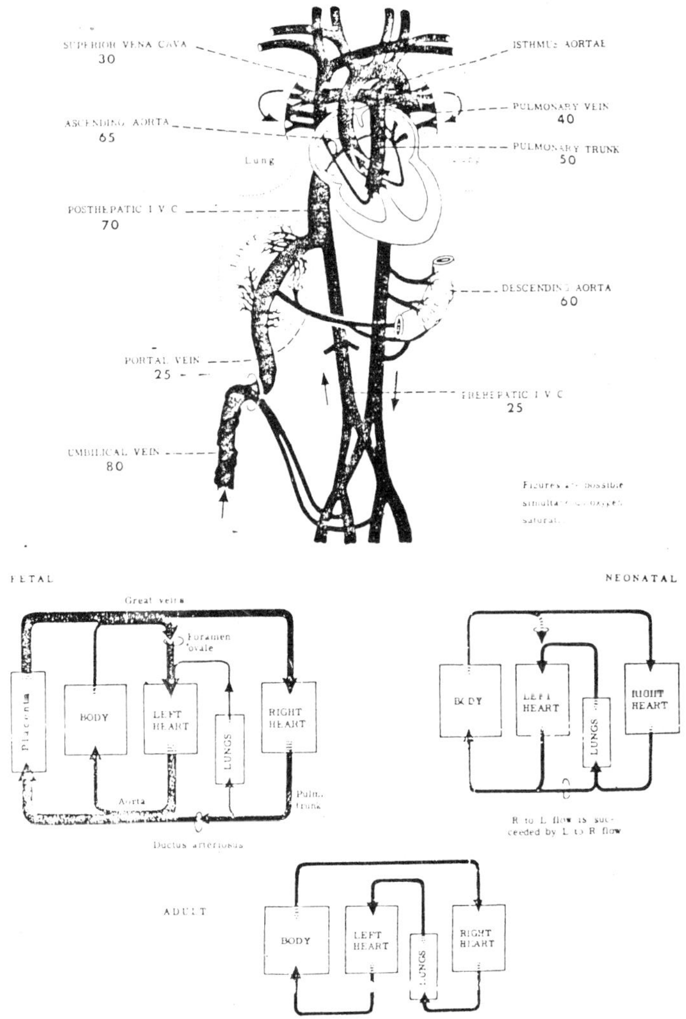

Fig. 7.77. Fetal Circulation

As the blood flow increases, the pulmonary arterial pressure falls and the left atrial pressure rises. Furthermore, when the venous return from the placenta ceases, the inferior vena caval pressure falls. These circumstances combine to press the valve of the foramen ovale against the limbus and produce *functional closure*. As this functional closure depends on relative pressures on the two sides of the valve, there may be an intermittant flow of blood from the inferior vena cava into the left atrium for some days after birth. Left atrial pressure, however, continues to rise, the shunt ceases and the valve begins to fuse with the limbus. During the following months the connective tissue of the valve increases, and structural closure usually occurs, forming the *fossa ovalis*.

The muscular wall of the *ductus arteriosus* contracts within a few minutes of birth, due to the direct action of oxygen or of catecholamines. For up to an hour the reduced shunt continues to be right-to-left in the fetal direction, but after this time the increase in systemic arterial pressure and the decrease in pulmonary arterial pressure combine to reverse the shunt, which may then produce a characteristic innocent heart murmur. These pressure changes are associated with a rapid adjustment of ventricular preponderance. This occurs mainly in the first month, as the right ventricle atrophies and the left ventricle hypertrophies. The degree of left ventricular preponderance characteristic of later life is reached by 6 months. Some days after birth (longer in premature infants) the ductus arteriosus shuts down completely. It becomes, occluded by overgrowth of the intima and media, subsequently undergoing fibrosis and contraction to form the *ligamentum arteriosum*. The umbilical arteries contract early, and before the umbilical vein and ductus venosus, so an appreciable amount of fetal blood is able to return from the placenta if the umbilical cord is not clamped immediately. Closure of the placental low-resistance circuit leads to a rise in systemic vascular resistance. The vessels ultimately become obliterated, forming the obliterated umbilical arteries, the ligamentum teres (umbilical vein) and the ligamentum venosum (ductus venosus).

Anomalies

As a group, malformations of the heart are the commonest serious congenital conditions and have an incidence of about 6 per 1000 total births. Of these, about 20 per cent each have a ventricular septal defect or a persistent ductus arteriosus; about 10 per cent eacg have an atrial septal defect, isolated puimonary stenosis. Fallot's tetralogy or transposition of the arterial trunks or coarctation of the aorta while 10 percent have other less common malformations of which more than a hundred varieties have been described.

In its commonest form an *isolated ventricular septal defect* occurs at the site of the membranous part of the interventricular septum and thus represents the failure of cushion tissue to bridge the gap between the bulbar and ventricular septa. In this case, the atrioventricular bundle lies in the fee edge of the septum and is vulnerable during surgical repair. Much less commonly the defect is in the muscular part of the interventricular septum and represents excessive outprouching and breakdown of the endocardium between trabeculae. In the case the defect may close spontaneously. In either from there is a left-to-right shunt between the ventricles.

Failure of the processes of muscular contraction and intimai proliferation which normally close it, results in a *persistent ductus arteriosus* and in persistence of the left-to-right shunt of neonatal life. Spontaneous closure may eventually occur but ligation in childhood, not only to close the shunt but also to reduce the risk of infective endarteritis, is preferred.

By far the commonest *atrial septal defect* is an *ostium secundum defect* in which the valve is inadequate to cover the foramen ovale or is fenestrated and thus is incompetent and permits a left-to-right shunt. In some normal individuals, structural closure may fail to occur but, provided the valve is competent, this is of no significance. *Endocardial cushion defects* and less common and occur in three degrees. In the first, the septum primum is underdeveloped of fails to fuse with the atrioventricular

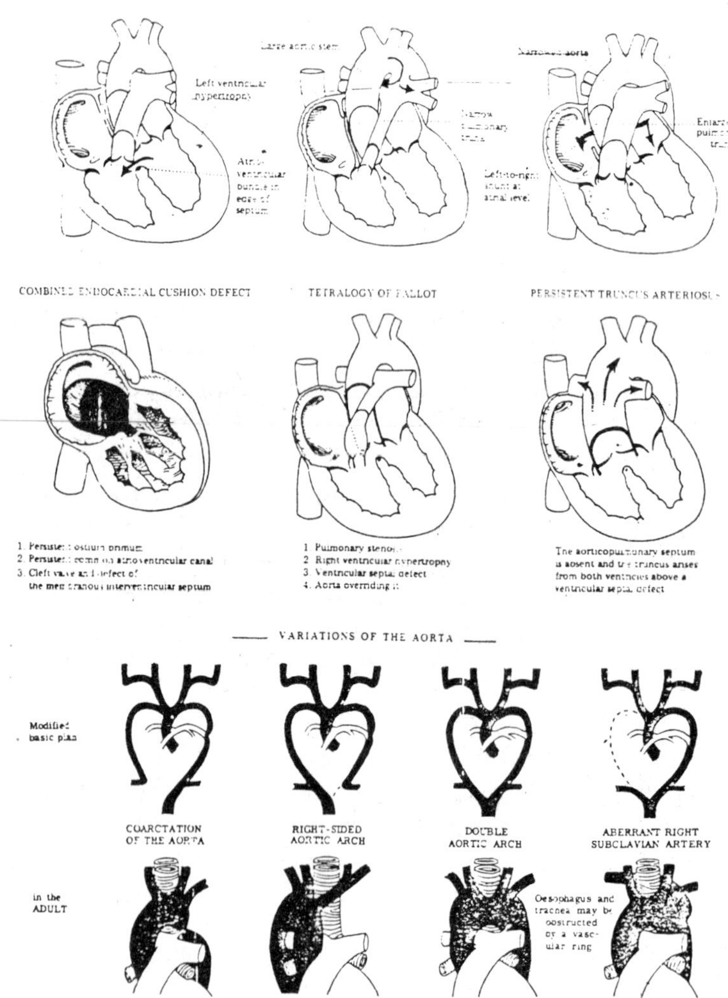

Fig. 7.78.   Variations of the Aorta

cushions, resulting in a *persistent ostium primum*. In the second, the atrioventricular cushions fail to fuse with each other resulting in a *persistent common atrioventricular canal*. In the third, the atrioventricular cushions also fail to fuse with the ventricular septum and result in a *defect of the membranous septum*. Thus in the third degree a continuous hiatus persists between the left and right sides of the heart.

Isolated pulmonary stenosis is commoner than *aortic stenosis* and may affect the pulmonary valve or the infundibulum. It is also common in combination with hypertrophy of the right ventricle and a ventricular septal defect, with the aorta over-riding it, in *Fallot's tetralogy*. This combination produces a right-to-left shunt and central cyanosis—the classic 'blue baby.'

In *transposition of the arterial trunks* the aorticopulmonary septum has failed to spiral, so the aorta arises from the right ventricle and the pulmonary trunk from the left. The condition is incompatible with postnatal life unless accompanied by a bidirectional shunt through a septal defect or a persistent ductus or by transposition of the whole or part of the heart so that the arterial transposition is 'corrected' and the connections are appropriate. Less commonly, the aorticopulmonary septum is absent in whole—*persistent truncus arteriosus*—or in part—aorticopulmonar window.

In *coarctation of the aorta* (L. *arctare* = to tighten) a constriction is present, usually near the ductus arteriosus, which may persist. Coarctation at this site may; represent a persistent isthmus aortae or involvement of the aorta in the closing mechanisms of the ducts. However, coarctation also occurs beyond the ductus and even, rarely, beyond the diaphragm, so other factors must operate. As there is a substantial pressure differential between the two sides of the coarctation, the blood pressure recorded in the upper limbs is higher than in the lower limbs and a collateral circulation develops.

Variations of the aorta other than coarctation are rare. Since flow factors, which can easily vary, are at least in part responsible for determining which blood vessels develop from a network at the expense of others variations in other blood vessels are common. Similarly, since venous pressure is lower, and thus less determinative, variations in veins are commoner than in arteries. However, in most cases variations are incidental findings and not of pathological significance.

Most cases congenital heart disease is multifactorial. The sex ratio is 1:1 and the incidence in twins is about twice that in non-twins; the incidence in the children and siblings of affected individuals may be 20-30 times higher than in the general population Mechanical factors may play a part and a whole gamut of anomalies can be produced by deformation of the heart tube. Altitude has effects, and viruses such as rubella may produce persistent ductus arteriosus, stenoses and other lesions.

Rarely, Mendelian inheritance operates. *Marfan's syndrome* is an autosomal dominant disorder of connective tissue which may be associated with congenital aortic valve disease. Autosomal recessive inheritance operates in *dextrocardia* as part of *situs inversus* and perhaps in some other conditions.

In *Down's syndrome* (trisomy 21) endocardial cushion defects are common, while with trisomies 13 and 15 ventricular septal defect, persistent ductus arteriosus and other anomalies occur. Aortic stenosis and less commonly coarctation of the aorta occur in the monosomic Turner's syndrome (45, X).

8

Development of Digestive System

The primary tissue of the entire digestive system is entoderm. This epithelial layer originally lines the whole yolk sac, but a region difference in the shape of the entodermal cells is apparent from the first. Those that underlie the embryonic disc (and serve as a flat roof to the early yolk sac) are taller than the rest; they are the ones that are destined to become gutentoderm. When, at the twentieth day, the rapidly expanding embryonic disc begins to fold into a cylindrical embryo, its gut-entoderm participates as a component layer. Folding first into the head end and then into the hind end of the elongating embryo, this entoderm necessarily takes the form of two internal, blind tubes. The open end of each tube, where it becomes continuous with the yolk sac, is called an *intestinal portal,* while the tubes themselves are named the *fore-gut* and *hind-gut.* An intermediate region, open ventrally into the yolk sac through the narrower yolk stalk, is sometimes termed the *mid-gut,* but its existence in man is brief since the yolk stalk constricts rapidly during the fourth week and detaches from the gut at the end of the fifth week. Both the fore-gut and the hind-gut elongate and broaden by interstitial growth, so as to keep pace with the growth of the embryo as a whole.

The primitive, tubular gut differentiates into the alimentary canal which has three chief segments: the mouth, pharynx and digestive tube. The latter division includes the esophagus, stomach, small intestine and large intestine; it lies mostly in the body cavity and is suspended or held in place by mesenteries. The fore-gut specializes into the mouth and pharynx, and into

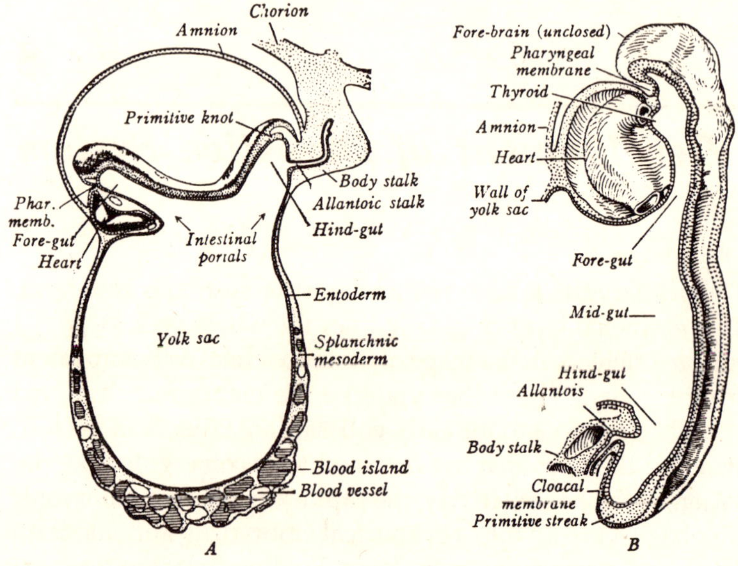

Fig. 8.1.   Entodermal tract of early human embryos, in sagittal section. A, At seven somites. B, At ten somites.

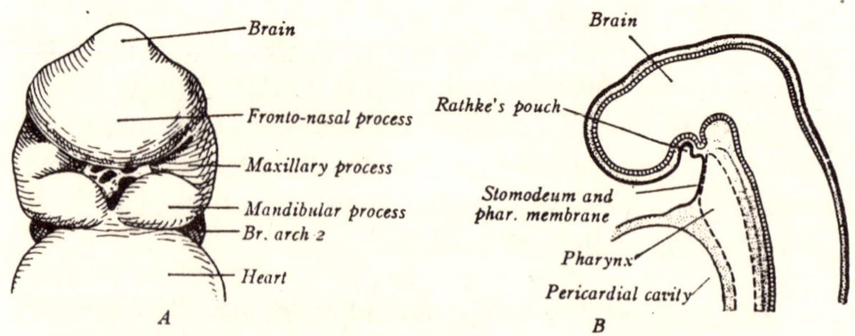

Fig. 8.2.   Human stomodeum and oral membrane. A, Boundaries of the stomodeum and a partly perforated pharyngeal membrane, shown in front view at 2.5 mm. B, Relation of ectoderm (full line) and entoderm (broken line) in this region, illustrated by a sagittal section at 2.5 mm.

the digestive tube to a point-far along the small intestine. The hind-gut becomes the rest of the small intestine and all of the colon and rectum. Throughout its length the entodermal digestive canal gives rise to numerous derivatives, chief of which are the respiratory tract and the thyroid, parathyroids, thymus, liver and pancreas. The entoderm furnishes merely the epithelial lining of the digestive and respiratory tracts, and the characteristic epithelial parts of the other organs. The various glands, both large (such as the liver and pancreas) and small (like the gastric and intestinal glands), are subordinate growths that push out from the lining epithelium. All of the accessory coats of the alimentary canal, such as muscle and connective tissue, are secondary investments. They differentiate from the splanchnic mesoderm nearby. The epithelium that lines the various hollow viscera is usually slimy. Together with its underlying connective tissue and glands, it comprises a unit named the *tunica mucosa* or mucous membrane.

At each end the primitive gut-tube comes ventrally into contact with the ectoderm. The fused plates, thus produced, are the *pharyngeal* (or *oral) membrane* and the *cloacal membrane*. The pharyngeal membrane makes a floor to an external depression known as the *stomodeum;* this pit is bounded by the fronto-nasal, maxillary and mandibular processes and is brought into existence by the overjutting of these parts as growth progresses. Midway in the fourth week (2.5 mm. embryos) the pharyngeal membrane ruptures and the stomodeum and fore-gut merge. The stomodeum develops into part of the mouth which is therefore, ectodermal.

The caudal end of the entodermal tube becomes the *cloaca*, or common vent. It communicates with the allantois (Fig. 9.), which preceded it in time of origin, and soon receives the urinary and genital ducts. Even before these latter connections are complete, the cloaca begins to sub-divide into a dorsal *rectum* and a ventral *bladder* and *urogenital sinu.* By the end of the seventh week the cloacal membrane has separated into an anal and urethral region; following this the two membranes

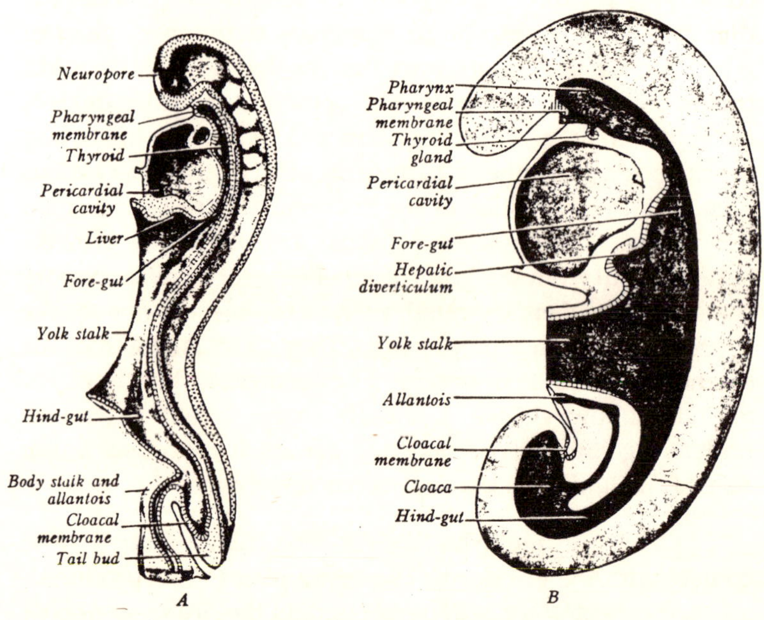

Fig. 8.3.   Entodermal tract, shown in hemisections of human embryos. A, At 2.5 mm., with eighteen somites. B, At 2.5 mm., with twenty-three somites.

rupture and disappear, thus concluding the division of the cloaca. Each of the new canals (rectum and urogenital sinus), so formed, acquires thus simply its individual opening to the outside. The end of the digestive tube is then lined for a short distance with ectoderm, and this portion (the *proctodeum*) constitutes some of the future anal canal. It will be noticed that the primitive entodermal tube extends caudad a little beyond the cloacal membrane; this *tail-gut* dwindles during the fifth week and soon disappears. How rapidly all these changes occur may be appreciated by comparing embryos of four weeks with those two or three weeks older.

The Mouth

After the loss of the oral membrane it is impossible to determine the exact junction of ectoderm and entoderm in the mouth. The inherent difficulties are increased by a considerable 'displacement' caudad of the dorsal line of union. The plane dividing ectoderm from entoderm is then a sltang one which passes forward from the roof of the pharynx to the floor of the mouth next the lower gum. This means that the roof and much of the sides of the mouth are ectodermal. More specifically, the nasal passages, palate, front of the tongue, and vestibule are considered to be lined with ectoderm; the enamel of the teeth and probably the salivary glands are likewise ectodermal derivatives. Although these various structures do not belong among the entodermal organs, it is simplest to describe them with the digestive and respiratory systems of which they are functional parts. A further derivative of the ectodermal stomodeum is a dorsal inpocketing, known as Rathke's pouch, which becomes the epithelial lobe of the hypophysis. Its point of origin marks the caudal extent of ectoderm in the completed mouth.

Causal Relations: Experiments on the amphibian neurula show that only when pharyngeal entoderm is in contact with the presumptive mouth ectoderm will a mouth, with teeth, form. Even entoderm from a mid-trunk level is unable to induce the formation of a mouth or mouth parts. The failure to provide a mouth opening is *astomia*.

Lips and Cheeks: Until the end of the sixth week the primitive jaws are solid masses which do not show any subdivision into lip and gum regions, as is also the permanent condition in animals below mammals. The separation of each lip from its respective gum is foreshadowed by the appearance of a thickened band of epithelium. This *labial lamina* grows from the ectodermal covering of the primitive jaw into the mesenchyme beneath. Following the contour of the jaw, it makes a long, curving band which deepens into a partitioning plate. Progressive disintegration of the more central cells causes each plate to split into two sheets. In this manner the *lips* become separate from the *gums* by the tenth week, and the epithelial-lined labial groove, so formed, deepens into the *vestibule*. In the midplane of each lip the splitting is not so deep, thus leaving a soft fold known as the *frenulum*.

The *cheeks* come into existence chiefly through a reduction in the extent of the originally broad mouth opening; this results from progressive fusion of the lips at their lateral angles. The labial and buccal muscles differentiate from mesenchyme of the second branchial arches which migrates between the epidermal covering and mucosal lining of these parts.

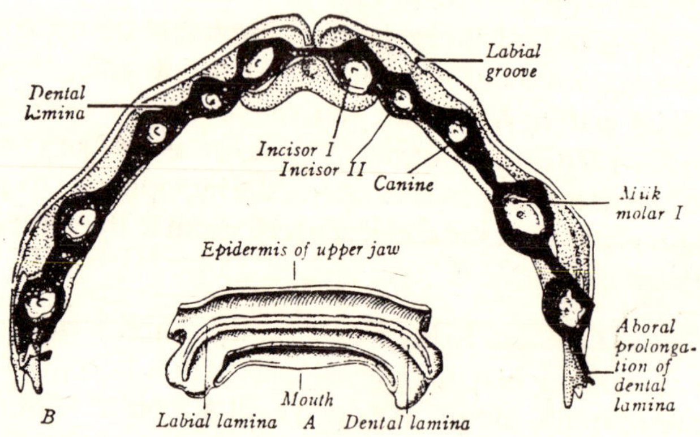

Fig. 8.4.   Isolated epithelium of the human jaws, showing the labial and dental laminae. A, At two months; B, at three months, with primordia of milk teeth.

The Teeth: Historically the teeth are products of the skin, and both the epidermis and derma (corium) contribute to their formation. A tooth is a greatly modified connective-tissue papilla that has both undergone a peculiar ossification into *dentine* and becomes capped by a hard *enamel* elaborated from the epidermis. In addition, the base encrusts with *cementum*, a bony deposit. The homology of a tooth primordium with a dermal papilla and its covering epithelium is somewhat obscured by an early ingrowing which results in the whole primordium coming to lie considerably beneath the surface of the gum.

The teeth have a double source of origin in the embryo: the enamel is from ectoderm; the dentine, pulp and cement are mesodermal. There are two generations of teeth in man and most other mammals, but no essential difference exists between the development of the temporary (milk) teeth and the permanent ones. Since the primordia of the *primary dentition* arise first, they will be described first and in greater detail.

The Dental Lamina and Early Tooth Buds: The first indication of on-coming tooth development is an epithelial plate, the *dental lamina,* which arises during the seventh week just gumward of the labial lamina, already described. The dental lamina soon becomes a horizontal shelf which projects perpendicularly from the labial lamina and extends well into the substance of the primitive gum. This is brought about as the gum, upgrowing, buries the dental lamina deeper and deeper. Each dental lamina thereby course alongside the curving labial groove and lies just gumward of it.

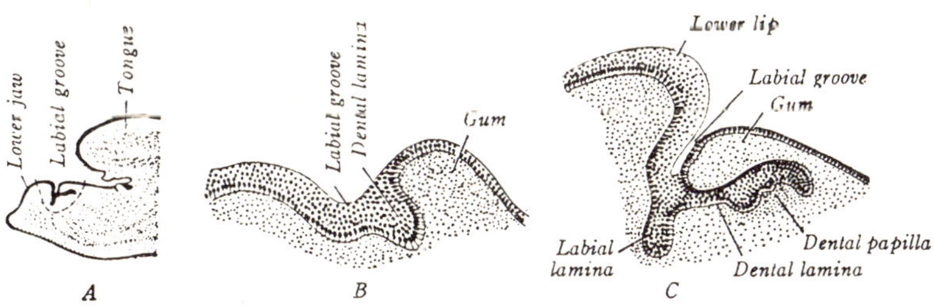

Fig. 8.5.   Development and relations of the labial and dental laminae, demonstrated by sections through the human lower jaw. *A,* Sagittal section, at nine weeks, to explain the areas included in *B* and *C. B,* Labio-dental lamina, at seven weeks. *C,* Detail at nine weeks, of the area set off in A by a broken line

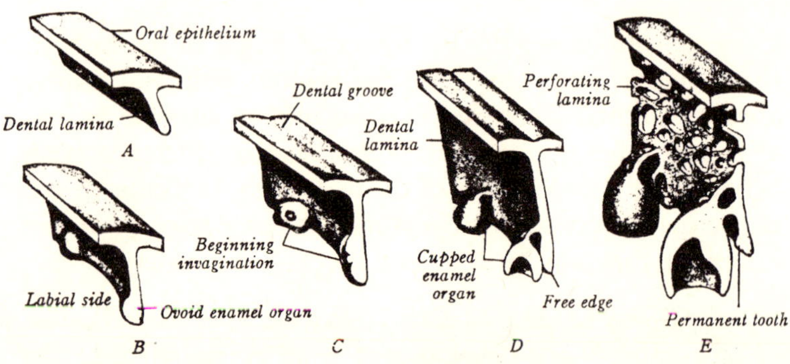

Fig. 8.6.   Development of the enamel organs, at two to four months, shown by models.

At intervals along the epithelial lamina there develops simultaneously a series of knob-like thickenings called the *enamel organs,* which will both produce the enamel and serve as the molds for the future teeth. Early in the third month the deeper side of each enamel organ presses against a dense accumulation of mesenchyme. The epithelial surface of contact both buckles inward (i.e., invaginates) and grows around the mesenchymal mound until the whole enamel organ is hollowed like a thick cup. The concavity, formed in this manner, is occupied by the condensed mesenchymal tissue of the *dental papilla* which is destined to differentiate into dentine and pulp. An enamel organ and its associated dental papilla are the developmental basis of each tooth. Ten such primordia of the *deciduous,* or *milk teeth* are present in each jaw of a ten-weeks fetus. Later the stalk connecting the enamel organ with the dental lamina, and most of the lamina itself, break down. However, the original free edge of the lamina persists longer and gives rise to the primordia of the permanent enamel organs.

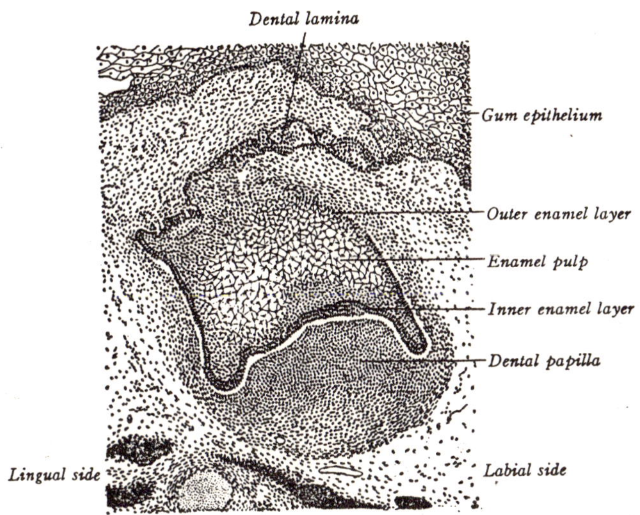

Fig. 8.7.   Primordium of a human tooth, at three months, shown in section.

The Enamel Organ: This primordium not only deposits enamel but also assumes the directing role in tooth development. During the third month the growing enamel organ becomes a double-walled sac, composed of an outer, convex wall (*outer enamel layer*) and an inner, concave wall (*inner enamel layer*). Between the two is a filling of looser cells which transform into a stellate reticulum named the *enamel pulp*. The enamel organ first encases the crown portion of the future tooth, molds its shape and deposits enamel there. Later the enamel organ elongates and similarly models the root portion of the dental papilla, which organizes in response to its influence. This extension is called the *epithelial sheath* of the root.

Neither the outer enamel cells nor the enamel pulp contributes directly to tooth development, although the building materials of enamel must pass from the nearby blood vessels through their loosely arranged tissue. By contrast, in the region of the future crown of the tooth the cells of the inner enamel layer become columnar and are designated *ameloblasts* (enamel formers), for they produce *enamel* at the ends that face toward the dental papilla. This deposit takes the form of parallel *enamel prisms*.

The enamel substance arises first as a cuticular secretion from the end of an ameloblast; calcification of this 'Tomes process' is secondary. Continued enamel formation produces elongate *enamel prisms*, one for each ameloblast, which become cemented together. As the enamel layer thickens, the ameloblasts retreat in an outward direction until finally the internal and external layers oft he enamel organ meet. The laying down of an enamel prism by an ameloblast can be compared to the band of toothpaste that is left behind when the tube is squeezed and at the same time drawn away. Enamel is first deposited at the apex of the crown; the process then spreads downward in a progressive manner so that the ameloblasts of the neck region are the last to become active. Molar teeth have a separate cap of enamel for each cusp; these eventually meet and merge into a compound crown. Long before a tooth cuts, its crown is finished and the enamel organ proper undergoes regression. The remains of the enamel organ constitute the transient *dental cuticula* (Nasmyth's membrane).

The *epithelial sheath* of the root is directly continuous with the enamel organ proper, but it differs from it both structurally and functionally. The inner layer of epithelial cells remains cuboidal and never produces enamel, while the pulp constituent of the typical organ is lacking. Perhaps the absence of the latter is significantly correlated with the failure of this region to form enamel.

The Dental Papilla: At the end of the fourth month the superficial cells of the dental papilla arrange themselves in a definite layer that simulates a columnar epithelium. These specialized, connective-tissue cells are named *odontoblasts* (tooth formers); actually, 'dentinoblasts' describes them better. Even so, it has not been demonstrated convincingly that odontoblasts are solely responsible for all the dentine substance. The fibrils of the early dentinal matrix are continuous with those coursing within the papilla as a whole. This soft, fibrillar *predentine* then calcifies into the definitive *dentine*, or dental bone. Whether the odontoblasts lay down the predentine fibrils is debatable; perhaps they are chiefly concerned with the subsequent deposit of calcium about them. In any region of the tooth, dentine formation precedes slightly the appearance of enamel.

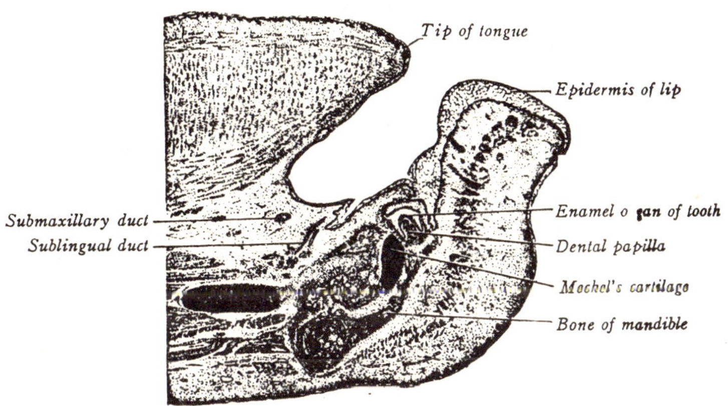

Fig. 8.8.   Differentiation of a human incisor, shown in sections. *A*, Deciduous and permanent primordia, at seven months.

The more central mesenchyme of the dental papilla, internal to the odontoblast layer, differentiates into a soft core-substance. This is composed of a framework of reticular tissue which binds together blood vessels, lymphatics and nerve fibers. Together with the odontoblast layer it constitutes the *dental pulp*, popularly known as the 'nerve' of the tooth.

As with enamel, the dentine layer is laid down first at the apex of the crown (or cusp of a molar) and then progressively toward the root. As the layer thickens, the odontoblast cells retreat before it and so always maintain a more central position. Yet, during the recession, a thread-like process of each odontoblast is spun out and remains behind in the dentine. This *dentinal fiber* (of Tomes) occupies a tiny *dentinal tubule*, essentially like any process of an osteoblast and its canaliculus in ordinary bone. The whole odontoblast layer persists throughout the life of a tooth and intermittently lays down dentine. The crowns of the various milk teeth are not completed until 2-11 months after birth, and only then is root development begun. As a preliminary the epithelial sheath elongates, and within this tube the primitive connective tissue is stimulated to condense and organize as it did in the crown. The epithelial sheath of a premolar or molar tooth branches and hence the root comes to have fangs.

The Dental Sac: The mesenchymal tissue surrounding the developing tooth is continuous with that of the early dental papilla. Outside the tooth it differentiates into ordinary connective tissue which constitutes the so-called *dental sac.* In the region of the future root the dental sac takes on three important functions: (1) Beginning at the time of eruption its inner cells differentiate into a layer of *cementoblasts.* With the progressive disintegration of the epithelial sheath in a downward direction, these cells deposit upon the dentine an encrustation of specialized bone, known as *cementum.* Deposition proceeds from the neck region downward. (2) During the period of tooth development there has been steady progress in the ossification of the jaw bone. In the region of the teeth the external surfaces of the dental sacs become active in producing bone. Thus each tooth comes to be surrounded with spongy bone, except over its top, and occupies its individual compartment or socket. As a tooth is cut and its root grows to full length, this bone-lined *alveolus* reaches a definitive state. (3) The fibrous sac itself consolidates into the thin *periodontal membrane* which holds the tooth in place by embedding some fibers in the cement and still others in the bony wall of the socket. It is a specialized periosteum.

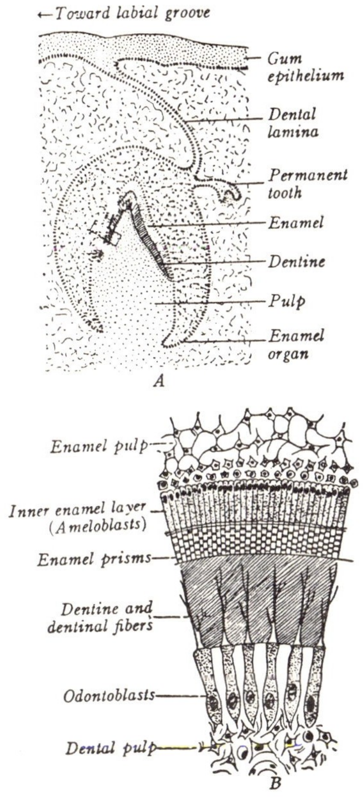

Fig. 8.9. Later tooth differentiation. *A*, Section of a milk incisor of a new-born dog *in situ*. *B*, Relation of pulp fibrils to the fibrillar matrix of dentine. *C*, Model of enamel prisms and their relation to ameloblasts.

Growth, Eruption and Shedding: Progressive growth of the root and other little-understood factors combine in pushing the crown of a milk tooth out of its bony socket, through the overlying portions of the dental sac and gum, and so to the outside. The periods of 'cutting' (*i.e., eruption*) of the various milk- or *deciduous teeth* vary with race, climate and nutritive conditions. The permanent teeth in their development eventually press the intervening tissues against the milk teeth. The roots of the latter then undergo partial resorption, whereupon their dental pulp is liberated. The combination of tissue loosening and pressure from the permanent teeth leads to the shedding of the milk teeth. Usually the milk teeth are begun, cut, completed and shed at the following times :

	Calcification of Crown Begins	*Tooth Erupts*	*Calcification of Root Ends*	*Tooth Sheds*
Central Incisors....	4 fetal months	6-7$^1/_2$ months	1$^1/_2$ years	7 years
Lateral incisors....	4• " "	7-9 "	1$^{*3}/_4$ "	8 "
Canines...........	5 " "	16-18 "	3$^1/_4$ "	12 "
First Molars.......	5 " "	12-14 "	2$^1/_2$ "	10 "
Second Molars.....	6 " "	2-24 "	3 "	11 "

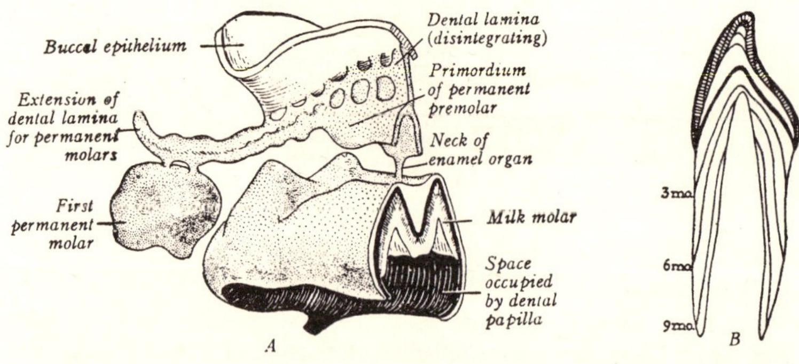

Fig. 8.10. *A*, Model showing the relation of a human milk molar and a permanent premolar to the dental lamina (adapted after Rose; X 7); at left, a permanent molar is developing from a backward extension of the dental lamina. *B*, The order of deposit of enamel and dentine in a milk incisor is shown at selected intervals; a heavy line encloses the portion of the tooth completed at birth.

The Permanent Teeth: During childhood and adolescence the jaws grow backward sufficiently so that they can accommodate 12 molar teeth that had no counterparts in the provisional set. Otherwise the *permanent dentition* develops essentially like the temporary set. The enamel organs of those permanent teeth that correspond to the milk dentition arise between the fifth and tenth fetal months in another series along the free edge of the disintegrating dental lamina. Located at the same intervals as the deciduous teeth, they come to lie on the lingual side of them and within the same dental sac. In addition, there are three molars not represented in the primary dentition, since the so-called milk molars really correspond to premolars. These permanent molars develops on both sides of each jaw from a backward-growing, free extension of the dental lamina.

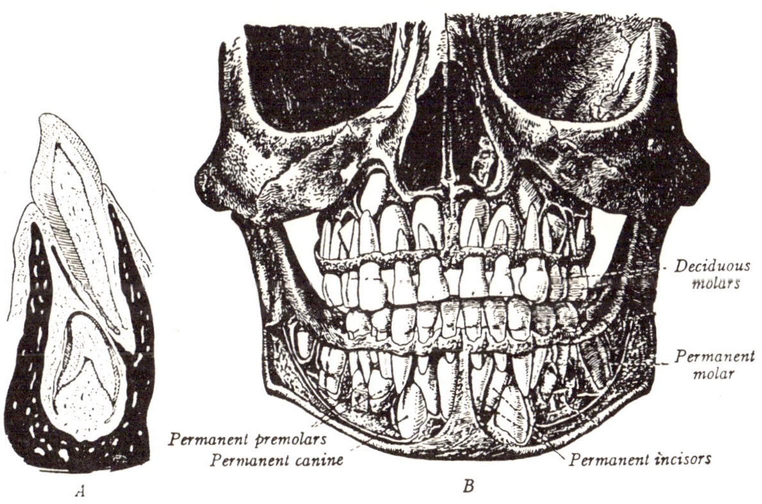

Fig. 8.11. Relations of the human deciduous and permanent teeth. *A*, Milk Canine, with eroding root, and a permanent canine, at four years, sectioned *in situ* (x 3). *B*, The two dentitions, at five years, in a partly dissected skull.

The primordia of the first permanent molars are present at four months, those of the second molars at birth, while indications of the third permanent molars, or 'wisdom teeth,' are not found until the fourth to fifth year. Provision for the permanent dentition of 32 teeth is then complete. The permanent teeth develop slowly for a while; later they progress more rapidly and, toward the sixth year, each jaw may contain 26 teeth. The enlarging, permanent teeth press against the corresponding milk teeth and aid in their shedding *(A)*. The permanent set is then free to remodel the alveoli as needed, to occupy them solely and to erupt.

Deposits of enamel and dentine are rhythmic; from time to time major disturbances such as birth, weaning, illness and puberty, leave visible records in a series of super imposed *growth lines*. Teeth, like bones, do not follow ordinary organs by continuing to grow after the developmental period ends. For example, a crown has attained full size before its eruption, and its enamel cannot increase further since the ameloblast layer is no longer present. By contrast, dentine is capable of continued development, but only in an inward direction at the expense of the pulp; such intermittent growth may eventually obliterate the root canal. Usually the permanent teeth calcify, erupt and are completed at the following times :

	Calcification of Crown Begins	Calcification of Crown Ends	Tooth Erupts	Calcification of Root Ends
Central Incisors....	3-4 months	4-5 years	6-8 years	9-10 years
Lateral incisors....	3-12 "	4-5 "	7-9 "	10-11 "
Canines...........	4-5 "	6-7 "	9-12 "	12-15 "
First premolars....	$1^1/_2$-2 years	5-6 "	10-12 "	12-13 "
Second Premolars..	2-$2^1/_2$ "	6-7 "	10-12 "	12-14 "
First Molars.......	At birth	$2^1/_2$-3 "	6-7 "	9-10 "
Second Molars.....	$2^1/_2$-3 "	7-8 "	11-13 "	14-16 "
Third Molars......	7-10 "	12-16 "	17-21 "	18-25 "

The teeth of vertebrates are homologues of the placoid scales of elasmobranch fishes (sharks and skates) and have a similar development. The teeth of the shark resemble enlarged scales, and many generations of such teeth are produced in the adult fish. In some mammals three, or even four dentitions occur. The primitive teeth of mammals were of the canine type, and from this conical tooth the incisors and molars have arisen. Just how the cusped tooth differentiated—whether by the fusion of originally separate units or by the development of cusps on a single primitive tooth—is debated.

SUMMARY OF RELATIONS IN TOOTH DEVELOPMENT

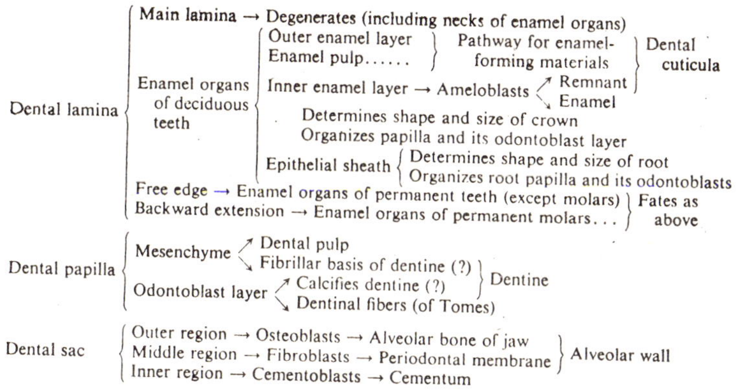

Causal Relations: The dependency of an ectodermal mouth, with teeth, on the formative influence of the associated entoderm has been mentioned. It is, however, the enamel organ that exerts the molding influence under which the entire tooth takes shape and gains size; through its inductive power the peripheral mesenchyme of the dental papilla is activated to produce the odontoblast layer and to differentiate dentine. For a time a halved primordium of a molar tooth, grown in culture, is able to regulate and develop as a whole tooth, even producing the normal cusp-pattern.

Anomalies: Developmental abnormalities comprise

irregularities in number, size, shape, structure, singleness, position and eruption. *Anodontia* designates a congenital absence of teeth; its mildest expression is the agenesis of a single tooth like the third molar; most extreme is the total suppression of tooth development which is very rare and then is associated with defective skin and its derivatives. Opposite in nature is *hyperdontia* which denotes a production of more than the normal number of teeth, either inserted into the regular series, added (an atavism?) as fourth molars beyond the wisdom teeth, or even occupying abnormal locations outside the dental arch. The two regular dentitions are exceeded when representatives of an extra set either precede the milk teeth or follow the permanent set. Persistence of more or less of the milk dentition may be associated with a corresponding failure in the development of the permanent dentition. *(A)*, berrations of form include distortion *(B)*, extra cusps or roots, branched or fused roots *(C)*, double crowns *(D)* and fused teeth *(E)*. Disturbances (acute diseases; rickets; syphilis) that interfere with enamel deposition produce pits, grooves or even irregular mulberry crowns *(F)*; total absence of enamel is known. *Enamel drops* formed in vesicles of dental-lamina or enamel-organ origin, may occur near a tooth or attached to its root *(G)*; *cysts*, located in the gums, have a similar origin.

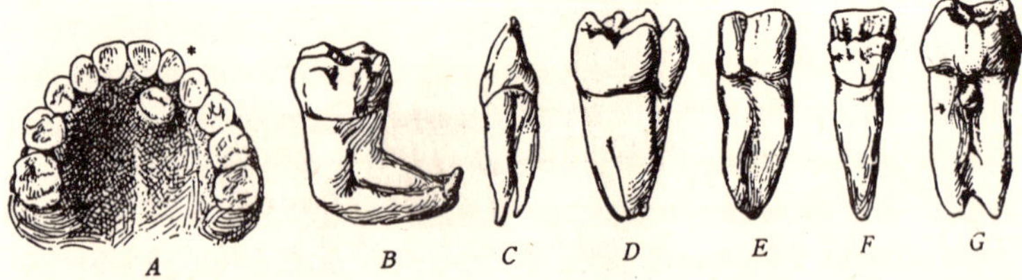

Fig. 8.12   Human dental anomalies. *A*, Supernumerary milk incisor (*), and an ectopic molar (on palate); *B*, distorted root; *C*, branched root; *D*, double crown; *E*, fused teeth; *F*, faulty enamel deposition; *G*, enamel drop (or pearl).

The Oral Glands: The glands of the mouth are distinctive characteristics of mammals, the only animals that chew their food. They are usually regarded as ectodermal derivatives, although the site of origin of the submaxillary and sublingual glands with respect to the vanished oral membrane is not surely known. All of the salivary glands have a common plan of origin and development. The primordium arises as an epithelial bud and grows by branching into a bush-like system of solid *ducts*, whose end-twigs round out into berry-like, secretory *acini*. Secondary hollowing of the whole system and specialization of the acinal cells complete the epithelial differentiation. A dense mass of mesenchyme, in which the epithelial primordium lies, furnishes as enveloping *capsule* and subdivides the gland into *lobules*. The major salivary glands attain this general plan of organization in the third month. Acinal cells and a canalized duct system are present at six months. As with other glands, acinal differentiation is not complete until sometime after birth even through the salivary enzyme is produced by the fetus.

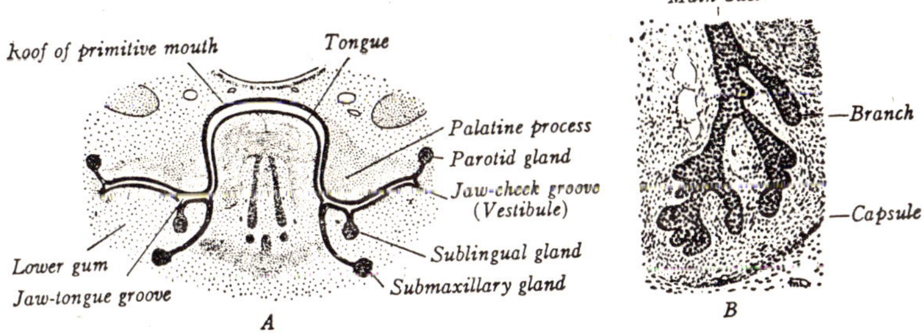

Fig. 8.13.   Origin and early growth of the human salivary glands. *A*, Sites of origin, shown by a diagrammatic frontal section across the jaws at about two months. *B*, Detail of the branching submaxillary gland, at two months.

The paired *parotid glands* are the first to appear. In the sixth week (10 mm) a keel-shaped, epithelial flange has been observed, near each angle of the mouth, growing away from the groove that will divided cheek from gum. The flange elongates and, in embryos of seven weeks, separates from the parent epithelium. A tube is then formed by hollowing and this grows backward toward the ear. It soon branches and organizes into the body of the gland, while the stem portion of the tube becomes the parotid duct opening into the vestibule.

Each *submaxillary gland* also arises at the end of the sixth week (12 mm.) as an epithelial ridge, located in the groove between the lower jaw and the tongue and at one side of the midplane. The caudal end of the ridge soon begins to separate from the epithelium and extend backward and ventrad in the mesenchyme beneath the lower jaw; here it enlarges and branches into the gland proper. The main stalk, separating in a rostral direction, persists as the submaxillary duct and opens at the side of the frenulum of the tongue.

Each *sublingual gland* appears during the eight week as a series of solid buds of epithelium growing downward from the groove between the lower jaw and tongue. This group, located just lateral to the submaxillary primordium consists of the sublingual proper, with its major duct (of Bartholin), and of about ten equivalent smaller glands, each with a minor duct. Growth is slower than in the sub-maxillary gland. The glands lie alongside the tongue and beneath it. The major duct opens just lateral to the submaxillary duct, or it may join it.

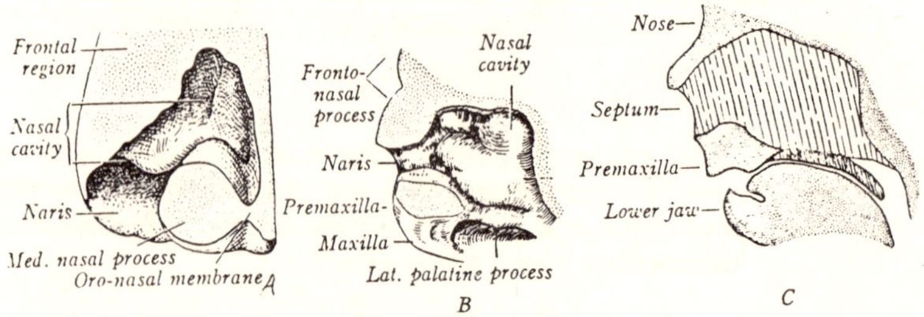

Fig. 8.14.   Separation of the human nasal and oral cavities by the palate. *A*, Medial half of the left nasal sac, at 12 mm. (after Shaeffer. *B*, Lateral half of the right nasal sac, at seven weeks. *C*, Relation of the fetal palate and nasal septum, demonstrated by a median section.

The smaller oral glands *(labial, buccal* and *palatine)* are aggregates that arise at about three months from multiple epithelial buds in their respective locations.

Anomalies: There may be an absence of any (or, rarely, all) of the salivary glands. Accessory glands occur, as do imperforate ducts that lead to retention cysts. Displacement and mutual fusion are more apt to involve the submaxillary and sublingual glands.

The Palate: The mammalian palate is a device for separating the mouth from the nasal respiratory passages, and thus making it possible for the young to such (and, later, to chew) and breathe at the same time. The two nasal cavities are at first represented by olfactory pits which quickly enlarge into blind sacs, as in adult sharks. The floor of each deepening sac then comes to overlie the roof of the front part of the primitive mouth, and is separated from it by an *oro-nasal membrane* only. The thinning membranes rupture during the seventh week and so create two internal nasal orifices, the *primitive choanae*. For a short time the two choanae open directly into the primitive oral cavity whose roof is merely the basal covering of the skull; this simulates the permanent condition in amphibia. Presently, however, the definitive nasal passages become separate from the mouth, and then open behind into the pharynx. This is accomplished by means of a horizontal partition that subdivides the primitive mouth cavity. The portion of the mouth, thus captured by the nose, extends the nasal passages greatly and they then communicate with the pharynx by secondary, definitive *choanae*. The horizontal plate, which thus divides mouth from nasal passages, is the *palate*; the details of its formation will next be described.

The primordia of the palate are two shelf-like projections that grow from each maxillary-process region of the upper jaw toward the midplane of the mouth cavity. In their growth mesad during the seventh and eighth weeks these *lateral palatine processes* encounter the tongue, which rises high at this period, and are forced to bend downward. A little later the tongue is withdrawn, due to growth changes, and the lateral palatine processes are then pushed upward to the horizontal plane *(B)*. The halves of the palate unite, first with each other and then with the nasal septum. Beginning in the ninth week, this fusion progresses rapidly from in front backward. Coincidently bone appears in the front part and forms the *hard palate*. More caudad (where union with the nasal septum does not occur; ossification fails. This region constitutes the *soft paiate;* the halves of its free apex, the *uvula*, are commonly still notched at birth.

Transverse ridges (to aid in the grinding of food) are developed in the mucosal covering of the hard palates of most mammals; their reduced state in man (more so in the adult than in the fetus) is perhaps correlated with the soft nature of his food. The folds of the soft palate are invaded from behind by tissue from the third branchial arches; this is responsible for those backward prolongations of the palate, known as the *palatine arches*, which delimit oral cavity from pharynx. From the same source comes the mesenchyme that differentiates into the muscles of the palate. The completed palate shows a median seam, or *raphe*, indicative of its bilateral origin.

The median nasal process, which participate so conspicuously in the formation of the face, also develop so-called *median palatine processes;* the latter do not contribute to the palate itself but become the premaxillary portion of the upper jaw. Fusion between the median palatine processes and the palate is incomplete, so that in the midplane there is a gap, the *incisive foramen*, flanked by the *incisive canals* (of Stenson). These ordinarily become covered with mucous membrane (*incisive papilia; C*) although they sometimes are still open at birth, as is the permanent condition is most mammals.

Anomalies: The lateral palatine processes occasionally fail to unite properly, thereby producing a malformation known as *cleft palate*, or *uranoschisis*. The extent of the defect varies considerably. In some persons it involves the soft palate alone and then is median in position. By contrast, clefts in the hard palate tend to lie at one side of the midline. Both the hard and soft palate may be involved in the same individual *(C)*. Cleft palate often accompanies cleft lip *(C)*; when double, the premaxilla (formed by the median nasal processes) tends to protrude prominently. Epithelial strands may become buried where the edges of the palatine processes fuse, and take the form of 'epithelial pearls' in the newborn or later.

The Hypophysis: The hypophysis or pituitary body, is an endocrine gland of double origin. One part, obviously glandular in nature, is epithelial; it develops from the ectodermal roof of the stomodeum, which in early stages is adherent to the floor of the fore-brain. The other component, not so plainly secretory, is a specialized extension from the brain wall. During the subsequent growth of these parts, and the filling-in of mesenchyme between them, both roof and floor become drawn out into hollow extension.

The stomodeal pocket, known as *Rathke's pouch*, is located originally just in front of the intact pharyngeal membrane. in embryos about 3 mm. long this pouch is a distinct, shallow sac which quickly enlarges and flattens against the tubular projection from the floor (infundibulum) of the fore-brain. The latter is the future *neural lobe* of the hypophysis. Meanwhile the connection of Rathke's pouch with the oral epithelium has elongated into a slender stalk which lags in development and vanishes by the end of the second month. The front wall of the pouch thickens greatly and becomes the important secretory mass known commonly as the *anterior lobe (B)*. The back wall of this epithelial pouch remains thin and has little functional significance. The cavity of the closed Rathke's pouch becomes the *residual lumen* of the mature gland; the final condition in man is unique since this cleft becomes reduced to cysts or even obliterates completely. The neural lobe differentiates secretory elements of functional value. Combined with the intermediate layer behind the cavity of the pouch it makes a complex commonly designated as the *posterior lobe*. During the third and fourth months the hypophysis attains its characteristic shape and arrangement. Somewhat later, cellular differentiation brings about the typical regional organization of this organ.

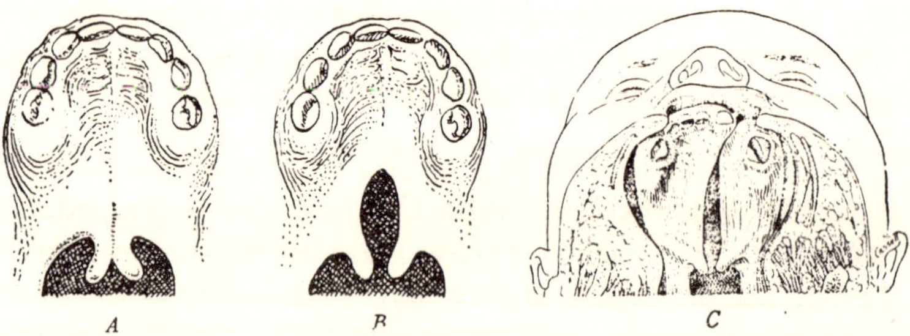

Fig. 8.15. Types of human cleft palate. *A*, Cleft uvula. *B*, Cleft involving total soft palate. *C*, Cleft involving total soft and hard palates; also combined with severe cleft lip.

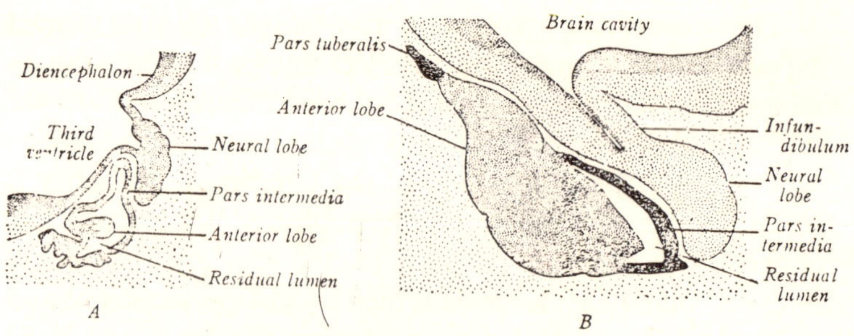

Fig. 8.16. Early development of the human hypophysis, shown by sagittal sections. *A*, At eight weeks; *B*, at eleven weeks.

The thick rostral wall of the pouch (technically, the *pars distalis*) differentiates into cords containing highly specialized cells; these cords are interspersed between abundant sinusoid. By the tenth week of fetal life at the specialized cell types are distinguishable. Later the growth- and then the gonadatropic hormone becomes detectible when extracted and used experimentally. The thin, caudal portion of the wall (originally the apex of the pouch), lying between lumen and neural lobe, is named the *pars intermedia;* in man it forms epithelial cysts but is not prominent. Another glandular region, the *pars tuberalis,* wraps about the infundibular stalk. It develops from the fusion of a pair of early wing-like lobes that bud off laterally from the main pouch. The mostly solid *infundibular stalk* connects the brain (diencephalon) permanently with the swollen end of the stalk which is officially designated the *pars nervosa.* This solid mass is composed of neuroglial tissue, nerve fibers and branching cells believed to be secretory in nature.

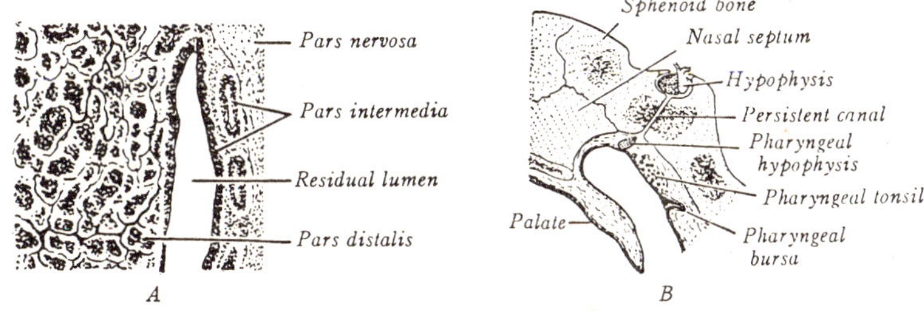

Fig. 8.17. The later hypophysis, shown in median section. *A,* Detail of the chief regions, at eight months. *B,* Relations of the main gland and pharyngeal hypophysis in the newborn.

Growth of the facial regions causes the jaws to jut forward, thus deepening greatly the originally shallow stomodeum and making its contribution to the mouth substantial indeed. In this merger (through the disappearance of the pharyngeal membrane) the original, virtually external site of origin of Rathke's pouch comes to be located well back on the roof of the primitive mouth. Still later, when the palate segregates the nasal passages, this point of origin lies at the dorsal and caudal border of the nasal septum.

Causal Relations: The cranial end of the notochord is said to induce the development of the neural lobe, while the neural lobe and the brain wall together induce the differentiation of the anterior lobe. The pars intermedia is determined as such by contact with the floor of the fore-brain (diencephalon).

Anomalies: The original course of the stalk of Rathke's pouch is sometimes marked by a canal in the sphenoid bone. This *cranio-pharyngeal canal,* however, is a secondary feature which develops after the stalk has degenerated; it usually obliterates before birth. Notable among the accessory glands that may occur along this pathway is a constant mass located between the nasal septum and the pharyngeal tonsil. It is known as the *pharyngeal hypophysis.*

The Pharynx

The fundamental importance of the early pharynx would scarcely be suspected from its adult simplicity and unspectacular role as a common corridor for the crossing pathways of air and food. Nevertheless, the primitive pharynx is the source of numerous organs and its developmental history is correspondingly complex. Most of these activities occur during

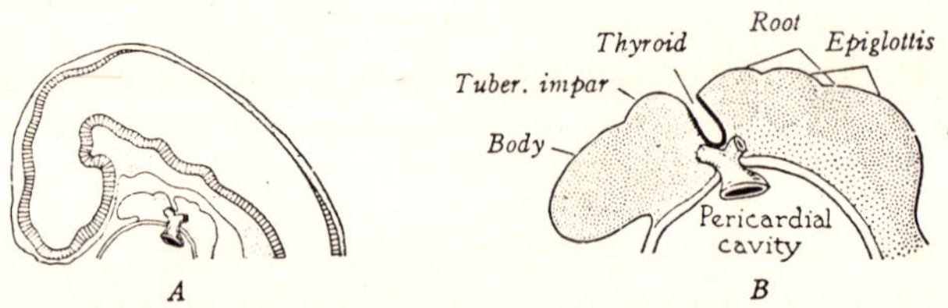

Fig. 8.18.   The tongue, at about four weeks, in diagrammatic sagittal section. *A,* General relations; *B,* detail of *A.*

the transitional period when the mammalian embryo passes from a stage in which the pharynx is arranged as for branchial respiration to a stage in which the breathing of air is anticipated. Naturally the parts altered most profoundly are the branchial arches and pharyngeal pouches themselves. In addition to the remodeling that is necessary to provide for the mammalian method of chewing and swallowing, various other conversions occur.

The Branchial Arches: In a previous chapter were described the general relations and significance of the branchial arches. These ancestral gill arches are converted into numerous things-including the neck, jaws, face and external ear, already discussed, and various arteries, muscles, cartilages and bones to be considered in later chapters. On the floor of the pharynx they contribute especially to the *tongue.* The nearby *larynx,* also of branchial origin, belongs to the respiratory system. Origins and fates in the branchial regions are tabulated; inspection of these lists will re-emphasize how important this territory really is.

The Tongue: This organ is primarily a pharyngeal derivative. Developmentally it is a mucous-membrane sac which becomes stuffed with voluntary muscle. The tongue arises from the ventral ends of the branchial arches. It consists of two different parts—one oral in origin, the other pharyngeal. The oral portion, comprising most of the *body* of the tongue, occupies the definitive mouth cavity. It arises from the mandibular arches, in front of the oral membrane, and hence is covered with ectodermal epithelium. This part of the tongue bears papillae and is concerned with mastication. The pharyngeal portion is the *root* of the tongue. It develops primarily from the union of the second branchial arches, but receives important contributions from the third and, apparently, the fourth arch as well. The entodermal-covered root becomes infiltrated with lymphoid tissue, but is mainly concerned with swallowing. The junction-line between ectoderm and entoderm is in front of the row of vallate papillae, whereas the body and root are demarcated by a V-shaped groove, the *terminal sulcus,* just behind these papillae.

In embryos of four weeks (5 mm.) the body of the tongue is indicated by three primordia. These are : (1, 2) paired *lateral swellings* of the first branchial arches, which latter have fused already as the mandible; and (3) the median, somewhat triangular *tuberculum impar* (i.e., unpaired tubercle) wedged in between the lateral swellings. At the same time, the future root is represented by a median elevation, brought into existence

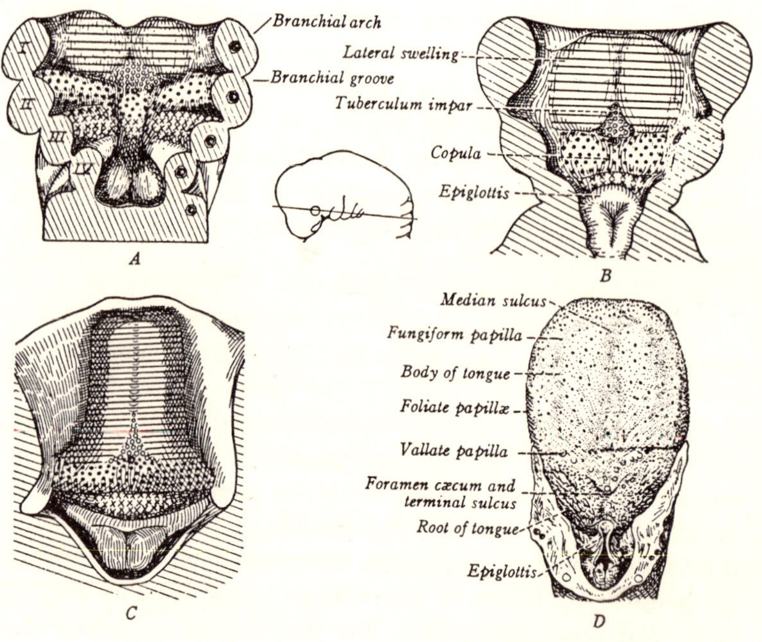

Fig. 8.19. Development of the human tongue, viewed from above. *A,* At 6 mm. (key figure shows section planes of *A-C*); *B,* at 9 mm.; *C,* at 15 mm.; *D,* at birth. Distinctive markings indicate the contributions of the branchial arches. The tuberculum impar is designated by circles, and at its base a larger circle indicates the foramen caecum.

by the union of the bases of the second branchial arches; for this reason it is named the *copula* (i.e., a yoke). Between the tuberculum impar and the copula is the point of origin of the thyroid diverticulum. This site becomes secondarily depressed by lingual growth-stresses into a prominent pit which occupies the apex of the terminal sulcus and constitutes the permanent landmark known as the *foramen coecum*. The progressive merger of the several lingual components can be followed in the stages shown as. Continued expansion of the tongue in both length and breadth brings into existence a deep, Ç-shaped furrow at the front and along the sides which will make this organ partly free and highly mobile. At the same time (seventh week), the tongue elevates and assumes prominence through the differentiation of voluntary muscle internally. The body of the tongue acquires *papillae* of several kinds, whereas the root is the site of the *lingual tonsil*.

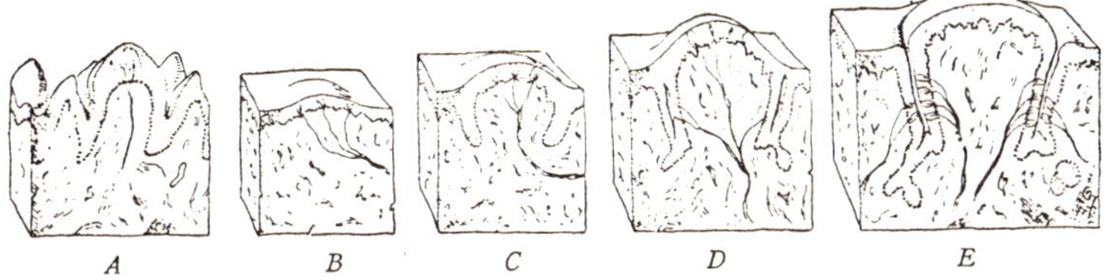

Fig. 8.20. Models of developing lingual papillae. *A*, Filiform and fungiform papillae, at eleven weeks; *B-E*, vallate papilla, at two to five months. Taste buds are present, and stages of Ebner's glands occur in *D,E*.

The individual components of the tongue are in process of fusion during the sixth week. The lateral swellings of the first arches increase rapidly in size, unite with the tuberculum impar, and nearly enclose it. In sharp contrast, the tubercle itself lags in development and becomes progressively less conspicuous. In the end, it furnishes little to the *body* of the final organ. The plane of union of the two latleral swellings is indicated superficially by the *median sulcus* and internally by the fibrous, median *septum*. The copula, together with the adjacent portions of the second branchial arches, enlarges greatly to produce the early *root* of the tongue *(B)*. Eventually, however, this territory seems not only to be encroached upon by the third and fourth branchial arches (which primarily form the epiglottis; *(C)*, but there is a slipping forward of their mucous membrane as well. This conclusion is supported circumstantially by the fact that the sensory portions of the trigeminal and facial eranial nerves (the nerves of the first and second branchial arches) ultimately supply the epithelium of the body of the tongue, while the glossopharyngeal and vagus nerves (the nerves of the third and fourth arches) supply the root.

The striated *musculature* of the tongue is innervated by the twelfth, or hypoglossal nerve, and both nerve and muscle belong ancestrally to the occipital region of the head, just caudad of the branchial arches. It is believed that during phylogeny the tongue migrated cephalad and invaded the branchial region, retaining its nerve supply the while. Such an invasion would also explain satisfactorily the forward dislocation of the mucosa, already mentioned. In present-day embryos, this migration is difficult to trace, although it has been demonstrated in the chick and kitten. Except for slight indications suggestive of migration, the muscles of the human tongue appear to arise *in situ* from the mesenchyme of the arches that make up the floor of the mouth. The temporary elevation of the tongue to a position between the developing halves of the palate is corrected by a withdrawal due to growth change in the lower jaw; the tongue is then able to broaden greatly *(B)*.

The early cuboidal epithelium that covers the tongue is changing to a stratified type at two months. Lingual papillae are confined chiefly to the oral, or masticatory part of the tongue. In fetuses of 9 and 11 weeks, respectively, the *fungiform* and *filiform papillae* may be distinguished grossly as elevations of the mucosa. The *vallate papillae*, which are entodermal, develop along a V-shaped epithelial ridge just in front of the terminal sulcus. At intervals there appear about nine elevation *(B)*. In the tenth week a thickened, epithelial ring delimits each elevation; this ring then grows downward and takes the form of an inverted, hollow cone *(C)*. During the fourth month circular clefts split each epithelial collar *(D)*, thus separating the sides of the vallate papilla from the surrounding wall and forming the trench from which this type of papilla derives its name *(E)*. At the same time lateral outgrowths arise from the bases of the epithelial cones and hollow into the *glands of Ebner*. The *foliate papillae* become distinguishable as parallel folds during the third month. The *taste buds* of the mouth begin to be indicated at about eight weeks; the details of their development are given. One taste bud precedes the appearance of each fungiform papilla and then occupies its top surface. Several buds occur on the dome of an early vallate papilla *(C,D)*; these vanish before birth and are replaced by definitive buds on the sides and on the trench wall *(E)*.

The *lingual tonsil* is foreshadowed in the fifth month by an infiltration of lymphocytes into the root of the tongue, whereas the pit-like crypts do not differentiate until the time of birth.

Fig. 8.21.   Anomalous trifid tongue.

Anomalies: The tongue may be abnormally large *(macroglossia)* because of its oversized muscle mass. Smallness *(microglossia)* or absence *(aglossia)* is referable to a failure of growth or development, respectively. Developmental arrest in the body of the tongue leads to a bifid organ, as in snakes; even a trifid tongue (unfused lateral components and tuberculum impar) has been known.

The Pharyngeal Pouches: The lateral walls of the entodermal pharynx give rise to a series of paired sacculations which extend outward toward corresponding ectodermal grooves. The early relations between these *pharyngeal pouches* and the *branchial grooves* (and branchial arches) are illustrated in. The pairs of pouches arise in succession in a caudalward direction. Toward the end of the fourth week (4 mm.) five sets have been formed, the last pair being atypical and attached to the fourth. Meanwhile the pharynx has flattened dorsoventrally and broadened at its cranial end; as a result, it is triangular in outline.

Each typical pouch develops a dorsal and a ventral wing. Also, in expanding, the pouch pushes aside the intervening mesenchyme and comes into contact with the ectoderm of the corresponding branchial groove. The two layers fuse and thus

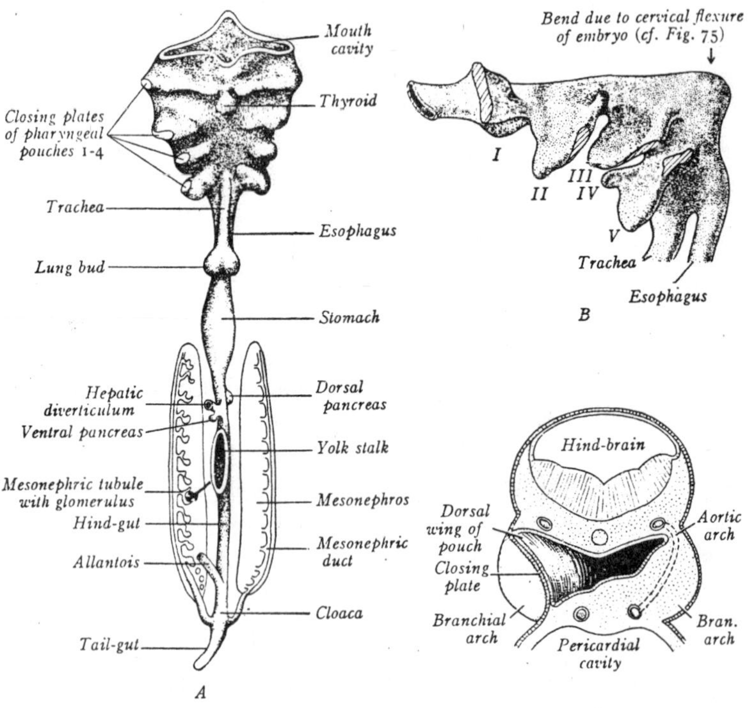

Fig. 8.22. Early human pharynx, shown as models. *A*, Entodermal tract, at 5 mm., in ventral view. *B*, Pharynx, at 7.5 mm., viewed from left side. *C*, Pharynx, at 8 mm., sectioned obliquely to show a pouch, at left, and an arch at right.

produce a *closing plate (C)*. Although the closing plates become perforate in human embryos only occasionally, each combination of pouch and groove, nevertheless, is homologous to a functional branchial cleft of fishes and tailed amphibia; their transitory appearance is an illustration of an unerased, ancestral imprint. The first and second pharyngeal pouches of each side soon extend off from a broad, lateral expansion of the pharynx. The third and fourth pouches, growing laterad, continue to communicate with the pharyngeal cavity through narrow ducts. The questionable fifth pouch is merely a blind diverticulum that shares a common duct with the fourth pouch.

The fates of the entodermal pouches are varied and spectacular. Although the several derivatives do not continue as parts of the digestive apparatus, their embryonic relations justify their inclusion in the present chapter. The first pouch retains its lumen and differentiates into the *auditory (Eustachian) tube* and the *tympanic cavity* of the middle ear. The second is greatly reduced and becomes the fossa and covering epithelium of the *palatine tonsil*. The third, fourth and fifth lose all trace of a lumen and give rise to a series of glandular organs; these are the *thymus, parathyroids* and *ultimobranchial bodies*.

The Auditory Tubes and Tympanic Cavity: The first pair of pouches is the only one that retains its embryonic relations in a recognizable manner. Each main pouch is drawn out during the eighth week into an *auditory tube*, whereas its dorsal angle expands into the *tympanic cavity* of the middle ear. An unconfirmed observation interprets the second pouch, as well, as becoming absorbed into this tract. Simultaneously with the transformation of the first entodermal pouch, the overlying ectodermal groove deepens to produce the *external acoustic meatus*. This canal leads inward to the closing plate which regains a middle layer of mesoderm and persists as the *tympanic membrane*.

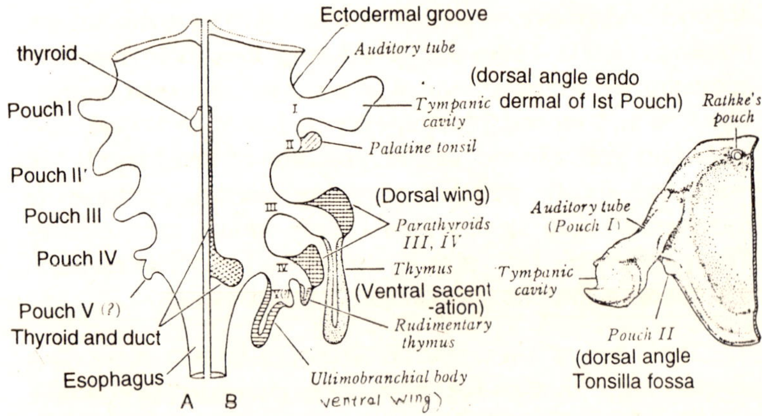

Fig. 8.23.   The human pharynx and its derivatives. *A*, Right half, in ventral outline at four weeks. *B*, Left half, in ventral outline, at six weeks. *C*, Model of left half, in dorsal view, at eight weeks.

The Palatine Tonsils. By the growth and lateral expansion of the pharynx, the second pouch of each side is largely absorbed into the pharyngeal wall. It would seem that the dorsal (some say ventral) angle of the pouch persists as the *tonsillar fossa*, while its entoderm furnishes the *epithelium* covering the tonsil and lining its crypts. The *crypts* arise progressively in fetuses of three to six months as solid ingrowths from the surface epithelium. They branch and hollow secondarily (B,C); many of the branches degenerate and reform after birth. Lymphocytes appear near the epithelium in the third month *(B)* and organize as nodules after the sixth month; the arrangement of *lymphoid tissue* makes the tonsil temporarily bilobed *(C)*. An enveloping *capsule* and an internal connective-tissue *framework* arise at five months from mesenchyme in the immediate vicinity of the developing organ. The permanent location and relations of the palatine tonsil.

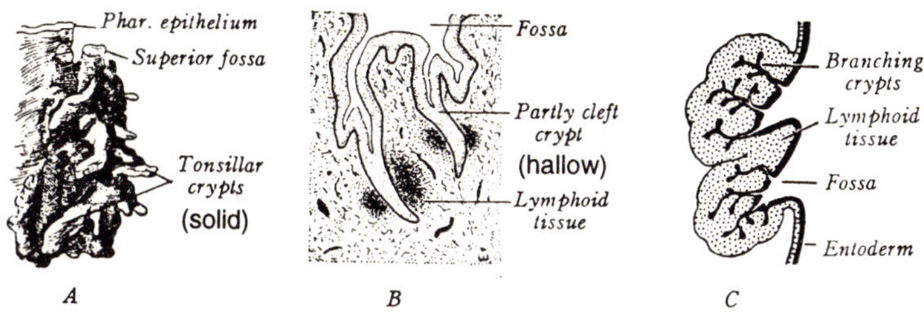

Fig. 8.24.   Development of the human palatine tonsil. A, Model of the crypt system at fourteen weeks. *B*, Vertical section, at fourteen weeks. *C*, Diagrammatic vertical section, at five months.

Some studies have cast doubt on the direct and continuous relation of the second pouch of the palatine tonsil. Rather, it is urged that the tonsil develops at the general site of the second pouch merely because this is a neutral, favourable position in a region of marked growth shiftings. Similarly, the anterior and posterior pillars of the tonsil represent merely the general positions of parts of the second and third branchial arches.

Anomallies: The palatine tonsils are highly variable, ranging from stalked, protruding organs to 'buried' types lying in a deep sinus. Cervical cysts, blind diverticula and complete fistulae occur. They are usually related to the second branchial clefts, and particularly to the cervical sinus, whose drawn-out portion commonly opens at the tonsillar fossa.

The Thymus: Toward the end of the sixth week each third pharyngeal pouch shows a pronounced ventral sacculation, and the entire pouch is set free in the week following *(B, C).* At first hollow, these thymic primordia rapidly become solid epithelial bars. The lower ends enlarge and unite superficially during the eight week to foreshadow the definitive organ; yet the thymus never loses wholly its paired nature. The two lower ends are attached to the pericardium and gradually sink with the latter to a permanent position in the thorax. During this descent the upper ends become drawn out and finally vanish.

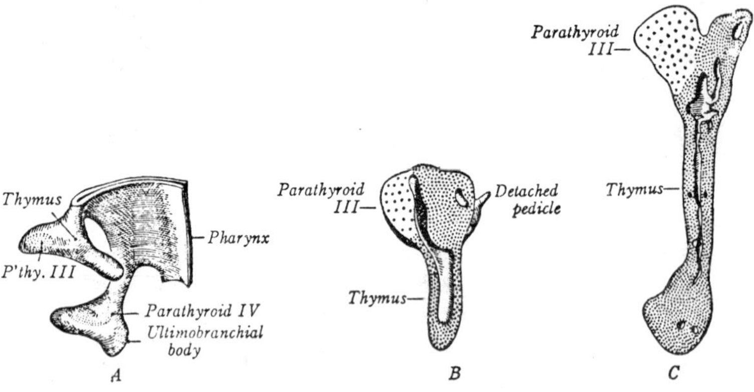

Fig. 8.25. The third and fourth pharyngeal pouches of human embryos, shown as models. *A*, At 10 mm., in ventral view. *B,C*, Detached third pouches, at 14 mm., hemisected lengthwise.

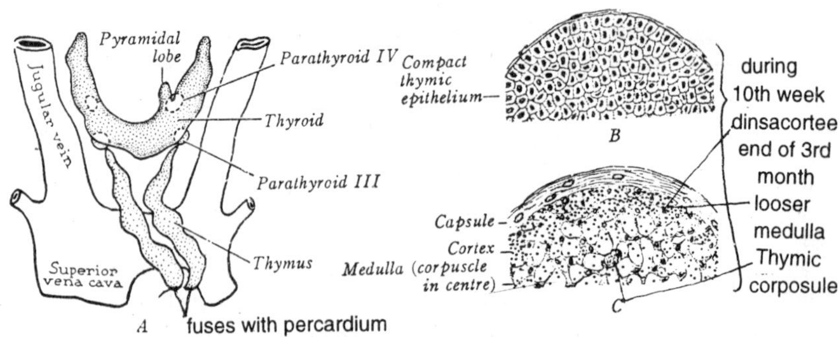

Fig. 8.26. Glandular derivatives of the human pharynx. *A*, Thyroid, parathyroids and thymi, at two months, in ventral view. *B*, Thymus, at seven weeks, in section. *C*, Thymus, at three months, in section.

By the tenth week the thymic epithelium is transforming into a stellate syncytium that resembles a sparsely fibrous reticular tissue. This spongework furnishes a delicate support for the entire organ. Some of the entodermal cells aggregate into tiny ball-like masses known as *thymic corpuscles* (of Hassal) *(C)*. Toward the end of the third month the thymus is becoming increasingly infiltrated with lymphocytes and is differentiating into a denser *cortex* and looser *medulla (C)*. The organization of a *capsule* and *trabeculae* from a surrounding mesenchyme and the partial sub-division of the organ into many *lobules* complete the process of development. The thymus enlarges steadily until puberty; it then begins to regress, although persisting in reduced form even into old age.

The epithelium of the cervical sinus attaches to the third pouch, but there is no clear evidence that this ectoderm contributes to the definitive organ. An origin of the characteristic small cells (so-called *thymocytes*) of the thymus from entodermal reticulum has been asserted frequently, yet most investigators view these elements as migratory lymphoeytes that invade the organ from without. It may be that the thymic epithelium actually attracts lymphocytes or even causes them to be produced. In addition to the main thymus, other small thymic masses are frequently encountered in embryos. They arise tardily in the region of the fourth pouches (ventral diverticula?), but their exact relations have not been surely traced.

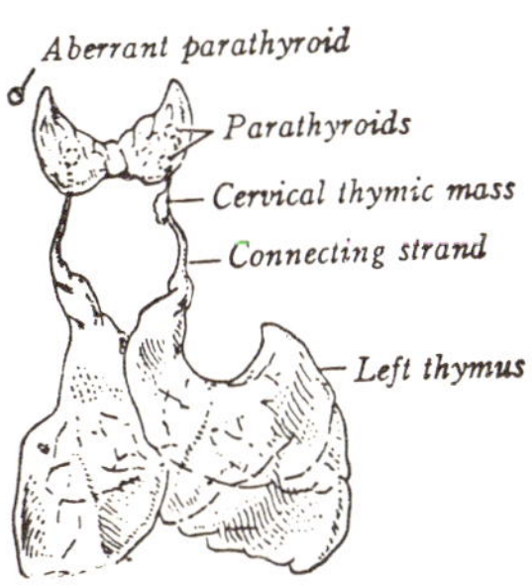

Fig. 8.27. Thyroid, parathyroids and thymi of a human newborn, in ventral view. An aberrant parathyroid and accessory thymic tissue also are seen.

Anomalies: The slender upper ends of the thymus sometimes persist. They either continue the thymus to the level of the thyroid gland or form separate accessory lobes there, depending on whether they persist wholly or in part.

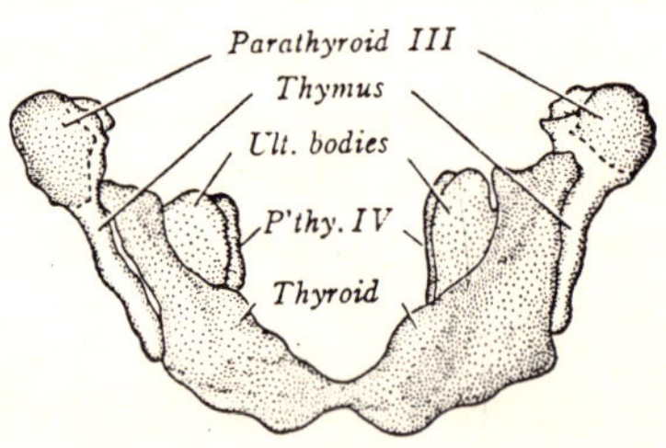

Fig. 8.28.   Glandular derivatives of the human pharynx, at 14 mm., shown as a model in ventral view.

The Parathyroid Glands: The dorsal wing of each third and fourth pouch thickens into a solid mass of cells which is prominent at 10 mm. Each is the primordium of a parathyroid gland. A few days later the two pairs of globular parathyroids (designated III and IV) are set free from the pharynx, although they still remain connected until the 20 mm. stage with the simultaneously detached thymi and ultimobranchial bodies, respectively. The pair from the third pouches is drawn down by the migrating thymic primordia to the level of the caudal border of the thyroid gland and become the *inferior parathyroids* of adult anatomy. The pair from the fourth pouches does not shift its position appreciably and remains at the cranial thyroid border as the *superior parathyroids.* Even at their earliest appearance, the glands are differentiating into distinctive, clear cells; acidophilic cells do not appear until late childhood. All four glands embed superficially in the thyroid capsule and gradually become well vascularized.

Anomalies: Variations in the number, size and location of the parathyroid glands are common. Both regular and accessory glands may locate at some distance from the thyroid;

parathyroids III sometimes even follow the thymus into the thorax.

The Ultimobranchial Bodies: In the fifth and sixth weeks these sacs are often classified as rudimentary fifth pouches, although there is some doubt as to their true status. At the beginning of the seventh week (13 mm.) each ultimobranchial body, joined with the adjacent parathyroid IV, is set free from the pharynx. Meanwhile growth of the thyroid brings its two lobes into contact with the ultimobranchial bodies. Each of the latter then loses its cavity and is incorporated into the thyroid. Some believe that the final fate of the ultimobranchial bodies is degeneration, other describe an actual conversion ('induction') of ultimobranchial into thyroid tissue, apparently through the dominating influence of a thyroid environment on a plastic, implanted tissue. Only to this degree, at best, can the ultimobranchial bodies qualify as 'lateral thyroid' primordia.

The Thyroid Gland: The thyroid is the earliest glandular structure to appear. Even an embryo 2 mm. long (six somites) shows a bulge on the ventral floor of its fore-gut that indicates the site of thyroid origin. A distinct entodermal pocket, the *thyroid diverticulum,* soon protrudes and lies between the first pair of pharyngeal pouches. This sac attaches to the pharynx by a narrower neck which is known as the *thyro-glossal duct.* It is so named because it is at first hollow and connects the primitive thyroid with the tongue which is organizing from the pharyngeal floor at the same time; here it opens at the aboral end of the tuberculum impar. The duct presently becomes a solid stalk and then breaks up in the sixth week, but its point of origin on the tongue is indicated permanently by an enlarged pit named the *foramen coecum.*

The thyroid sac quickly becomes a solid mass which lies against the primitive aortic sac. It is bilobed at an early stage and, when set free by the atrophy of its stalk, the thyroid begins to be converted into an irregular mass of epithelial plates *(C).* Early in the seventh week the gland becomes crescentic in shape and settles to a transverse position, with lobe on each side of the trachea. Actually its shift is illusory and is caused by the forward growth of the pharynx, which leaves the aortic trunk and thyroid behind. During the seventh week the enlarging ultimobranchial bodies come in contact with the thyroid primordium and fuse with it, thus forcing the thyroid to part company with the descending aorta and pericardium. As discussed in a previous paragraph, the ultimobranchial bodies rapidly lose their original identity and possibly transform into thyroid tissue. In the eighth week discontinuous cavities begin to appear in swollen or beaded portions of the solid thyroid plates *(D,E)*; these represent the beginning of *follicles* which acquire colloid in the third month and soon afterward become functional. By the end of the fourth month this conversion into follicles ends; thereafter new follicles arise only by the budding and subdivision of those already present. A *capsule* and vascular *stroma* differentiate from the local mesenchyme.

Anomalies: Persistent portions of the thyro-glossal stalk give rise to accessory thyroids, cysts or even a median fistula opening on the neck. *Accessory thyroids* may also be derived from detached portions of the main primordium. The variable pyramidal lobe of the thyroid, leading upward from the gland, results from the retention and growth of the lower end of the thyro-glossal stalk. Failure to descend properly leaves the thyroid located in the base of the tongue.

The Pharynx Proper: This funnel-shaped passage is the residual product after the transformation of its roof, floor and side walls is finished. The original epithelium becomes pseudostratified to stratified. As in the mouth and upper esophagus, the neighbouring mesenchyme differentiates a coat of striated (voluntary) muscle. Besides the organs already described, there are a few additional derivatives.

The Pharyngeal Tonsil: The entrance to the definitive pharynx is in a sense encircled by a lymphoid ring. In addition to the lingual tonsil below and the paired palatine tonsils at each side, there is still another tonsil mass that lies in the dorsal wall. This *pharyngeal tonsil* starts development in the fourth month; its lymphoid accumulation is a response to local vascularity and freedom from growth tensions. The so-called crypts are merely epithelial folds, wrinkled by the stresses of this region, and peculiarly dilated ducts of mucous glands.

The *pharyngeal bursa* is a pit located just below the pharyngeal tonsil. It results from the ingrowth of epithelium along the course of the degenerating tip of the notochord. Until the end of the second month the latter is fused with the epithelium of the pharynx at this point. *Seessel's pouch* is merely the dorsal, blind end of the entodermal fore-gut which, after the disappearance of the oral plate, persists for a short time as a sort of pit. It has no further significance. The lateral *pharyngeal recess* (of Rosenmuller) is a secondary formation, not related to the second pharyngeal pouch, as formerly claimed. The *piriform recess*, at each side of the entrance to the larynx, marks the site of the earlier third and fourth pouches.

THE DIGESTIVE TUBE AND ASSOCIATED GLANDS

The Digestive Tube

The digestive canal proper (esophagus, stomach and intestine) exhibits a rather uniform developmental history, except in such details as size, shape, position and glandular specialization. It consists originally of : (1) an internal tube of entoderm, which is the primary tissue that becomes the *epithelial lining* (including glandular ingrowths); and (2) an investing layer of splanchnic mesoderm that specializes into the thick, supporting wall. In accordance with the general progress of development in a cranio-caudal direction, higher levels of the digestive tube begin specialization sooner than lower levels and maintain this advantage for sometime. The mucosal lining expands faster than the outer wall and so becomes thrown into fold (*C,D*); these anticipate a future distention by food and also provide additional secretory and absorptive surface.

The mesenchymal investment about the lining entoderm differentiates into the *lamina propria, submucosa, muscularis* and *serosa* (or *adventitia*). Of the two chief muscle coats, the circular layer uniformly develops earlier (6-10 weeks) than the longitudinal layer (10-14 weeks).

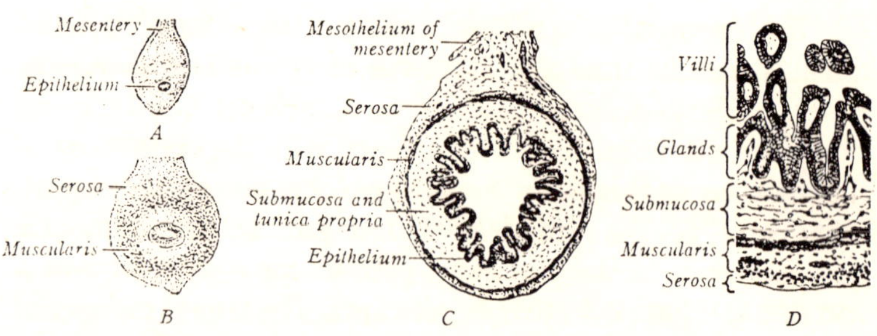

Fig. 8.29. Organization of the human intestine, shown in transverse sections. *A*, At six weeks; *B*, at eight weeks; *C*, at three months; *D*, at four months.

The Esophagus: Embryos of about 2.5 mm lack a definite esophagus, as in fishes, and at four weeks it is still a short tube extending from pharynx to stomach. However, the esophagus soon elongates rapidly, keeping pace with the differentiating neck and the growing heart and lungs alongside. A stratified epithelial lining is acquired slowly.

At five weeks the epithelium has acquired two layers of nuclei, and in the seventh week vacuoles begin to appear in it so that presently there is an increase in the size of the lumen. In this way the lining becomes channeled for a while, but at no time is it totally occluded like the fetal esophagus of reptiles and birds. The epithelium begins to acquire cilia at ten weeks and it is not until into the fifth month that a stratified squamous epithelium starts replacing it. At birth the epithelium numbers ten layers, but may still include some ciliated patches. *Superficial glands* are developing in the fifth month, whereas the *deep glands* arise mostly after birth. As a component of the mediastinum, the esophagus never acquires a typical mesentery or serosal tunic.

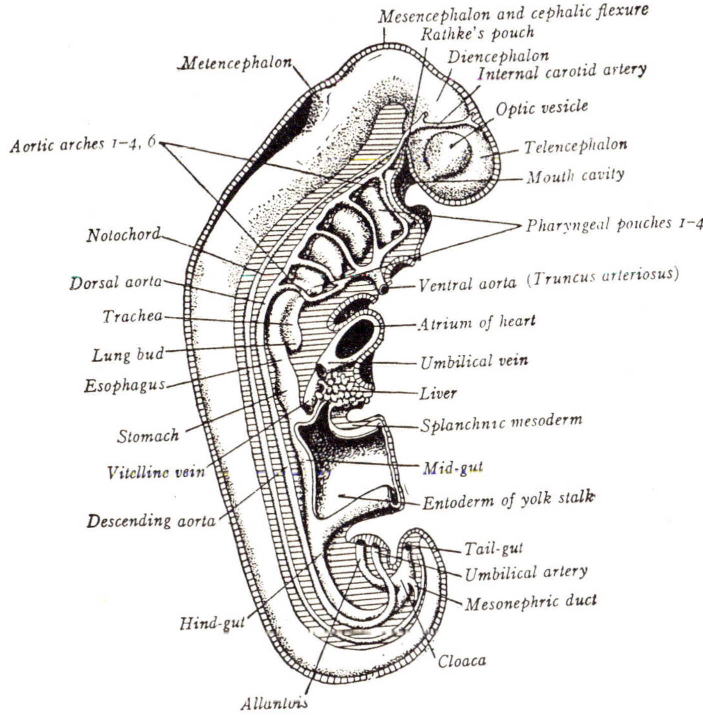

Fig. 8.30. Entodermal tract of a 4 mm. human embryo, exposed in a lateral dissection.

Anomalies: There may be *stenosis* (narrowing) or local *atresia* (no cavity). As the latter condition usually involves the trachea as well, it will be easier to explain after that organ has been discussed. The partial epithelial occlusion, normally a transient feature, predisposes toward all these abnormalities.

The Stomach: The stomach is discernible in embryos of 4 mm. as a spindle-shaped enlargement of the fore-gut, somewhat flattened on its lateral surfaces. Originally the stomach lies in the future neck region, but by the end of the seventh week a 'descent' has been completed through a distance of 16 segments to the permanent location in the abdomen. During the period of descent (4-7 weeks) the stomach undergoes certain changes in shape and orientation; (1) the entire organ increases in length; (2) the dorsal border grows faster than the ventral wall and so produces the convex *greater curvature* in contrast to the passively concaved *lesser curvature;* (3) the *fundus* arises as a local bulge near the cranial end; (4) the stomach rotates 90 degrees about its long axis until the greater curvature (primitive dorsal wall) lies on the left and the lesser curvature (primitive ventral wall) is on the right; and (5) the rotating stomach is displaced by the enlarging liver until it extends obliquely from left (above) to center (below).

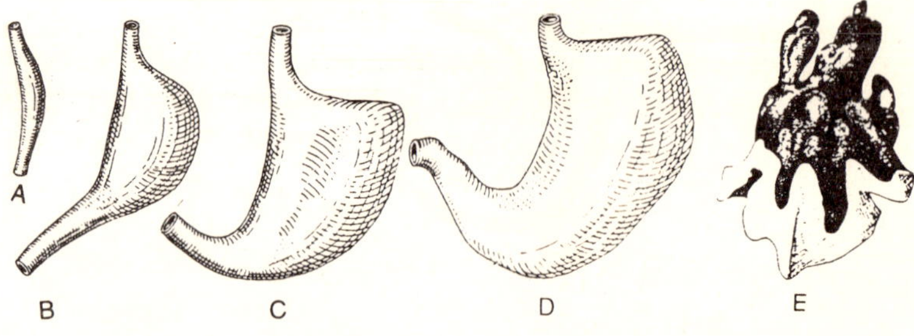

Fig. 8.31.   Models of the human stomach. *A,* At 5.5 mm.; *B,* at 9 mm. *C,* at 15 mm.; *D,* at 23 mm.; *E,* gastric glands, at seven months.

The epithelium differentiates distinctive *gastric glands* and the mesenchyme produces three incomplete layers in the muscular coat. Rotation causes the original right surface of the stomach to become dorsal and the original left surface to become ventral. This explains why the vagus nerves, coursing down the right and left sides of the esophagus, shift to dorsal and ventral positions in the region of the stomach.

Although usually described as a shift caudad, either by active growth or secondary crowding, 'descents' of the stomach, heart, lungs, and diaphragm are probably better interpreted as relative rather than actual. They result from the forward growth of the head-end of the embryo which leaves these organs behind. Especially are the more dorsal regions of the body concerned in this forward overgrowth, owing primarily to the rapid elongation of the neural tube. The simultaneous, but passive, transport cephalad of the dorsally located somites would explain the apparent descent of other relatively fixed organs, since the somites are customarily used as reference points.

The dorsal mesentery of the stomach, active at the time of stomach-rotation in forming the sacculated omental bursa, expands rapidly, whereas the ventral mesentery grows slowly. These factors assist in causing the rotation. Further more, the enlarging liver displaces the freely movable cephalic end of the stomach to the left, whereas the caudal end is relatively anchored by the short ventral mesentery and bile duct; the vitelline artery also aid by acting as a block. All of the factors involved in the rotation and displacement of the stomach are not understood but, as with intestinal placement and liver positioning, as they are at least partly intrinsic. In short, the movements of all these organs are not wholly passive, although neighbouring influences can serve as effective agents in a chain of sequential events.

A pyloric region becomes distinguishable in the third month, but the *pyloric sphincter* is still weakly developed at birth. The mucous membrane shows two early folds that course along the lesser curvature from esophagus to pylorus. These ridges delimit a groove, recognizable permanently as the *gastric canal*. The early epithelium is a simple sheet. Pits, or *foveolae*, are indicated in the mucosa at seven weeks, and at 14 weeks *gastric glands* begin to bud off from them. Both continue to increase many-fold between birth and maturity until they reach a total in the millions. Enzymes are secreted by the fifth month, but the secretion of hydrochloric acid is perhaps delayed until considerably later. Gastric motility occurs at four months.

Anomalies: Transposition to the right side of the abdomen, as in a mirror-image, sometimes occurs in conjunction with the reversal of other asymmetrically placed organs. Location of the stomach above the diaphragm (*thoracic stomach*) is due to an arrested descent. Stenosis results from an overdevelopment of the pyloric sphincter.

The Intestine: In embryos of four weeks (5 mm.) The intestine is a simple tube, beginning at the stomach and ending in the cloaca. It occupies the median plane and parallels the curving neural tube; midway along its course the intestinal tube bends slightly ventrad and becomes continuous with the now slender yolk stalk. For convenience the segments of the early intestine above and below the attachment of the yolk stalk are designated as the cranial and caudal *limbs of the intestinal loop (B).* At this stage the location of the future *duodenum* is recognizable by its relations to the stomach and the primordial liver and pancreas; the remainder of the cranial loop will become the *jejunum* and much of the *ileum,* while the caudal loop represents the lower ileum and all of the *colon* and *rectum.* The intestine is supported from the dorsal body wall by the *dorsal mesentery;* a *ventral mesentery* exists only in the duodenal region.

In the fifth week (5-8 mm.) the intestine elongates faster than the trunk, and the intestinal loop becomes a prominent flexure. By the end of this period the yolk stalk detaches from the apex of the loop. The loss of this marker on the intestinal wall is offset by the acquisition of a more useful one; a bulge in the caudal limb indicates the *caecum,* and consequently denotes the boundary between the *small* and *large intestine.* Succeeding gross changes involving the intestine include its torsion, herniation into the umbilical cord and coiling; next follow a withdrawal into the abdomen, final placement and fixation. Structurally, the differentiation of *villi* and *glands* is a characteristic feature.

Rotation: While the intestinal loop is developing it undergoes a torsion; in this rotational displacement the superior mesenteric artery, which courses in the mesentery between the two limbs, may be considered as the axis. The twisting is so executed that the cranial limb is carried from the midplane to the right; conversely, the caudal limb shifts to the left. In other words, there has been anticlockwise rotation as one views the embryo from the ventral side. The torsion is said to result from a change of position of the enlarging left umbilical vein which forces the cranial limb of the bowed gut to the right, but the primary factors stem from a symmetry determination that is established much earlier.

Herniation and Coiling: During the sixth week the elongating intestinal loop can no longer be contained within the slower growing abdomen and it begins to escape into the umbilical cord, still retaining the rotated positions of the two limbs. This protrusion constitutes a temporary (but normal) umbilical hernia, the intestine then occupying a cavity continuous with the embryonic coelom. Thickenings of the mesentery at the kinked lower end of the duodenum and at the site of the future (splenic) flexure of the colon prevent the duodenum and descending colon from entering the cord, but the loop proper (future jejunum, ileum, ascending and transverse colon) is not so restrained. Continued elongation of the herniated small intestine leads to extensive coiling; by contrast, the large intestine and its associated mesentery grow relatively little at this period.

Re-Entry and Placement: In embryos of ten weeks the abdominal cavity has increased sufficiently in size both absolutely and relatively (the latter due particularly to a decline in the growth rate of the liver), so that the intestine can again be accommodated. The exact cause of withdrawal from the umbilical sac is not well understood. But whatever, the underlying forces may be, it is plain that the return, once begun, is completed quickly and the coelom of the cord promptly obliterated. The small intestine is the first to re-enter the abdomen. It does this in a progressive manner, the proximal portions leading. The returning coils first fill the available space on the left side of the abdomen, thereby pressing the non-herniated (future descending colon also to the left, whereas the later coils to return locate in the right half of the abdominal cavity.

Because of the prominent caecal swelling, the large intestine is the last to leave the umbilical cord and re-enter the abdominal cavity. Its tendency to straighten then carries this limb slantingly

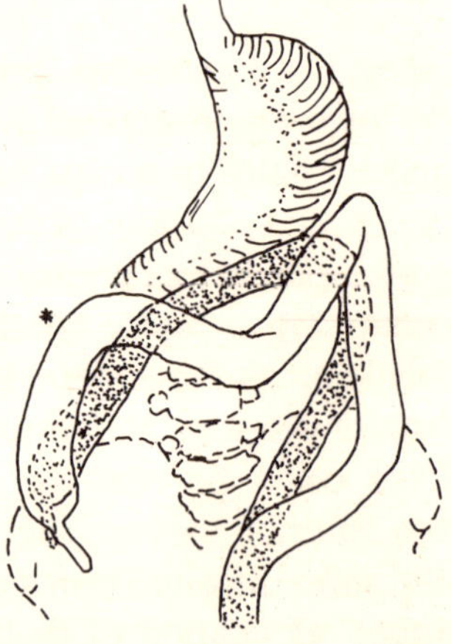

Fig. 8.32.   Position of the colon in human fetuses. In stipple, the relations at ten weeks on return from the umbilical cord; in full line, the relations during the seventh month with the ascending colic limb and hepatic flexure (*) evident.

across to the right side, thus completing a rotation that totals 270 degrees. in this placement the caecum lies close to the crest of the ileum and here it becomes fixed in its permanent position. From this point the colon passes obliquely upward to the left of the stomach, where it recurves sharply (*splenic flexure*) into the future descending colon which maintains its displaced position on the left side (in stipple). For several months these conditions persist and there are no definite ascending and transverse limbs of the colon.

Completion: The *duodenum* acquires an early, characteristic curve in relation to the growing pancreas. It loses its mesentery by becoming fixed against the dorsal body wall. The remainder of the small intestine, thrown into loops since its elongation while in the umbilical cord, is given the somewhat arbitrary names of *jejunum* and *ileum.* The original caecal bulge grows and makes a definite, blind sac which extends the large intestine beyond its junction with the ileum. The distal end of this sac lengthens rapidly for a time *(B)*, but it fails to keep pace in thickness with the rest of the sac *(C)*. As a result, the characteristic *vermiform process* of the higher apes and man becomes distinct from the *coecum (D).*

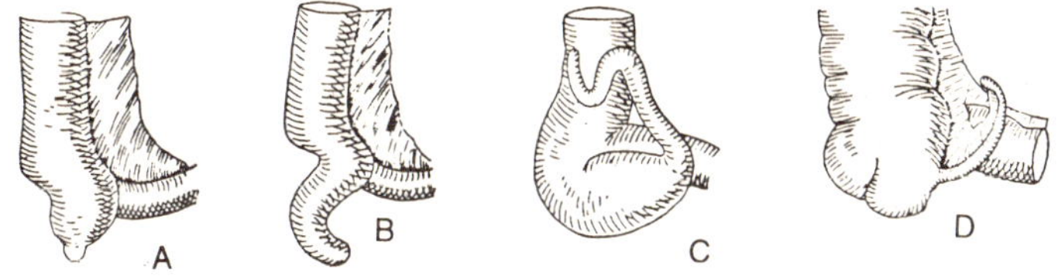

Fig. 8.33.   Development of the human caecum and vermiform process. *A,* At two months; *B,* at three months; *C,* in the newborn; *D,* at five years.

As the liver decreases in relative size and accordingly 'retreats' cephalad, as *hepatic flexure* appears in the originally oblique proximal limb of the colon and becomes increasingly sharper. This flexure progressively demarcates an ascending from a transverse colic limb. The *ascending colon*, beginning to elongate as such in the middle of fetal life, is not completed until early childhood. The *transverse colon* necessarily courses in front of (i.e., ventral to) the duodenum, which never left the abdomen, and hence remains suspended by its mesocolon *(B)*. Quite different is the *descending colon*, which, like the ascending colonic limb, is applied against the body wall; each then loses

its free mesentery and becomes fixed in a way to be explained
(C,D). A relatively early, pronounced elongation of the most
caudal portion of the colon is retained as the *sigmoid colon*. The
terminal portion of the intestine, or *rectum*, is derived from the
subdivision of the cloaca; illustrate the process of separation,
which is described in full. After the anal membrane ruptures
at the end of the eighth week, a short ectodermal pit, or
proctodeum, is added to the entodermal digestive tube. This
anal canal results from the encircling growth of several anal
hillocks. The junction between ectoderm and ectoderm is
somewhat below the so-called anal valves.

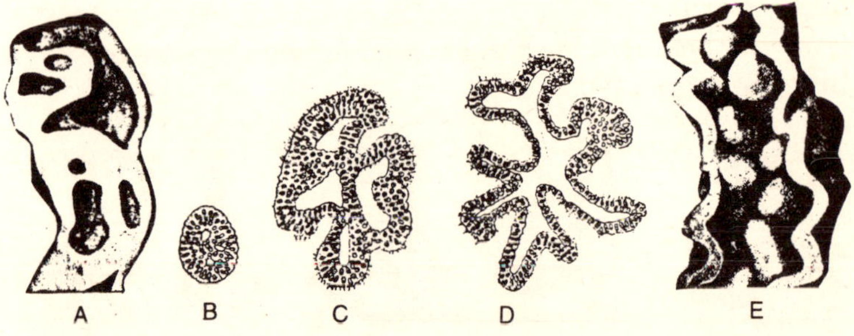

Fig. 8.34. Differentiation of the epithelium of the human small intestine.
A, Duodenal epithelium, reconstructed and cut lengthwise
to display the temporary occlusion and vacuolation at eight
weeks. *B-D*, Transverse sections of the duodenal epithelium
at six, eight and nine weeks. *E*, Jejunal epithelium, reconstructed
and cut lengthwise to display the villi at eight weeks.

Between the fifth week and birth the intestine increases its
length 1000 times, while the small intestine becomes six times
the length of the large intestine. The small intestine is original
thicker than the large intestine; it is not until the fifth month
that the large intestine becomes greater in diameter. Proliferation
of the epithelial lining of the duodenum leads to its occlusion
in the sixth and seventh weeks, but vacuolation soon restores
a continuous lumen *(C,D)*. All of the small and large intestine
shows a similar phenomenon, but in lesser degree; the entire

intestinal tract is finally clothed with a single-layered epithelium. *Villi* begin to appear at eight weeks as independent, rounded elevations of the epithelium. *Intestinal glands* (of Lieberkuhn) arise as tubular ingrowths of the epithelium about the bases of the villi. They first appear toward the end of the third month and are closely followed by the compound *duodenal glands* (of Brunner). Both villi and glands increase greatly in number during childhood. *Lymph nodules* and *Peyer's patches* are present at five months. The colon bears villi in the middle third of fetal life; the *teniae* are linear thickenings of the longitudinal muscle layer. At ten weeks the caecum makes a sharp bend with the colon proper, and this flexure is responsible for the production of the *colic valve*. Peristalsis of the intestine has been observed at 11 weeks.

Meconium begins to collect in the intestine after the third month. This mass is a pasty mixture of mucus, bile and cast-off epithelial cells, to which are added lanugo hairs, desquamated epidermal cells and sebaceous secretion swallowed with amniotic fluid. It is green in colour and is wholly voided by the third or fourth day after birth—a fact sometimes of medicolegal value in proving how long an infant has lived. At birth the intestine and its contents are perfectly sterile, but a bacterial flora is acquired promptly.

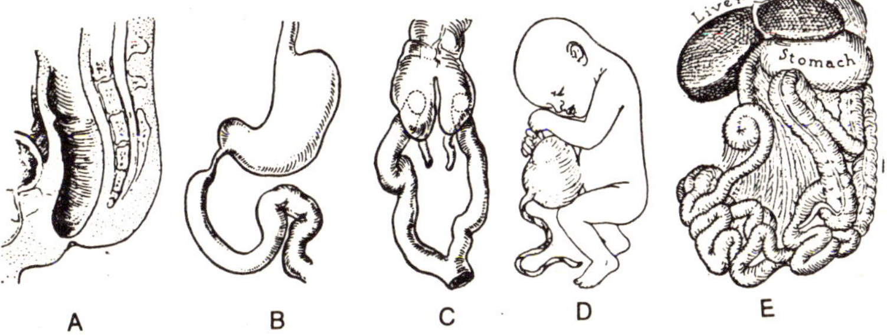

Fig. 8.35. Anomalies of the human intestine. *A*, Imperforate anus; *B*, atresia of the duodenum; *C*, duplication of the lower ileum and caecum; *D*, congenital umbilical hernia; *E*, non-rotation of the intestine.

Causal Relations: Transplantation experiments on amphibians show that the fore-gut has been irreversibly determined at the neurula stage, not only as a whole but also regionally (mouth; pharynx; esophagus; stomach; intestine) Other experiments prove that the factors related to the rotation of the stomach and intestine, and the asymmetrical displacement of these organs and the liver, are already present in the gastrula stage (and probably earlier). This conclusion follows since reversal of the archenteron-roof leads to the subsequent transposition of the stomach, liver and intestine as in a mirror-image. The invagination of ectoderm to form the proctodeum is induced by the presence in that region of ventral-lip mesoderm of the amphibian gastrula. But it is the entoderm of the hind-gut, thus brought into contact with the floor of this pit, that is responsible for the perforation of the ectoderm which establishes the anus.

Anomalies: The failure of the cloacal membrane to rupture results in an *imperforate anus;* it may be combined with atresia of the rectum. More or less of a permanent cloaca follows the incomplete separation of rectum from urogenital sinus. The intestine may undergo stenosis or atresia; this occurs most often, in relation to the length of the various regions, in the duodenum. It represents a partial or complete retention of the temporary fetal occlusion. Two per cent of all adults show a persistence of the proximal end of the yolk stalk which forms a pouch, Meckel's *diverticulum of the ileum;* this may extend even to the umbilicus *(B),* and when patent constitutes a fecal *umbilical fistula (B,C).* Duplication of the digestive tube occurs at all levels. This may result from longitudinal splitting and take the form of parallel, communicating tubes. Another duplicating method is sacculation, in which persistent diverticula or separate sacs range in shape from the commoner spheroid to elongate, blind tubes.

Congenital *umbilical hernia* may be due either to the perpetuation of the transitory fetal condition or to a secondary protrusion of intestinal loops after the primary withdrawal. Ordinary cases are secondary herniations results from muscular and fascial defects where the body wall has been pierced by umbilical blood vessels. Most severe is *omphalocoele*, in which intestine and other viscera occupy a thin sac of amnion and peritoneum located at the base of the umbilical cord. Other hernias of the bowel are explained. Rarely there is *non-rotation* of the returning intestine; the jejuno-ileum then lies on the right side, the colon on the left *(E)*. Reversed torsion of the colon, after re-entry, can result in the transverse colon passing behind the duodenum without any other relations being disturbed. The caecum may have a high position because it does not become fixed but ascends with the liver. Transposition of the digestive tract right for left, as in a mirror image, is one feature of the more general conditions known as *situs inversus*; the participation of the intestine in this process is characterized by a complete reversal of the normal course of rotation.

The Liver

The liver is a ventral outgrowth from the gut-entoderm in the region of the anterior intestinal portal. In embryos with 17 somites (2.5 mm.) its shallow primordium lies between the pericardial cavity and the attaching yolk stalk. Here is the floor of the future duodenum which continues to give rise to the definite sacculation named the *hepatic diverticulum*. This consists of a cranial portion, which will differentiate into the glandular tissue and its bile ducts, and a caudal portion which becomes the gall bladder and cystic duct *(C)*. The hepatic diverticulum forces its way ventrad into a mass of splanchnic mesoderm that will furnish most of the substance of the diaphragm; at this stage the primitive diaphragm is named the septum transversum *(A,B)*. A little later, the region of the septum occupied by the liver becomes drawn out as the ventral mesentery, and the final relation of the liver is then more intimately related to this mesentery than to the diaphragm proper.

Straightway after its appearance, the cranial portion of the hepatic diverticulum buds off epithelial cords which invade the septum transversum farther and continue to proliferate there into a rapidly expanding sponge-work. From the first the diverticulum lies close to the paired vitelline veins which flank the gut, and these veins send branches into the region of proliferation. The result is a mutual, intimate intergrowth of tortuous liver cords and sinusoidal channels (C). Perhaps it is because of its rich blood supply that the hepatic mass enlarges so rapidly. In any event, the liver of a 5 mm. embryo is a large crescentic mass with a wing extending upward on each side of the gut. While these changes have progressed, the original diverticulum is elongating and differentiating into the duct system.

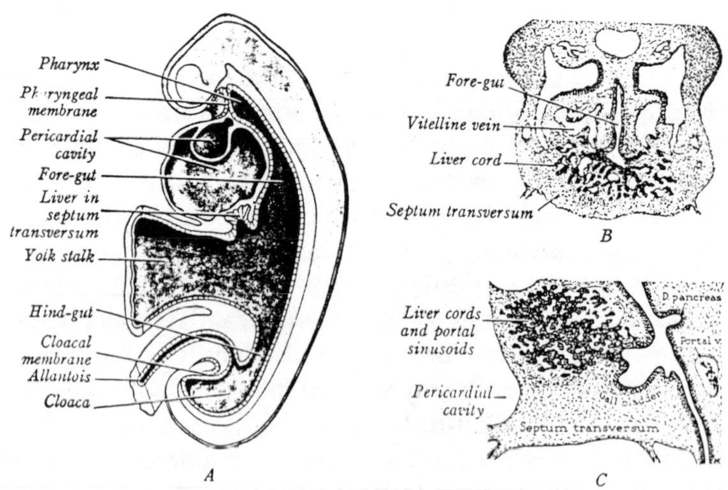

Fig. 8.36. Origin and relations of the human hepatic diverticulum. A, At 3 mm., in sagittal section. B, At 3.5 mm., in transverse section. C, At 5 mm., in sagittal section.

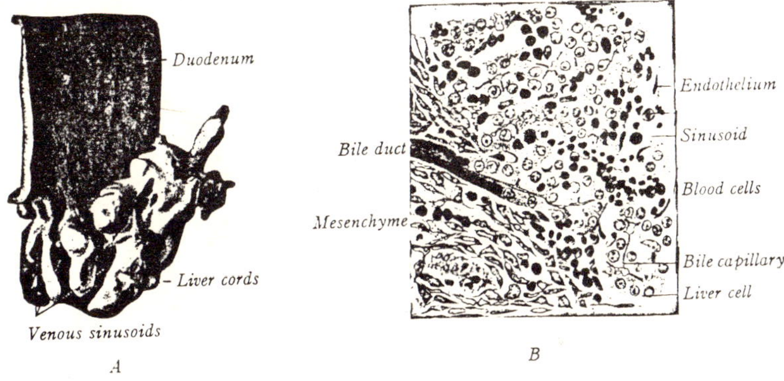

Fig. 8.37. Differentiation of the human liver. A, Model of half of the duodenal wall and liver, at 4 mm. B, Section, at 16 mm.

Glandular Tissue and Blood Vessels: The early epithelial strands are the forerunners of the definitive *liver cords* (actually thin plates) around which the endothelium of the broad sinusoids becomes closely applied. In its early growth upward around the gut, the wings of the liver come to enclose and interrupt the nearby vitelline veins. After this occurs, only *sinusoids* interconnect the supplying (portal) and draining (hepatic) vessels. At first relatively far apart, these two venous trees grow steadily as the liver expands and thus progressively 'approach' each other in an alternating (or dovetailing) manner. The regularity of the system of branching that is employed is responsible for the creation of the characteristic *hepatic lobules* from the epithelial tissue and sinusoids.

From the second to the seventh month of fetal life blood cells are actively differentiating between the hepatic cells and their covering endothelium, but only small foci remain at birth. This potential ability remains latent throughout life, and blood formation can be resumed in the liver whenever, the need for replacement is sufficiently urgent. The sinusoidal lining partially transforms into large macrophages which becomes the so-called *Kupffer cells*. Typical bile is secreted by the hepatic cells in fetuses five months old and colours the meconium.

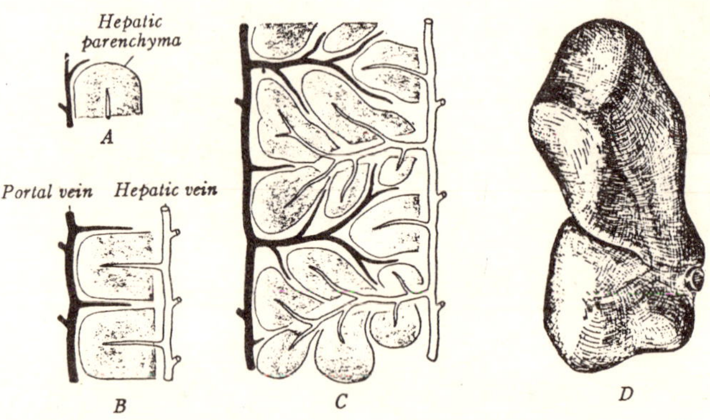

Fig. 8.38. Method of origin of the hepatic lobules. *A-C*, Diagrams of successive stages of growth and subdivision. *D*, Bifurcating lobule of the postnatal pig's liver.

In a 4 mm embryo the whole liver is a single, complete lobule; at 7.5 mm it is bilobed and has two lobules; at 11 mm there are six lobules, while a late fetus has many thousands. Each lobule is surrounded by several terminal branches of the portal vein and is drained by a single hepatic vessel. Toward the end of the fetal period, but mostly after birth, these primary lobules subdivide into smaller, secondary units. Each central (hepatic) vein takes the initiative and bifurcates or gives off a side (branch. New lobules then arise by the simple splitting (i.e., through connective tissue invasion of such a lobule which has thus acquired two central veins *(C,D)*. The portal veins at the periphery branch correspondingly as they push in between the new lobules to keep the vascular relationship unchanged. A clear demarcation of the definitive lobules, some 500,000 in number, is not seen until early childhood.

The Ducts and Gall Bladder: The main portion of the hepatic diverticulum elongates into the *ductus choledochus* (common bile duct) and its direct continuation to the liver, the *hepatic duct*. The original ventral site of origin of the common duct shifts, during the seventh week, through torsion of the duodenum, until its attachment becomes dorsal *(B,C)*. The *bile ducts* within the liver, which are tributary to the hepatic duct, arise in a secondary manner beginning at eight weeks. Wherever, the liver cords come under the influence of connective tissue that grows in with the branching portal vein, they transform into *interlobular ducts*. The liver cords are said to be microscopically hollow at their earliest appearance and hence the tiny *bile capillaries* of the permanent cords are primary lumina and not secondary acquisitions.

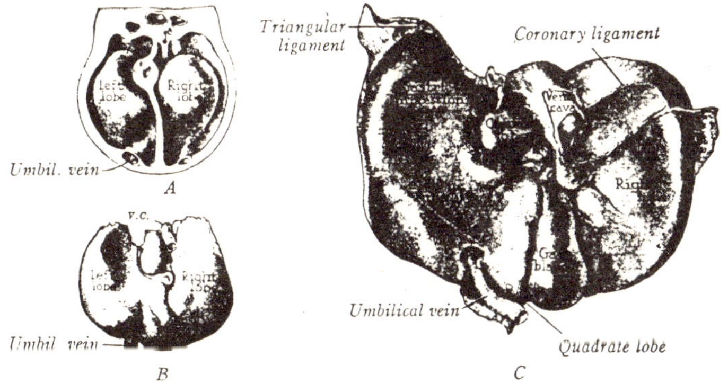

Fig. 8.39.   External form of the human liver, viewed from the dorsocaudal surface. *A*, At 8 mm.; *B*, at 16 mm.; *C*, at birth.

The *gall bladder* constitutes a separate, caudal region of the originally shallow hepatic diverticulum. In a 5 mm embryo it is a solid, epithelial cylinder which is carried away from the duodenum by the elongating common duct. A distinct stem, or *cystic duct*, is then recognizable *(B,C)*, and in the seventh week a lumen has been established throughout most of the tract which then appears like an offshoot from the main biliary passage.

Accessory Tissues: The growing liver expands greatly that particular portion of the septum transversum (soon ventral mesentery) within which it lies. Superficially the mesentery becomes a *capsule*, with a peritoneal covering, about the liver. More internally the mesenchyme differentiates into the connective-tissue framework (*Glisson's capsule*) about the lobules and into the muscular walls of the larger ducts and gall bladder.

The Liver as a Whole: The primary attachment of the liver to the septum transversum causes it to 'descend' with the latter organ from a cervical level of origin. The liver soon outgrows its original location in the septum transversum and at four weeks bulges caudad into the abdominal cavity. The continued progressive separation of liver from septum occurs at the time when the gut is also drawing away from the septum to produce a definite ventral mesentery. This is the reason why the later liver is intimately associated with both the septum and ventral mesentery. Such relations and the development of the *hepatic ligaments* will be described. The history of the vitelline and umbilical veins with respect to the liver may be found. The subdivision of the liver into its characteristic *lobes* is largely the result of unequal growth, while its final shape is a passive response to adjacent pressures.

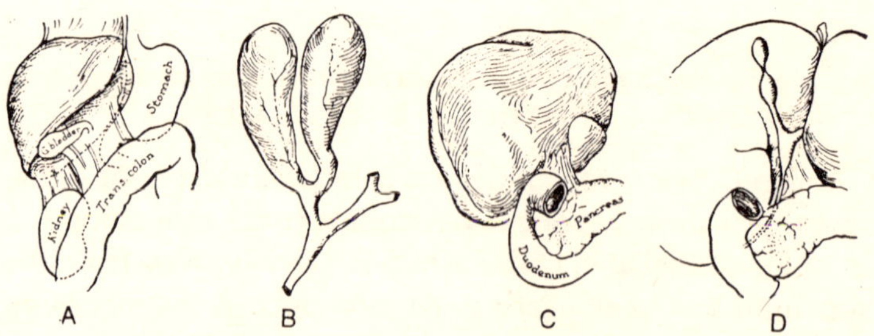

Fig. 8.40. Anomalies of the human liver and gall bladder. *A*, Absence of the left lobe of the liver. *B*, Double gall bladder. *C*, Absence of the gall bladder, and the consequent lack of a quadrate lobe. *D*, Stenosis of the gall bladder and common duct.

The primary gross swellings are the paired *right* and *left lobes*. Originally these are of equal size, but the right lobe becomes larger after the third month. In part this asymmetry is due to intrinsic growth factors, although the greater available space on that side plays a practical role while the vitelline and umbilical veins are usually credited with some influence as well. Parts of the early right lobe become set off as subordinate lobes. Thus, at six week the *caudate lobe* is recognizable, bounded by the ventral mesentery and inferior vena cava *(B)*. The *quadrate lobe* originates later with the atrophy of the liver tissue overlying the intrahepatic portion of the umbilical vein; it lies between that vein and the gall bladder *(C)*.

The developing liver is spongy and highly plastic, so that it tends to occupy the available space not used by firmer, neighbouring organs; this accounts largely for its general final shape. In certain regions the hepatic tissue undergoes degeneration (due to pressure atrophy?) and especially is this true peripherally in the left lobe. For a time the liver grows at a faster rate than other organs or the body as a whole. Its maximum relative size is attained at nine weeks when it constitutes 10 per cent of the body volume and occupies most of the belly cavity.

Causal Relations: Suitable transplants of the ventral yolk-mass of an amphibian neurula show that the liver and pancreas, although structurally indistinguishable from the future gut-entoderm of this period, are already irreversibly determined as to their respective fates. Even further, the more dorsal portion of the actual liver bud is a region that is already determined as gall bladder. These results prove that both glands are not secondary specializations, budding out from a previously established gut-entoderm, as sections seem to imply, but that they merely occupy for a time, along with gut-entoderm, a common medial strip on the floor of the archenteron. At a still earlier stage every cell of the gut is presumably pluripotent in these respects. This is supported by abnormal human development, since nests of stomach or pancreas cells can occur in a Meckel's diverticulum or elsewhere in the digestive tube. In the chick it appears that contact of the cardiac primordium with the gut floor is a prerequisite to liver formation.

Anomalies: A reduction or an increase in the external lobation of the liver is a rare occurrence. An increase sometimes results in lobation resembling that in lower mammals. The main ducts and the gall bladder are subject to duplication as the result of early splitting, subdivision or sacculation *(B)*. Absence of the gall bladder (as occurs normally in the horse and elephant) is well-known *(C)*. A congenitally narrowed, solid or interrupted condition of the gall bladder or of the chief ducts is related to the temporary embryonic occlusion *(D)*.

The Pancreas

Two outpocketings from the entodermal lining of the gut represent the earliest indications of the future pancreas. These buds arise on opposite sides of the duodenum in embryos of 3 to 4 mm.. One pushes out from the dorsal wall, just cephalad of the level of the hepatic diverticulum; it is the *dorsal pancreas*. The other, probably originally paired, appears ventrally in the caudal angle between gut and hepatic diverticulum and consequently is designated the *ventral pancreas*. These two primordia meet and unite, thus producing a joint organ *(B-E)*.

Of the two pancreatic primordia, the dorsal one grows more rapidly. In the sixth week it is an elongate, nodular structure, which extends into the dorsal mesentery. Since it arises near the mouth of the developing omental bursa, it continues its growth within the dorsal layer of that mesenterial sac. The ventral pancreatic bud remains smaller; it is carried away from the duodenum by the lengthening common bile duct and then arises directly from the latter. Unequal growth of the duodenal wall shifts the bile duct dorsad and thereby brings the ventral pancreas into the dorsal mesentery, near the stem of the dorsal pancreas *(B,C)*. During the seventh week the two primordia interlock intimately *(D)*. Grossly the dorsal pancreas forms all of the mature gland except most of the head and the uncinate process, which come from the ventral primordium *(E)*.

The Ducts: Both pancreatic buds have an axial duct. The dorsal duct arises directly from the duodenal wall, but the base of the ventral duct is carried upward onto the elongating common bile duct and shares a common stem with it. When duodenal torsion brings the two pancreatic primordia side-by-side, the short ventral duct taps the dorsal duct *(C,D)*. Thereafter the long distal segment of the dorsal duct plus the entire ventral duct will serve as the chief line of drainage *(E)*. This combined tube is known in adult anatomy as the *pancreatic duct* (of Wirsung). The proximal, or stem, segment of the dorsal duct constitutes the so-called *accessory duct* (of Santorini). It becomes tributary to the main duct, but commonly retains its duodenal outlet as well *(E)*.

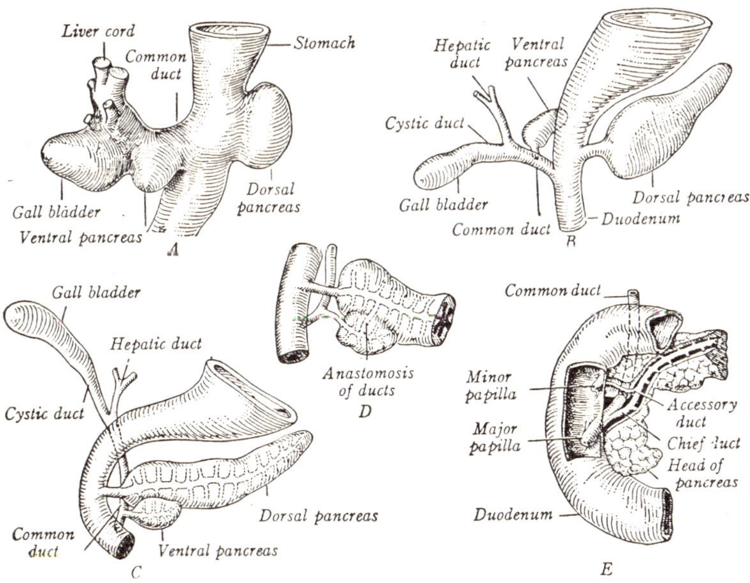

Fig. 8.41. Development of human pancreas, shown by models viewed from the left side. *A,* At 6 mm.; *B,C,D,* at 8,12 and 16 mm., respectively; *E,* at birth.

The occurrence of a permanent common outlet into the duodenum for bile and pancreatic juice is a direct consequence of the close relationship between the bile and ventral pancreatic duct. The region of the common outlet is the *ampulla* (of Vater) which opens at the major *duodenal papilla*. This joint duct gains a circular sheath of smooth muscle (*sphincter of Oddi*) in the seventh week. A final arrangement of ducts similar to that in man occurs in sheep, while the hog and ox reverse the relation and use the dorsal duct as the chief stem; less specialization occurs in the horse and dog, which retain both ducts as functional outlets into the intestine.

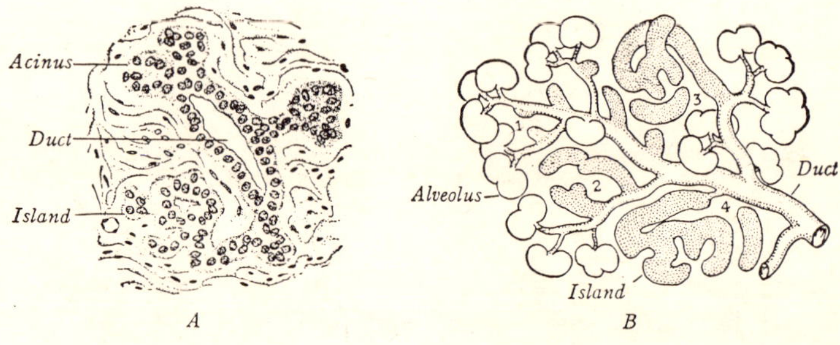

Fig. 8.42.   Differentiation of the human pancreas. *A*, Section, at fourteen weeks, demonstrating the origin of acini and islands from ducts. *B*, Diagram showing four progressive stages (1-4) in the organization of islands.

The Glandular Tissue: Secretory *acini* begin to appear in the third month as terminal and side buds from the ducts. *Pancreatic islands* (of Langerhans) also are differentiating from the ducts at about the same time (*A*). They are composed of distinctive cells which take the form of single sprouts, but later through growth (and, it is claimed, through union) become complex island masses (*B*). In all about a million islets are formed, some of which retain their original (but soon impervious) connections with the parent ducts.

No histological distinction exists between the acini of the dorsal and ventral pancreatic masses, but probably the dorsal pancreas alone differentiate islands. The divergent specialization pursued by acini and islands and their subsequent inability to convert from one type to the other are no more astonishing than the unlike differentiation of serous and mucous cells in the submaxillary gland or of peptic and acid cells in the stomach. Trypsin has been detected at five months and insulin seems to be present still earlier. The mesenchymal bed, in which the gland develops, furnishes a connective-tissue *capsule* and subdivides the organ into *lobes* and *lobules*.

Causal Relations: A pancreas-forming potency is irreversibly determined before the primordium can be recognized as an entity in the gut-lining of amphibians. Moreover, experiments can prove the existence at this time (neurula stage) of separate dorsal and ventral primordia; only the former is able to differentiate islands. Pancreatic primordia, like liver and gut, will self-differentiate in culture medium even to the extent of producing secretion.

Anomalies: Accessory pancreases are common. Many of these lie within the wall of the intestine and stomach; others are associated with the spleen and omentum. The development of supernumerary primordia and the displacement of parts of the diffuse, early pancreas are responsible for these several conditions. An annular pancreas encircling the intestine, bile duct or portal vein sometimes occurs *(B).* It is apparently due to the ventral primordium failing to rotate as a wromic. The ventral pancreas, and accordingly the main duct of the adult gland, may arise directly from the duodenum. Absence of that part of the gland derived from either primordium, failure of union between the dorsal and ventral pancreatic components, completely independent ducts, and a single duct *(C)* are all well-known.

THE RESPIRATORY SYSTEM

The nose and naso-pharynx belong to the respiratory apparatus, but since this relation is a secondary adaptation their development is described in other chapters. As with all hollow viscera, the larynx, trachea and smaller respiratory passages are lined with an epithelium (in this instance, entoderm) which is strengthened and supported by other layers differentiated from the surrounding mesenchyme. In addition, the lungs expand into the coelom (pleural cavities) and in so doing gain a covering of splanchnopleure (visceral pleura) whose free surface is mesothelium.

The earliest indication of the future respiratory tree is in embryos of 3 mm., with 20 somites. It constitutes the so-called *laryngo-tracheal groove* which runs lengthwise in the floor of the gut, just caudal to the pharyngeal pouches. In surface view the entoderm projects as a ventral ridge. This primordium is destined to become in order the *larynx* and the *trachea*, while its more rounded caudal end will shortly give rise to the *bronchi* and *lungs (B)*. A lateral furrow appears on each side, along the line of junction between ridge and esophagus *(C)*; becoming progressively deeper and extending cephalad, the furrows join, thereby splitting off first the lung bud and then the trachea. At the upper end, the laryngeal portion of the tube actually advances slightly cephalad until it lies between the fourth branchial arches. At the 4 mm stage the lung bud begins to bifurcate and the respiratory organs are then represented by: (1) a *laryngeal slits;* (2) the tubular *trachea;* and (3) two *primary bronchi (D)*. The latter buds are potentially more than bronchi, since by growth and branching they will ultimately produce all the subdivisions of the respiratory tree (bronchial branches, bronchioles and air sacs).

The Larynx

This organ develops somewhat differently in its upper and lower halves. The lower portion organizes around the cranial end of the trachea, whereas the part above the vocal folds rises out of the pharyngeal floor, in the region of the primitive glottis, as a sort of vestibule.

The *epiglottis* is peculiar to mammals. Embryos of 5 mm. show a rounded prominence that elevates midventrally from the bases of the third and fourth arches. This soon alters it shape *(B-D)* and consolidates into the transverse flap that guards the entrance to the larynx during swallowing *(E)*. It becomes concave on its laryngeal surface and in the middle of fetal life differentiates cartilage internally *(F)*.

The slit that opens from the floor of the pharynx into the trachea is the primitive *glottis*. Presently it is bounded on each side by a rounded eminence, of fourth and fifth arch origin, known as an *arytenoid swelling (B)*. These two swellings straightway begin to grow in a tongueward direction. On meeting the primordium of the epiglottis they arch upward

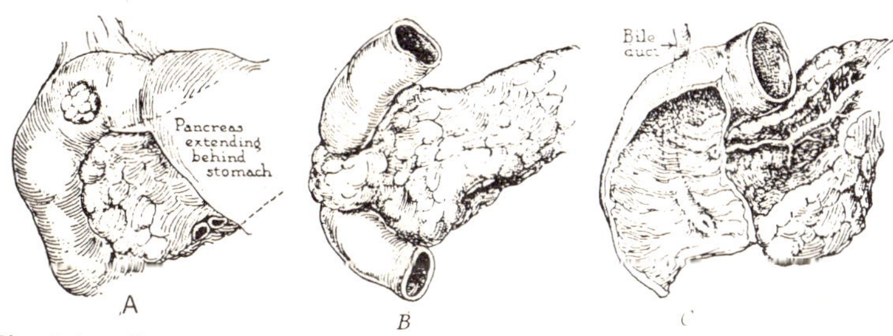

Fig. 8.43.   Early stages of the human respiratory primordium. *A*, At 2.5 mm., in ventral view; *B*, *C*, at 3 mm., in ventral and lateral view; *D*, at 4 mm., in ventral view.

and forward against its caudal surface *(C)*. In the seventh week this results in the original, sagittal slit adding a transverse groove to its upper end, so that the laryngeal orifice becomes T-shaped *(D)*. However, the entrance to the larynx ends blindly for sometime because fusion of the epithelium in the upper larynx has obliterated the lumen. When the epithelial union is dissolved (10 weeks) the entrance becomes more oval in contour *(E)* and a pair of lateral recesses *(laryngeal ventricles)* is evident in the restored cavity. Each is bounded cranially and caudally by a projecting, lateral shelf. The caudal pair, lying at the same level as the primitive laryngeal slit, is the *vocal folds (F)*; they later differentiate elastic tissue.

The epithelial lining of the larynx is supported by dense mesenchyme derived from the fourth and fifth branchial arches. Early in the seventh week this mass shows localized condensation that foretell the *laryngeal cartilages*. These belong to the skeleton and will be treated more in detail in a later chapter. The *laryngeal muscles* also originate from the same branchial arches and consequently continue to be innervated by the vagus nerves which supply those arches.

The Trachea and Primary Bronchi

The early 'lung bud' promptly bifurcates (at 4 mm) into the two *primary bronchi*, and the tubular system then becomes

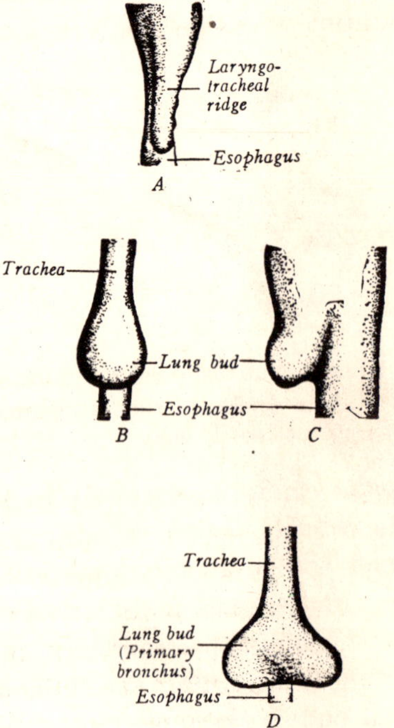

Fig. 8.44. Development of the chief bronchi of the human lung, in ventral view. *A*, at 5 mm.; *B*, at 7 mm.; *C*, at 8.5 mm.; *D*, at 10 mm.

l-shaped. The tracheal tube elongates rapidly for a time and the point of bifurcation ultimately 'descends' (like the lungs) a distance of eight body segments. The right bronchus extends more directly caudad than does the left bronchus. The difference is maintained throughout life and accounts for the more frequent aspiration of foreign bodies into the right bronchial tube.

The *epithelial lining* changes but little from its early columnar form to the final pseudostratified ciliated type. Smooth-muscle fibers and incomplete cartilaginous *'rings'* are differentiating from the surrounding condensed mesenchyme at the end of the seventh week. The *glands* develop as ingrowths from the epithelium after the fourth month.

The Lungs

In an embryo 7 mm. long the right primary bronchus gives rise to two side buds, or *lateral bronchi,* while the left bronchus forms but one; the end of each main tube continues

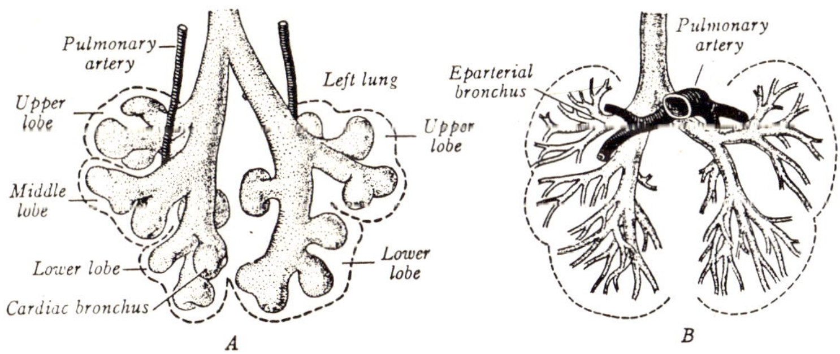

Fig. 8.45.   Developmental plan of the human lungs, in ventral view. *A,* At 14 mm.; *B,* at birth.

as the *stem bronchus.* Even at this early stage the plan of the future lobed lung is begun, as slightly older embryos make clear *(C,D).* On the right side, the upper bud is small and is called the *apical bronchus,* since it presages the *upper lobe;* the other lateral bud is the axis for the *middle lobe,* whereas the stem bronchus will form the *lower lobe.* On the left side the single lateral bud identifies the future *upper lobe* and the stem bronchus is the forerunner of the *lower lobe.*

The development of the lungs proceeds much like the branching of a compound gland into a bush-like set of tubules. Their development is in three phases which occupy consecutive periods, as follows :

1.   Establishment of the larger conducting tubes—*bronchi* and *bronchioles* (five weeks to nearly four months).

2.   Laying down of the *respiratory bronchioles* (four to six months).

3. Extension into a system of *alveolar ducts* and the differentiation of early *alveoli* (six months to term). The lung loses its glandular appearance and becomes highly vascular.

Epithelial Growth and Differentiation: As the bronchial buds continue to grow and branch, the tubular system in each pulmonary lobe becomes increasingly bush-like with dorsal, ventral, lateral and mesial rami. In the fifth month the epithelial lining of the terminal buds of the respiratory tree is cuboidal. Early in the sixth month the adjoining capillaries begin to push through the epithelium so that the lining soon becomes discontinuous and the epithelium largely disappears *(B).* At this time 18 generations of pulmonary branchings have formed; they present the fully prenatal plan of the lung in which the terminal buds, or *alveolar sacs,* appear as irregular spaces bordered by capillary networks. According to some investigators this is the permanent structure of the alveolar sacs. Such an interpretation contrasts sharply with the traditional description of a continuous epithelial lining which merely flattens in the later months to assume the character of a thin but intact layer.

After birth the branching of the pulmonary tree resumes; it continues at least through middle childhood, thus producing the final number of about 24 generations. There is a progressive budding-out of new alveolar sacs with a corresponding

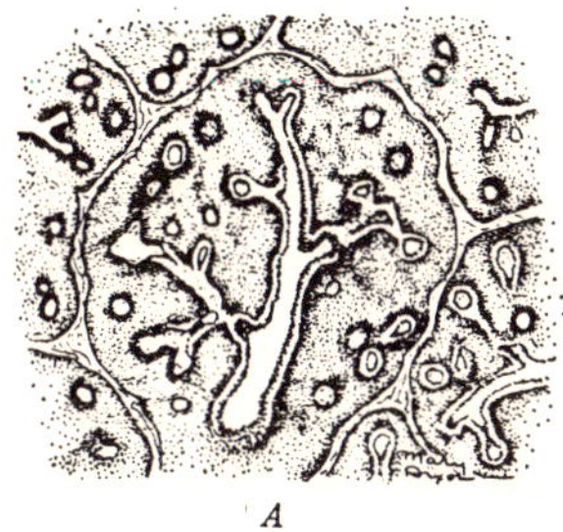
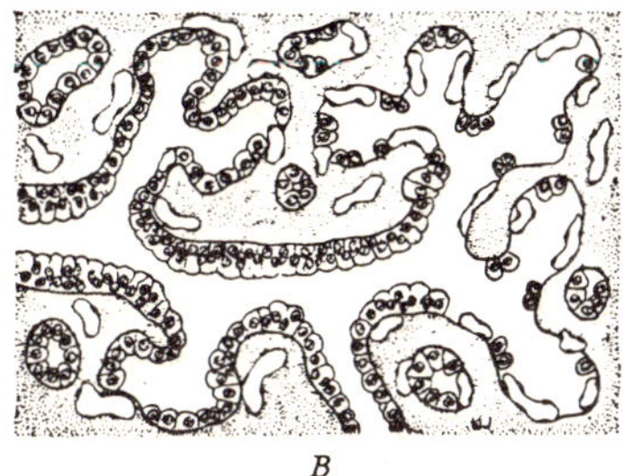

A   *B*

Fig. 8.46. Sections of the human lung. *A*, Developing lobules, at four months. *B*, Loss of epithelium in the terminal air passages, at eight months.

transformation of the parent sacs into air tubes. This explains the occurrence of permanent alveoli on the walls of the *alveolar ducts* and *respiratory bronchioles*.

The early apical bronchus of the right upper lobe comes to be called the *eparterial bronchus* because it alone lies upon the pulmonary artery. Originally this bronchus is dorsal to the artery, as the name implies. Later, when the heart descends, it is cranial with respect to this vessel *(B)*; yet, on the basis of upright posture, the designated meaning 'upon the artery' is again wholly appropriate. It is commonly stated that this bronchus was anciently a secondary branch in what was then the upper lobe of the lung; in the course of evolutionary advance it is supposed to have migrated upward onto the main stem and induced the formation of a new lobe about it. Others view this bronchus as an entirely indpendent replacing outgrowth, at a higher level, that became selected as the basis for a new lobe.

The left upper lobe seems to contain a bronchial branch that is the equivalent of the entire apical bud on the right side. Since, however, this branch remains shall and fails to induce the formation of a separate lobe, the upper lobe of the left lung is homologous to both the upper and middle lobes of the right side. The suppression of the upper left lobe has been interpreted as an adaptation to facilitate the normal caudal recession of the aortic arch. An alternative explanation stresses the lessened opportunity for pulmonary expansion on the left side consequent on the results of the rotation of heart and esophagus in opposite directions. As a contributory factor, the more caudal position of the left common cardinal vein is perhaps significant. Also on the left side an important branch is suppressed in the lower lobe, owing to the position of the heart and pulmonary vein; this, however, affords opportunity for an excessive development of the corresponding right ramus which then projects into the space between the heart and diaphragm as the *cardiac bronchus*.

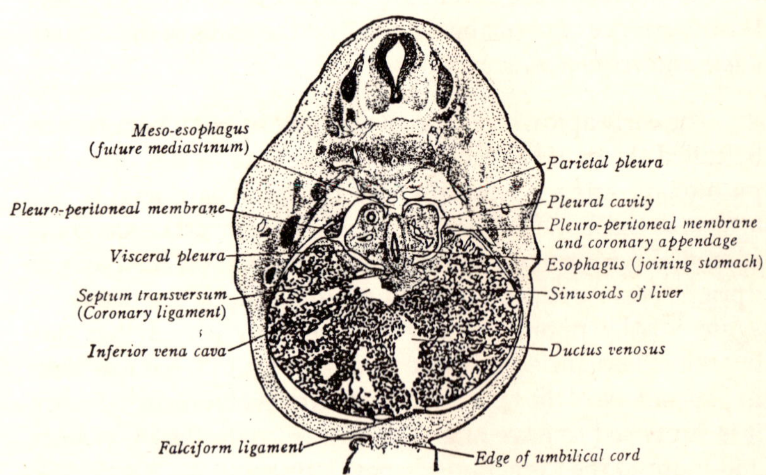

Fig. 8.47. Mediastinum, lungs and pleural cavities of a 10 mm human embryo, in transverse section.

It has been much discussed whether the primary method of pulmonary branching is by simple forking of a growing tip, by side-branching proximal to the non-bifurcating tip, or by a combination of these two methods. Both styles apparently occur, but not always typically enough to avoid differences in interpretation. The method of postnatal growth at the periphery of the pulmonary tree is likewise disputed. A continuation of the prenatal type of branching, retrograde splitting of air passages already present, and a combination of both methods have all been advocated.

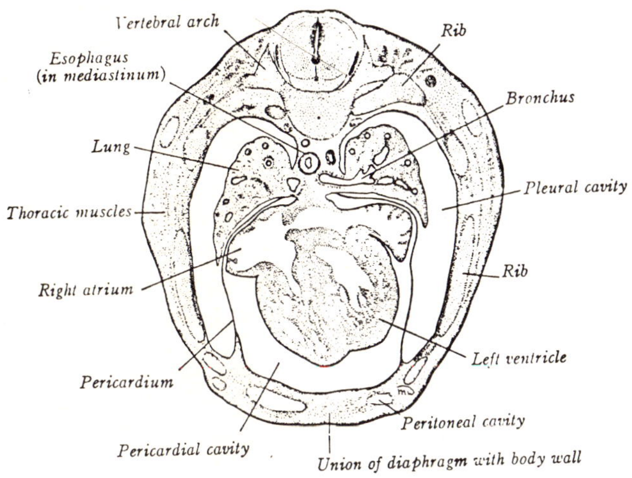

Fig. 8.48. Growth of the human lungs and pleural cavities (and the consequent extension of the pericardium), shown in a transverse section at nearly eight weeks.

Relations to Mesenchyme and Coelom: The entodermal lining of the early respiratory primordium develops within a median mass of mesenchyme, located dorsal and cranial to the man peritoneal cavity. This tissue resembles a broad mesentery; it is later named the *mediastinum.* The original right and left bronchial buds grow out laterally into their respective pleural cavities, carrying before them dome-shaped investments of mesenchyme surfaced with mesothelium. The subsequent branching of the bronchial buds takes place with this simultaneously growing tissue-mass. The mesoderm adapts itself to the shape of the two bronchial trees and gradually the external lobation of the lungs takes form Internally each *lobe* becomes subdivided into *lobules*. The mesenchyme actually encasing the respiratory tree ultimately differentiates into the muscle, connective tissue and cartilage plates of the walls of the air tubes and the supporting tissue of the alveolar sacs. Into it grow blood vessels and nerve fibers.

As the lungs enlarge, they make room at the expense of the spongy tissue of the adjacent body wall. This burrowing advance splits off an increasingly extensive *pericardium* from the thoracic wall and allows the lungs more and more to flank the heart on each side. When the pleural cavities are completed, the mesothelial and connective-tissue covering of the lungs becomes the permanent *visceral pleura*. The corresponding layers lining the thoracic wall constitute the *parietal pleura*. These two pleural layers are derived respectively from the visceral (splanchnic) and parietal (somatic mesoderm of the embryo.

Birth Changes: Respiratory-like movements of the chest, which tend to aspirate amniotic fluid into the lungs, sometimes occurs in fetuses; they are probably due to oxygen want. Nevertheless, until normal breathing distends the lungs with air, these organs are relatively small; in particular they leave vacant the ventral and caudal portions of the pleural cavities. With the onset of respiration after birth the lungs gradually expand and occupy better the space allotted them. The pulmonary tissue, which was previously compact and resembled a gland in structure, becomes light and spongy owing to a great increase in the size of its alveoli and blood vessels *(D)*. The alveolar sacs then interlock and press against each other until their arrangement is highly intricate. Evidence of such expansion (the lungs float in water) has medicolegal value in determining whether respiration ever occur. When inflation has been completed and the amniotic fluid in the lungs has been absorbed (at three days after birth), the lungs are considerably larger in every diameter and have more rounded margins. Because of the greater amount of blood admitted to the lungs after birth, their absolute weight increases considerably.

Premature babies, born before the alveolar sacs have developed properly, appear to accomplish their respiratory exchanges through the terminal ducts which function well enough to sustain life.

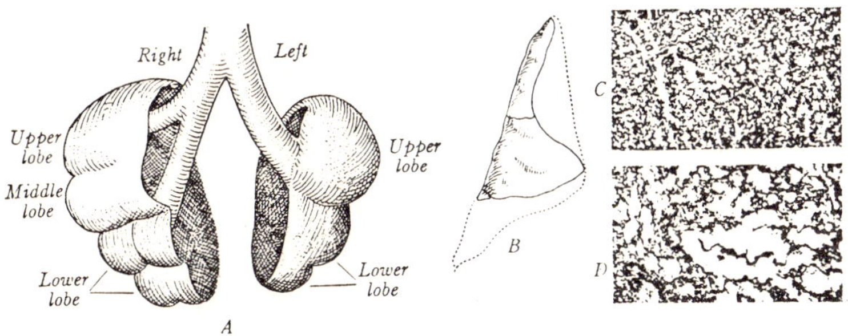

Fig. 8.49. The lungs as a whole and their expansion. *A,* Human lungs, at 13 mm., in ventral view. *B,* Human right lung, at birth, in ventral view broken lines indicate the unfilled extent of the pleural cavity before breathing begins. *C, D,* Sections demonstrating the appearance of the lungs of guinea pigs, just before breathing begins and after eleven minutes of breathing.

Anomalies: Absence of one or both lungs has been recorded, as has the presence of accessory lungs. Variations occur in the size and the number of the major lobes. Rarely there is an eparterial bronchus, or even a third lobe, on the left side. The right eparterial bronchus at times arises directly from the trachea, as in the sheep, pig and ox; contrariwise, it may imitate the relations on the left side. The presence of a distinct cardiac lobe of the lung, though infrequent, is interesting since it occurs regularly in some mammals, including certain primates. A serious anomaly is presented when there is a fistulous connection between the trachea and esophagus; the esophagus usually is atretic and divided transversely, the trachea opening into its lower segment while the upper portion ends as a blind sac. The cause lies in an incomplete separation of the early laryngo-tracheal groove from the gut.

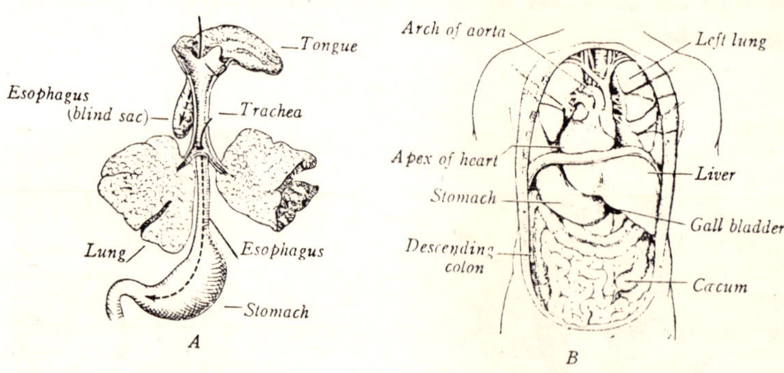

Fig. 8.50. A, Atresia in the esophagus of a newborn, with a fistulous opening of its lower segment into the trachea (X •). B, Complete transposition of the adult viscera.

A striking malformation of the viscera in general is *situs inversus,* in which the various organs are transposed in position, right for left and left for right, as in a mirror image. This reversal may affect all the internal organs, or an independent transposition of the thoracic or abdominal viscera alone may occur. Positive knowledge of the cause is lacking, but this reversal is merely a part of the larger problem of how bilateral symmetry and asymmetry are established normally. It is said that in the production of bilateral asymmetry an activity gradient exists between the two halves of the embryo. A determining factor for the asymmetries of the viscera can be shown to lie in the gut of amphibians, since a 180 degree rotation of the archenteron roof that corresponds to the later duodenum brings about situs inversus. Transposition of the viscera is an extreme type of symmetry reversal; left-handedness and counter-clockwise hair whorl on the crown are other more familiar but milder expressions of the same tendency. The relation of these reversals to identical twinning is discussed.

9

Hormones in Development

The normal human menstrual cycle is dependent upon a complex system containing several components, including higher brain centers, the hypothalamus, the pituitary gland (hypophysis), the ovaries, and the uterus. The cohesive function of these components is integrated by positive and negative feedback signals, and comprises the endocrine system in the female.

In the male the same system operates, only the testes take the place of the ovaries as the major steroid-hormone-secreting organs. Briefly, the reproductive endocrine system functions as an integrated unit in which a releasing hormone (known as gonadotropin releasing hormone [GnRH], or luteinizing hormone releasing hormone or factor [IRF or LHRH] is produced and released by the hypothalamus under the stimulatory influence (positive feedback) of steroid hormones produced by the target organs (ovaries and testes). LRF then stimulates the anterior pituitary gland to release the gonadotropic hormones—follicle-stimulating hormone (FSH) and/or luteinizing hormone (LH)—which stimulate production of steroidal hormones from the target organs. In the case where steroid hormones inhibit release of hypothalamic releasing hormone, this is known as negative feedback (see the section on Integration).

Recent technological advances which have facilitated the measurement of hormones in biologic fluids, coupled with advances in the study of the interaction between hormones and target organs, have opened the way to a better understanding of the regulatory systems. We now enjoy a reasonable complete understanding of the endocrinology of the menstrual cycle. This chapter will first discuss the individual components of the system, then their integrated function, and finally the effects of the hormones on their target organs.

THE COMPONENTS

The Central Nervous System

Although there are conflicting data and hypotheses, the majority of evidence suggests that transmitter substances such as dopamine and norepinephrine, in the brain (particularly in the hypothalamus) affect the specialized nerve cells which contain the hormone-releasing and inhibiting factors. Current data support the existence of a dual system: dopamine inhibits, and norepinephrine stimulates, gonadotropin releasing factor secretion. The cell bodies of the neurons containing norepinephrine are largely located in the medulla and pons, whereas those containing dopamine are found in highest concentration in the arcuate and periventricular nucleii. This type of neuronal activity is largely regulated directly by changes in the circulating levels of the steroid hormones—estrogen and progesterone.

The hypothalamus secretes a releasing factor for each of the tropic hormones produced by the anterior pituitary gland except the gonadotropins, which are probably regulated by a single releasing hormone. The structures of two of these hypothalamic releasing hormones, thyrotropin releasing factor or hormone (TRF or TRH) and gonadotropin releasing factor or hormone (LRF, GnRh, or LHRH), have been elucidated. Thyrotropin releasing factor is a tripeptide that causes release of thyroid-stimulating hormone (TSH), as well as prolactin (PRL)—another trophic hormone which has a major action on the mammary gland. However, it may not be the primary prolactin releasing factor. Gonadotropin releasing factor is a decapeptide. Although it causes synthesis and secretion of both follicle-stimulating hormone (FSH) and luteinizing hormone (LH), it is a much more potent stimulator of the latter.

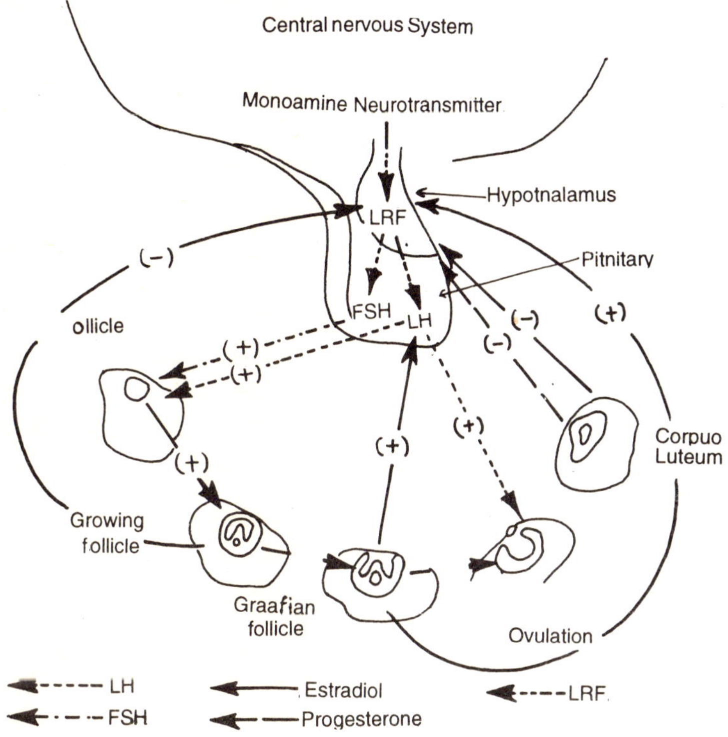

Fig. 9.1. Positive (+) and negative (-) feedback (long-loop) pathways in the central nervous system and the ovary. Arrows point in direction of feedback. See text for explanation of the pathways.

In addition to its inhibitory action on release of gonadotropin releasing factor, dopamine has an inhibitory effect on the synthesis and release of prolactin, probably through synthesis of an as yet uncharacterized prolactin inhibiting factor (PIF).

The LRF neuronal system has been described in the monkey and in the human. The cell bodies of LRF-containing neurons are found in the anterior hypothalamus and in the tuberal hypothalamus. LRF is synthesized in these neurosecretory cells of the hypothalamus, and then released into the hypothalamic-hypohyseal portal system, in which they are transported to the anterior pituitary. At this site, they stimulate synthesis and release of LH and FSH.

The Pituitary

Three gonadotropic hormones of the anterior pituitary have been identified: follicle-stimulating hormone (FSH), luteinizing hormone (LH), and prolactin (PRL).

FSH and LH are glycoprotein hormones containing alpha (α) and beta (β) subunits. They are closely related, structurally and chemically. The α subunits of these hormones are essentially identical. The immunologic specificity of FSH and LH is invested in the B subunit. FSH and LH are synthesized in basophilic cells, called *gonadotropes*, which are scattered throughout the anterior pituitary.

There has been debate about the source of FSH and LH. The critical question was whether both hormones are made in the same gonadotropes, or whether there are two separate populations of gonadotropes, one for LH synthesis and one for FSH synthesis. This question remains incompletely settled, but recent evidence strongly supports the contention that FSH and LH are located in the same gonadotropes in the human pituitary. LRF seems to control both FSH and LH synthesis and release.

The third hormone produced by the pituitary, prolactin, is a single-chain polypeptide composed of 198 amino acid residues. It is synthesized and released by acidophilic cells of the anterior pituitary. The role of PRL in the normal human menstrual cycle has been less precisely defined than the roles of FSH and LH. Recently Vekemans and his coworkers reported a midcycle PRL peak, with continued elevated levels during the luteal phase. However, other investigators have found no cyclic variation in PRL. Current opinion holds that prolactin is necessary for ovarian and testicular steroidogenesis.

FSH and LH exert two types of influence on the reproductive cycle. Their primary effect is stimulation of target cells in the gonad. Although there may be some overlap of activity, because of the structural similarities of the two hormones, each gonadotropin has a specific role in the regulation of gonadal function. FSH stimulates follicular growth and maturation in the female, and initiates spermatogenesis in the male. In addition, FSH catalyses the aromatization of testosterone to estradiol in both sexes. LH stimulates the conversion of cholesterol to pregnenolone in the gonads of both sexes. LH stimulation is indispensable to the complete maturation of germ cells in male and female gonads. However, steroidogenesis proceeds to functional levels under the stimulation of relatively low levels of LH, FSH alone produces neither complete germ cell maturation nor complete steroidogenesis. Thus, it appears that LH, even if FSH is present only at very low levels, can induce steroidogenesis, but that both trophic hormones must be present in adequate amounts, in order for reproduction to be normal. Thus, the gonadotropins are ultimately responsible for the production of the ovarian hormones, which, in turn, control LRF release through the "long-loop" positive or negative feedback system.

In addition to their primary effects on the gonads, the pituitary gonadotropins exert a negative feedback effect on the hypothalamus, through a retrograde "short-loop" vascular system connecting the pituitary and the hypothalamus. Recent animal experiments have demonstrated a neurohypophyseal capillary network common to the median eminence and the infundibular stalk and neural lobe of the pituitary. Three vascular routes have been demonstrated: fenestrated portal vessels to the anterior lobe of the pituitary; capillary connections to the medial basal hypothalamus, with orientation of the capillary loop toward the arcuate nucleus; and an internal plexus to the ependyma of the median eminence (3,4,17). Additional data have been reported demonstrating retrograde transport to the hypothalamus of LH and prolactin, in the rat. Thus, there is strong evidence for short-loop feedback of the gonadotropins to the hypothalamus.

The Ovary

The human menstrual cycle is under ovarian control, in the sense that the ovarian steroids profoundly influence the entire complex. For clarity of exposition, it is convenient to begin with a description of the hormonal events of the menstrual cycle.

The average length of the menstrual cycle is 28 days. By convention, the day of initiation of bleeding is designated day 1. The new follicle for a given cycle starts to develop during the luteal phase of the preceding cycle. Just before and during menses, the cells of the developing follicles proliferate (the follicular phase), and the follicles increase in size, soon to exceed 1 mm in diameter. Many do not attain this size, but become atretic. The follicle that is most sensitive to FSH stimulation gets a lead that it never relinquishes. The rising FSH stimulates the follicle to grow, by inducing mitotic proliferation of the granulosa cells and formation of the theca. The early follicular phase ends when a definite rise in plasma estrogen is noted.

The locally produced estrogen also stimulates the follicle to grow, so that the mature follicle, i.e., the Graafian follicle, reaches a diameter of 6 mm. The next largest follicles become atretic at a size of 1 to 3 mm. The concentrations of estradiol (the major biologically active estrogen) in the general circulation rise from 50 pg./ml. on day 1 to about 75 pg./ml. on day 6 of the cycle. Estradiol concentrations increase more sharply to reach levels of about 150 pg./ml. on day 9. A sharp rise, usually called a "surge," then occurs, to reach a peak of about 350 pg./ml. on day 11. Estradiol levels decline rapidly, to values of about 250 pg./ml. on day 14. The estradiol concentration then gradually rises again.

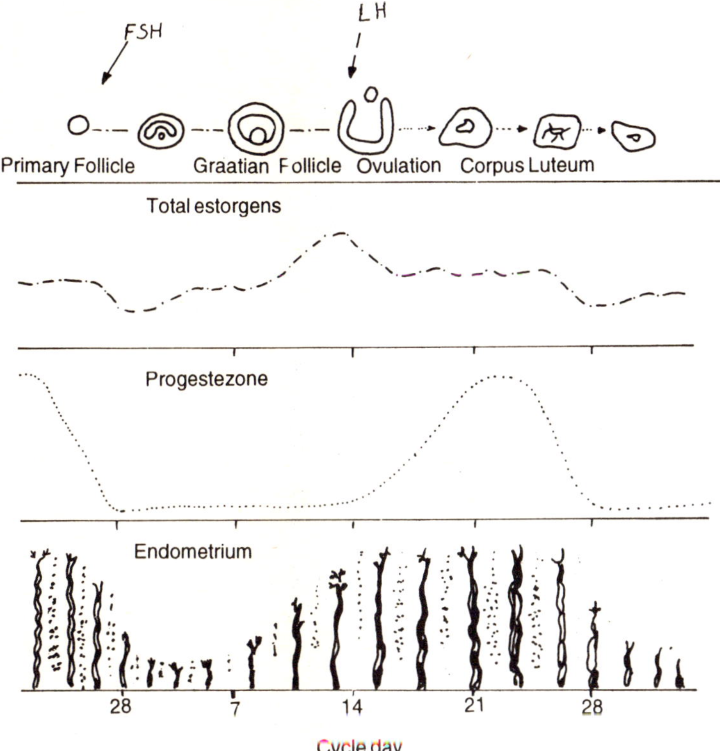

Fig 9.2. Relationships among pre- and post-ovulatory changes in the ovary, plasma levels of estrogen and progesterone, and endometrial response.

The central nervous system-hypothalamic-pituitary axis becomes increasingly sensitive to the positive feedback action of estradiol as the concentration of estradiol in the bloodstream increases during follicular development. An acute surge of LH and FSH at mid-cycle is induced by estradiol. This positive feedback requires a circulating level of estradiol of 100 to 200 pg./ml., which is sustained for 36 to 42 hours (8,9, 23,26,29). Current evidence suggests that estradiol is required for the initiation, but not for the maintainence, of the LH surge. In primates, including man, the interval between the estrogen peak and the LH peak is 14 to 27 hours. Ovulation follows the LH peak within 11 to 24 hours.

Shortly before ovulation, plasma levels of 17-hydroxyprogesterone, produced by the ovary, increase. Just after the onset of the LH surge, levels of progesterone, coming from the ovary, also begin to rise.

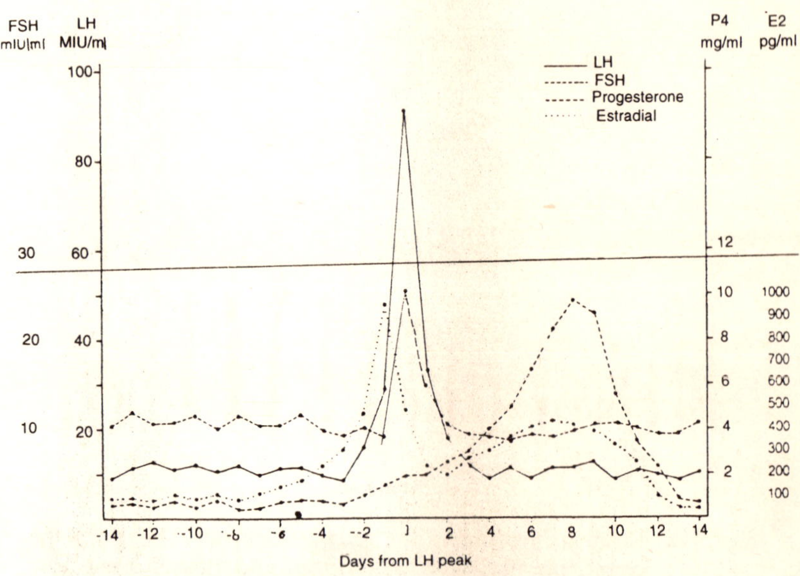

Fig. 9.3. Patterns of estradiol, progesterone, LH, and FSH in the normal menstrual cycle. See text for explanation.

Around the time of ovulation, the granulosa cells luteinize. Luteinization is initiated by LH. For the first three postovulatory days (cycle days 15 to 17) the granulosa cells proliferate, and the corpus luteum reaches a size of 1 cm. During postovulatory days 4 to 9 (cycle days 18 to 23) capillaries, previously found only in the thecal cells, grow to reach the granulosa, and peak vascularization begins. After day 23, regression begins. The highest levels of plasma progesterone are reached on cycle days 18 to 23.

To summarize, the following endocrinologic events are hallmarks of the normal menstrual cycle. The estrogen peak precedes the LH peak; 2) estrogen secretion attains appropriate levels; 3) the LH peak occurs at least 13 days prior to the onset of menses; 4) initial rises in plasma 17-hydroxy-progesterone and progesterone levels occur, coincident with the LH rise;-5) a second rise in plasma 17-hydroxyprogesterone occurs later, coincident with the marked rise in plasma progesterone; 6) plasma progesterone begins to rise with the LH surge, and reaches a maximum 6 to 8 days after the LH peak.

The Endometrium

The cyclic changes in the ovary elaborating first estrogen, and then estrogen and progesterone, induce, in the endometrium, first, proliferation, and then differentiation. These changes are reflected in the endometrial morphology, and the functional state of the ovaries can be inferred with reasonable accuracy from the histologic appearance of the endometrium.

During the **first** half of the cycle, endometrial glandular and stromal **cells** undergo replication. The endometrium thickens, and the glands begin to elongate. The epithelial cells become taller, and mitotic activity is seen in glands and stroma. The first morphologic change after ovulation is the formation of subnuclear vacuoles in the glandular epithelium. This is first seen on the 15th to 16th day of a 28 day cycle. The vacuoles are caused by the accumulation of glycogen, which on the 19th or 20th day is deposited in the glandular lumen. From the 19th to the 23rd day, no mitoses are seen in either the glands or the stroma, suggesting that progesterone causes cessation of hyperplastic growth. The earliest stromal change seen under progesterone influence is progressive loosening of the stroma, with accumulation of fluid between the cells. This is due to a change in the ground substance, presumably related to preparation for implantation. At about the 23rd day, evidence that progesterone causes hypertrophic growth and differentiation can be inferred from the appearance of changes in the stromal cells surrounding the blood vessels which will allow for implantation. These decidual changes progress over the next few days, until the stroma has undergone generalized decidual transformation. If pregnancy does not occur, the stroma becomes infiltrated with inflammatory cells.

The premenstrual endometrium consists of thousands of tiny units composed of large numbers of stromal cells and exhausted glands surrounding markedly coiled arterioles. The first event in menstruation is vasoconstriction of these arterioles. The resulting ischemia results in dissolution of the architecture of the endometrium. Foci of swelling and distortion appear, the stromal cells begin to clump, hemorrhage occurs within the stroma, and the endometrium is shed as the menstrual discharge or flow. The raw, denuded surface of the endometrium is rapidly covered by epithelium, so that within three days healing is complete.

Thus, the changes caused by progesterone in a non-fertile cycle are peak secretion, followed by peak edema in the endometrium, followed by decidual formation. If conception occurs, instead of each of these endometrial changes peaking and waning in sequence, all continue and peak together. This lends a characteristic appearance to the endometrium, which has been described as "gestational hyperplasia."

INTEGRATION

Integration of the components of the endocrine system is controlled by the hormones described in the first section of this chapter. The same hormones are synthesized by both men and women, but the patterns of secretion are different in the two sexes.

The gonadotropins are secreted in pulses, at intervals of about 90 minutes. Pulse amplitude is greater for LH than for

FSH. In males, the pulses are maintained in a tonic pattern, whereas in females the cyclic gonadotropic peak that occurs before ovulation is superimposed upon the tonic pulsatile pattern.

Tonic secretion of LH and FSH is regulated by a feedback loop involving central nervous system components (i.e., dopamine, norepinephrine, and LRF), ovarian steroids, and the gonadotropins. Estradiol is the most potent gonadotropic inhibitor. A quantitative relationship between the negative feedback action of ovarian steroids and gonadotropin release can be demonstrated by interrupting the negative feedback loop by withdrawing estradiol through ovariectomy; this results in significant increases in circulating LH and FSH concentration. The rise in gonadotropins continues until a plateau is attained at approximately ten times preoperative levels about 3 weeks after the operation. The increase is greater for FSH than for LH. This differential effect may reflect preferential inhibition of FSH by estradiol (25,27,28) and possibly, by follicular inhibin (6).

There is some evidence that inhibin, a non-steroidal gonadal factor, is involved in the feedback control of gonadotropic secretion. A protein which inhibits FSH secretion has been identified in the testes and in testicular secretions (1,7,11). Because the increase in FSH levels in menopausal women is disproportionate to the increase in LH levels, an inhibin factor has also been postulated to be present in women. Some evidence for the presence of inhibin in ovarian follicles has been obtained in animals (6,12,22). Although incomplete, the inhibin story may prove to be an exciting breakthrough, because the presence of a specific inhibitor of FSH secretion and release would explain the differential levels of FSH and LH observed under certain circumstances.

Other observations provide further evidence of the feedback effects of estradiol on gonadotropin output: 1) during the first week after ovariectomy, a greater rise in LH and FSH is seen in women operated upon during the follicular phase of the cycle than in those operated upon during the luteal phase; 2) withdrawal of estradiol results in elevated secretion of hypothalamic LRF and thus of pituitary gonadotropins (5,10); 3) there is also a rapid decline of circulating gonadotropin levels in response to estradiol administration to ovariectomized and postmenopausal women.

Moderate levels of estradiol, as seen in the early follicular phase of the menstrual cycle, maximally inhibit gonadotropin output. In the normal menstrual cycle, a change in estradiol levels in either direction reduces inhibition and stimulates the hypothalamic-hypophyseal axis. As mentioned above when the hypothalamic-pituitary axis is maximally stimulated through ovariectomy or the menopause, then the action of estradiol will be solely inhibitory. The negative and positive feedback actions are not interrupted phenomena, but rather a continuum. The positive feedback effect of estrogen on gonadotropin release is always preceded by a phase of negative feedback.

To summarize, in response to a nadir in circulating levels of estrogen, gonadotropin releasing hormone is secreted by the hypothalamus. The pituitary, in turn, releases FSH, which stimulates follicular growth. LH, secreted by the pituitary in basal amounts, stimulates synthesis of estradiol by the theca interna of the developing follicles. Estradiol synthesized by the Graafian follicle has a local trophic effect on the follicle and an inhibitory effect on pituitary release of FSH. As FSH declines, the remaining follicles regress and become atretic. Estradiol synthesis in the ovulatory follicle increases dramatically. Peak levels of estradiol are reached approximately 2 days prior to ovulation. In response to the sustained estradiol peak, a surge of stored LH is released from the anterior pituitary, presumably due to increased secretion of gonadotropin releasing factor by the hypothalamus. A smaller increase in FSH concentration is noted at the same time. Thus estradiol inhibits the hypothalamus in the follicular phase, but stimulates the hypothalamus at mid-cycle.

After ovulation, the granulosa cells become vascularized as the corpus luteum forms. Basal levels of LH stimulate secretion of estradiol and progesterone by the corpus luteum. In combination, these steroids are potent inhibitors of the synthesis of gonadotropin releasing factors. Progesterone may also inhibit further follicular maturation. In the absence of a luteotrophic factor (hCG), the corpus luteum ceases to function after 12 to 14 days. The cells of the corpus luteum show evidence of degeneration. Although the yellow colour is maintained for months, the structure becomes smaller and is eventually replaced by hyaline tissue. The resultant structure is called a corpus albicans.

In response to falling levels of estradiol and progesterone, FSH secretion increases, and a new crop of follicles begins to grow and synthesize estradiol, thus perpetuating the cyclic pattern. The nadir of steroid synthesis occurs approximately 2 days before the onset of menses. The rise in FSH begins shortly before this nadir is reached.

Variations in cycle length are generally considered to take place at the expense of the follicular phase. Thus, in normally ovulating women, the length of the postovulatory phase was thought to remain constant from cycle to cycle, whereas the duration of the preovulatory phase was thought to be variable. However, recent data suggest that the luteal phase may also vary in duration. These variations may be a reflection of very fine adjustments between FSH secretion and follicular response which take place until one follicle becomes large enough to assume major estradiol synthesis. Once this critical size has been achieved, only modest amounts of FSH and LH are required for continued growth of the follicle.

THE TARGET ORGANS

The Testis

The fetal testis differentiates earlier than the fetal ovary and secretes testosterone and dihydrotestosterone by the end of the first trimester of pregnancy, at which time differentiation of the external genitalia also occurs. Fetal gonadal steroidogenesis is thought to occur in response to hCG secreted by the placenta. This mechanism is substantiated by the identification of Leydig cells in the fetal testis from the end of the first trimester. They revert to mesenchymal cells shortly after birth.

From birth until shortly before puberty, the testis is inactive. The germinal epithelium, although present, displays no active spermatogenesis. Levels of FSH and LH rise perceptibly as puberty approaches and induce increased testosterone synthesis and spermatogenesis.

As adult males reach middle age, dihydrotestosterone levels may increase, testosterone levels decrease, androgen-binding globulin levels increase, and spermatogenesis becomes less efficient. However, there is no sudden cessation of reproductive function in the male such as occurs in the female.

Medullary elements dominate in the adult testis. Each testis is composed of specialized cell lines, the Sertoli cells, which nourish and support spermatogenesis, and the germinal epithelium, from which the spermatozoa develop. Scattered about the stroma surrounding the seminiferous tubules are interstitial (Leydig) cells which synthesize and release testosterone. The seminiferous tubules converge into a secondary collecting tubular system called the rete testis, and from there merge into a single long convoluted tubule, the epididymis, in which spermatozoa complete maturation and are stored. Testosterone secreted by the Leydig cells diffuses directly to the Sertoli cells. It does not become absorbed into the bloodstream until it reaches the region of the rete testis. An effective blood-testis barrier appears to exist up until this point.

FSH is necessary for the initiation of spermatogenesis, and testosterone is required for the later steps in spermatogenesis. Both FSH and testosterone stimulate Sertoli cells to elaborate androgen-binding protein (ABP). The exact role of this protein in spermatogenesis is uncertain. Sertoli cells also convert testosterone to 5α-dihydrotestosterone and estradiol. Testosterone, rather than 5α-dihydrotestosterone, appears to be important in spermatogenesis. Estradiol may be the primary mediator of feedback inhibition from the testis to the hypothalamus. However, as noted with respect to the ovary, inhibin may, either alone or in concert with estradiol, be the mediator of testicular negative feedback.

The spermatogenic cycle in man takes 74 ± 6 days. All stages of spermatogenesis can be observed simultaneously in the same seminiferous tubule. Spermatogenesis involves production of haploid germ cells (spermatids) from diploid stem cells (spermatogonia). Spermiogenesis is the process by which the spermatids mature into cells which are capable of independent motility and transport (spermatozoa).

The spermatogonia are divided into two groups: those which constantly divide to maintain the stem cell population and those which undergo growth preparatory to meiotic division, producing two secondary spermatocytes. Each secondary spermatocyte, in turn, undergoes reduction division to produce two spermatids. Development of head, midpiece, and tail then occurs, resulting in mature spermatozoa.

Spermatozoa are released into the lumen of the seminiferous tubules and transported to the epididymis, where they undergo further maturation and storage. At ejaculation, they are propelled as a bolus through the vas deferens. Seminal plasma secreted by the seminal vesicles and prostate is added to the bolus of sperm as it traverses the prostatic urethra, thus completing the composition of the ejaculate.

As noted above, the testis produces testosterone and other androgens, which are secreted in pulsatile fashion. The testis also produces estrogen, which may be involved in the feedback mechanisms at the hypothalamic and pituitary levels.

Thus FSH is important for spermatogenesis and LH for production of testosterone. The regulation of the components of the hypothalamic-pituitary-testis system is similar to that in the female, with testosterone and testicular inhibin playing major roles in the long-loop feedback system. The short-loop system of pituitary tropic hormones (FSH and LH) feeding back on the hypothalamus and its releasing hormone (LRF) is also probably operative. Since the production of LH in the male is only tonic, and not both tonic and cyclic as in the female, the regulatory system is simpler and less susceptible to disruption.

Testosterone has two major roles in the male. It stimulates development of male secondary sex characteristics, such as enlargement of the genitalia, increase in laryngeal size with concomitant deepening of the voice, increase in muscle mass, and growth of body hair. It also is indispensable to normal spermatogenesis. The effects of androgens on other tissues and organs are discussed later in this chapter.

The Ovary

The fetal ovary does not demonstrate the hormonal activity seen in the fetal testis. Ovarian steroid synthesis is not necessary for differentiation of female external or internal genitalia. However, the fetal ovary actively produces germ cells. The full complement of ova is achieved by 20 weeks of intrauterine life, and the process of follicular maturation and regression is evident by the third trimester of pregnancy.

The ovary is composed of numerous follicles, in varying stages of maturation, surrounded by stromal cells. Follicular units consist of an ovum surrounded by a layer of granulosa cells, and internal and external layers of thecal cells. The maximum number of follicles (millions) is reached at approximately 20 weeks of intrauterine life. From this time, follicles become involved in an inexorable cycle of partial maturation followed by atresia. Most follicles become atretic; only about 500 develop to the stage of ovulation in a lifetime.

Gonadotropin levels in females are low in infancy and

rise gradually as puberty approaches. Levels of FSH are slightly higher than those of LH, but the ratio is nearly one to one. In contrast, during reproductive life LH is always higher than FSH, except early in the follicular phase of the cycle, when growth of a new crop of follicles is stimulated.

The rising gonadotropin levels of late childhood stimulate estrogen synthesis in the ovarian stroma. The secreted estrogen causes maturation of the external genitalia and stimulates growth of the breasts. Eventually, when there has been sufficient estrogen stimulation of the endometrium, uterine bleeding occurs. The first episode of bleeding is called the menarche. The initial endometrial shedding probably occurs in response to a drop in circulating levels of estrogen. The first several "menstrual" episodes may occur at irregular intervals because ovulation has not yet been initiated. Once an ovulatory pattern has been established, it recurs with consistent periodicity in an individual woman. Toward the end of the reproductive age, ovulation and menstrual cycles again become irregular. At menopause, estradiol synthesis decreases drastically, and gonadotropin levels become markedly elevated.

The pattern of ovarian steroid secretion fluctuates throughout the menstrual cycle. The ovulatory follicle and corpus luteum are the major sources of estradiol synthesis and secretion. Average production rates range from 0.07 mg./day in the early follicular phase to 0.8 mg./day at the time of the preovulatory peak. Estrone is also secreted by the ovary, but most of the circulating estrone is derived from peripheral conversion of androstenedione. About half of circulating androstenedione is of ovarian origin and half is of adrenal origin; it is secreted by the ovary in a cyclic pattern which parallels that of estradiol and estrone. Mean testosterone levels in the normal woman are about 40 ng./ml. of serum. Approximately half is secreted by the ovary and the remainder by peripheral conversion of androstenedione and of dehydroepiandrosterone, an androgen mostly produced (in normal circumstances) in the adrenal gland.

About 50 per cent of the estrone and estradiol secreted daily undergoes 16-hydroxylation to form estriol. Estriol is then conjugated with sulfate or glucuronide to facilitate excretion.

Progesterone and 17-hydroxyprogesterone are predominantly secreted by the corpus luteum during the luteal phase of the cycle. Low levels of progesterone found in the follicular phase of the cycle are presumably of adrenal origin. As noted earlier, ovarian progesterone and 17-hydroxyprogesterone secretion become detectable in the periovulatory period concomitant with the LH surge. Peak secretion of progesterone and 17-hydroxyprogesterone in the mid-luteal phase is about 24 mg./24 hours and 4 mg./24 hours, respectively. Progesterone gradually rises, beginning in the periovulatory period, whereas 17-hydroxyprogesterone rises sharply before ovulation, and again during the mid-luteal phase.

Mechanisms of Hormone Action

Gonadotropins and steroids select their target cells because of the presence of specific receptors on, or in, these cells. The mechanism of hormone receptor interaction differs for each class of hormones.

The protein hormones, FSH and LH, attach to receptors located on the plasma membrane. The resulting membrane receptor-hormone complex, in turn, activates a membrane-bound enzyme, adenylcyclase. In the presence of intracellular magnesium, adenyl-cyclase acts on ATP to produce cyclic AMP. Cyclic AMP then activates protein kinases, which mediate the biologic responses of the cell. In the case of LH, steroidogenesis is initiated, and in the case of FSH, specific proteins are elaborated which affect germ cell growth and development.

In contrast, steroid hormones combine with specific receptors in the cytoplasm of target cells and are carried by these receptors to the nucleus of the cells, where they interact with specific sequences of DNA to effect messenger RNA synthesis and protein synthesis.

Steroids secreted by gonads are transported to target tissues in the bloodstream in free and bound forms. Testosterone and estradiol bind to a specific binding globulin known as TEBG (testosterone-estradiol binding globulin) or sex steroid binding globulin. Progesterone has no specific binding globulin of its own but appears to bind preferentially to corticosteroid binding globulin (CBG). Only free steroids are biologically active in target organs and in the feedback mechanism.

In addition to feedback relationships with the hypothalamus and pituitary, gonadal steroids affect a variety of target organs including the skin and its appendages, subcutaneous fat, muscle, breast, larynx; uterus, cervix, vaginal mucosa, external genitalia, and the gonad itself.

Androgens exert their primary tropic effect on skin, muscle, laryngeal tissue, and male external genitalia. Androgens are responsible for the laryngeal enlargement which produces the deep male voice and for the increased muscle mass characteristic of the male. Androgens stimulate penile and testicular growth and are responsible for rugations of the scrotum. They stimulate growth of body hair and secretory activity of sebaceous glands and, paradoxically, are also responsible for the temporal recession of the hair line seen in males. At the peripheral level, the aetive androgen is 5α-dihydrotestosterone, which is usually derived by peripheral conversion of testosterone through an irreversible reaction mediated by 5α-reductase.

Estrogens are responsible for growth of the breast and development of the ductile system, and for the deposition of fat in the breasts, abdomen, hips, and things which produces the characteristic female body habitus. Estrogens also induce hyperplastic growth of the endometrium, increased production of cervical mucus, and thickening and rugation of the vaginal mucosa. Estrogen induces the growth of estrogen and progestrone receptors in target tissues, whereas progesterone eliminates these receptors.

Progesterone is a complement, as well as an antagonist, to estrogen. In the breast progesterone induces growth of the lobule-alveolar complexes after estrogen-induced growth of the ductile system has taken place. Estrogen causes increased water content of cervical mucus, resulting in a copious, clear mucoid secretion, and progesterone induces increased viscosity and production of an opaque, gummy mucoid material.

The endometrium is probably the single most important tissue that the responsive to the effects of both estrogen and progesterone. Estrogen induces mitotic growth of the endometrial glands and stroma, and increased endometrial blood flow. Progesterone halts proliferation of endometrial tissue and induces hypertrophic growth and differentiation of the existing tissues. The interaction of these two steroids is finally responsible for the cyclic bleeding referred to as "menstrual periods."

Gonadal steroids also affect target tissues that are not related to sexual or reproductive function. The anabolic action of testosterone results in positive nitrogen balance and retention of potassium, phosphorus and calcium. Testosterone also stimulates hemoglobin synthesis. Both testosterone and estradiol are vasoactive and abrupt depletion, as experienced in loss of gonadal function, produces vasomotor instability, with accompanying "hot flashes."

Estradiol increases serum levels of trigly-cerides and hormone binding globulins, including those for sex steroids, cortisol, progesterone and thyroxine. Estrogen increases insulin response to carbohydrate and causes sodium retention, through interaction with the renin-angiotensin system; it also affects the ability of the liver to secrete substances such as bromosulfophthalein. Progesterone induces natriurisis, probably by competing with estradiol.

All three major sex steroids may cause changes in mood. Estradiol and testosterone are alleged to enhance libido and contribute to a feeling of optimism and well-being. Progesterone is said to contribute to depression. Evidence for these effects is largely inferential, and requires further documentation.